MODELING ONLINE AUCTIONS

STATISTICS IN PRACTICE

The texts in the series provide detailed coverage of statistical concepts, methods, and worked case studies in specific fields of investigation and study.

With sound motivation and many worked practical examples, the books show in down-to-earth terms how to select and use an appropriate range of statistical techniques in a particular practical field. Readers are assumed to have a basic understanding of introductory statistics, enabling the authors to concentrate on those techniques of most importance in the discipline under discussion.

The books meet the need for statistical support required by professionals and research workers across a range of employment fields and research environments. Subject areas covered include medicine and pharmaceutics; industry, finance, and commerce; public services; the earth and environmental sciences.

A complete list of titles in this series appears at the end of the volume.

MODELING ONLINE AUCTIONS

WOLFGANG JANK
GALIT SHMUELI
Department of Decision, Operations, and Information Technologies
Robert H. Smith School of Business
University of Maryland
College Park, MD

A JOHN WILEY & SONS, INC., PUBLICATION

Published by John Wiley & Sons, Inc., Hoboken, New Jersey
Published simultaneously in Canada

For general information on our other products and services or for technical support, please contact our Customer Care Department within the United States at (800) 762-2974, outside the United States at (317) 572-3993 or fax (317) 572-4002.

Wiley also publishes its books in a variety of electronic formats. Some content that appears in print may not be available in electronic formats. For more information about Wiley products, visit our web site at www.wiley.com.

Library of Congress Cataloging-in-Publication Data:

ISBN: 978-0-470-47565-2

Printed in the United States of America

10 9 8 7 6 5 4 3 2 1

To our families and mentors who have inspired and encouraged us, and always supported our endeavors:

Angel, Isabella, Waltraud, Gerhard and Sabina

– Wolfgang Jank

Boaz and Noa Shmueli, Raquelle Azran, Zahava Shmuely, and Ayala Cohen

– Galit Shmueli

CONTENTS

PREFACE

Our fascination with online auction research started in 2002. Back then, empirical research using online auction data was just beginning to appear, and it was concentrated primarily in the fields of economics and information systems. We started our adventure by scrutinizing online auction data in a statistically oriented, very simple, and descriptive fashion, asking questions such as "How do we plot a single bid history?," "How do we plot 1000 bid histories without information overload?," "How do we represent the price evolution during an ongoing auction?," "How do we collect lots and lots of data efficiently?". During that period, we started to collaborate with colleagues (primarily non-statistician) who introduced us to auction theory, to its limitations in the online environment, and to modern technologies for quickly and efficiently collecting large amounts of online auction data. Taking advantage of our "biased" statistical thinking, we started looking at existing questions such as "What auction factors affect the final price?" or "How can we quantify the winner's curse?" in a new, data-driven way. Then, after having been exposed to huge amounts of data and after having developed a better understanding of how the online auction mechanism works, we started asking more "unusual" and risky questions such as "How can we capture the price dynamics of an auction?" or "How can we forecast the outcome of an auction in a dynamic way?" Such questions have led to fruitful research endeavors ever since. In fact, they have led to several rather unique innovations. First, since we posed these questions not only to ourselves but also to our PhD students, it has led to a new PhD specialization. Our graduating students combine a strong background in mathematics or statistics with training in problem solving related to business and electronic commerce. This combination is essential for competing and advancing in the current business analytics environment, as described in the recent monograph

Competing on Analytics by Davenport and Harris or many other articles and books on the same topic.

Our research has also led to the birth of a new research area called *statistical challenges in eCommerce research*. Our interactions and collaborations with many colleagues have led to a new annual symposium, now in its sixth year, that carries the same name (see also `http://www.statschallenges.com`). The symposium attracts researchers from different disciplines who all share the same passion: using empirical techniques to take advantage of the wealth of eCommerce data for the purpose of better understanding the online world.

There are several aspects that make online auction research a topic to be passionate about. Looking long and hard at data from online auctions, we have discovered surprising structures in the data, structures that were not straightforward to represent or to model using standard statistical tools. Two examples are the very unevenly spacing of event arrivals and the "semicontinuous" nature of online auction data. To handle such data and tease out as much information as possible from them, we adapted existing statistical methods or developed new tools and methodology. Our statistical approach to online auction research has resulted in a wide variety of methods and tools that support the exploration, modeling, and forecasting of auction data. While this book primarily focuses on auctions, the lessons learned can be applied more generally and could be of interest to any researcher studying online processes.

The book is intended for researchers and students from a wide variety of disciplines interested in applied statistical modeling of online auctions. On the one hand, we believe that this book will add value to the statistician, interested in developing methodology and in search of new applications—this book carefully describes several areas where statisticians can make a difference and it also provides a wealth of new online auction data. On the other hand, the book can add value to the marketer, the economist, or the information systems researcher, looking for new approaches to derive insights from online auctions or online processes—we are very careful in describing our way of scrutinizing online auction data to extract even the last bit of (possibly surprising) knowledge from them. To allow researchers to replicate (and improve) our methods and models, we also provide the code to our software programs. For this purpose, we make available related data and code at a companion website (`http://ModelingOnlineAuctions.com`) for those who are interested in hands-on learning. Our own learning started with data, and we therefore encourage others to explore online auction data directly.

Wolfgang Jank and Galit Shmueli

January 2010

ACKNOWLEDGMENTS

There are many people who helped and influenced us and we would like to thank them for making this book come to fruition:

First of all, we would like to thank our following students who have worked (and sometimes suffered) with us while developing many of the ideas and concepts described in this book: Phd students Shanshan Wang, Valerie Hyde, Shu Zhang, Aleks Aris, and Inbal Yahav, as well as Masters' students Brian Alford, Lakshmi Urimi, and others who participated in our first *Research Interaction Team* on online auctions.

We also thank our many co-authors and co-contributors. The generation of new knowledge and ideas, and their implementation and application would have never seen the light of day without our close collaborations. A special thanks to Ravi Bapna (University of Minnesota), who introduced us early on to the research community studying online auctions and to the immense potential of empirical methods in this field. We thank our University of Maryland colleagues Ben Shneiderman and Catherine Plaisant from the Human-Computer Interaction Lab, PK Kannan from the Marketing department, and Paul Smith from the Mathematics department. We also thank Ralph Russo (University of Iowa), Mayukh Dass (Texas Tech University), N. D. Shyamalkumar (University of Iowa), Paolo Buono (University of Bari), Peter Popkowski Leszczyc (University of Alberta), Ernan Haruvy (University of Texas at Dallas), and Gerhard Tutz (University of Munich).

We express our deepest gratitude to our many colleagues who have helped shape ideas in our head through fruitful discussions and for their enthusiasm and support for our endeavors: Paulo Goes (University of Connecticut), Alok Gupta (University of Minnesota), Rob Kauffman (University of Arizona), Ramayya Krishnan (Carnegie Mellon University), Ram Chellapa (Emory University), Anindya Ghose (NYU), Foster Provost (NYU), Sharad Borle (Rice University), Hans-Georg

Mueller (University of California at Davis), Gareth James (University of Southern California), Otto Koppius (Erasmus University), Daryl Pregibon (Google), and Chris Volinsky (AT&T).

We also greatly thank our statistician colleagues Ed George (University of Pennsylvania), Jim Ramsay (McGill University), Steve Marron (University of North Carolina), Jeff Simonoff (NYU), David Steinberg (Tel Aviv University), Don Rubin (Harvard University), David Banks (Duke University), and Steve Fienberg (Carnegie Mellon University), who supported and encouraged our interdisciplinary research in many, many ways.

Finally, we thank Steve Quigley and the team at Wiley for their enthusiasm and welcoming of our manuscript ideas.

1

INTRODUCTION

Online auctions have received an extreme surge of popularity in recent years. Websites such as *eBay.com*, *uBid.com*, or *Swoopo.com* are marketplaces where buyers and sellers meet to exchange goods or information. Online auction platforms are different from fixed-price retail environments such as *Amazon.com* since transactions are negotiated between buyers and sellers. The popularity of online auctions stems from a variety of reasons. First, online auction websites are constantly available, so sellers can post items at any time and bidders can place bids day or night. Items are typically listed for several days, giving purchasers time to search, decide, and bid. Second, online auctions face virtually no geographical constraints and individuals in one location can participate in an auction that takes place in a completely different location of the world. The vast geographical reach also contributes to the variety of products offered for sale—both new and used. Third, online auctions also provide entertainment, as they engage participants in a competitive environment. In fact, the social interactions during online auctions have sometimes been compared to gambling, where bidders wait in anticipation to win and often react emotionally to being outbid in the final moments of the auction.

Online auctions are relatively new. By an "online auction" we refer to a Web-based auction, where transactions take place on an Internet portal. However, even before the advent of Internet auctions as we know them today, auctions were held electronically via email messages, discussion groups, and newsgroups. David Lucking-Reiley (2000) describes the newsgroup `rec.games.deckmaster` where

Modeling Online Auctions, by Wolfgang Jank and Galit Shmueli

Internet users started trading "Magic" cards (related to the game *Magic: the Gathering*) as early as 1995. He writes

> By the spring of 1995, nearly 6,000 messages were being posted each week, making `rec.games.tradingcards.marketplace` the highest-volume newsgroup on the Internet. Approximately 90 percent of the 26,000 messages per month were devoted to the trading of Magic cards, with the remaining 10 percent devoted to the trading of cards from other games.

Lucking-Reiley (2000) presents a brief history of the development of Internet auctions and also provides a survey of the existing online auction portals as of 1998. The first online auction websites, launched in 1995, went by the names of *Onsale* and *eBay*. Onsale (today *Egghead*) later sold its auction service to Yahoo! and moved to fixed-price retailing. Yahoo! and Amazon each launched their own online auction services in 1999. Within 10 years or so, both shut down their auction services and now focus exclusively on fixed-price operations. (At the time of writing, Yahoo! maintains online auctions in Hong Kong, Taiwan, and Japan.) Thus, from 1995 until today (i.e., 2010) the consumer-to-consumer online auction marketplace has followed the pattern of eCommerce in general: An initial mushrooming of online auction websites was followed by a strong period of consolidations, out of which developed the prominent auction sites that we know today: eBay, uBid, or Swoopo (for general merchandize), SaffronArt (for Indian art), or Prosper (for peer-to-peer lending).

Empirical research of online auctions is booming. In fact, it has been booming much more compared to traditional, brick-and-mortar auctions. It is only fair to ask the question: "Why has data-driven research of online auctions become so much more popular compared to that of traditional auctions?" We believe the answer is simple and can be captured in one word: data! In fact, the public access to ongoing and past auction transactions online has opened new opportunities for empirical researchers to study the behavior of buyers and sellers. Moreover, theoretical results, founded in economics and derived for the offline, brick-and-mortar auction, have often proven not to hold in the online environment. Possible reasons that differentiate online auctions from their offline counterparts are the worldwide reach of the Internet, anonymity of its users, virtually unlimited resources, constant availability, and continuous change.

In one of the earliest examinations of online auctions (e.g., Lucking-Reiley et al., 2000), empirical economists found that bidding behavior, particularly on eBay, often diverges significantly from what classical auction theory predicts. Since then, there has been a surge in empirical analysis using online auction data in the fields of information systems, marketing, computer science, statistics, and related areas. Studies have examined bidding behavior in the online environment from multiple different angles: identification and quantification of new bidding behavior and phenomena, such as bid sniping (Roth and Ockenfels, 2002) and bid shilling (Kauffman and Wood, 2005); creation of a taxonomy of bidder types (Bapna et al., 2004); development of descriptive probabilistic models to capture bidding and bidder activity (Shmueli et al., 2007; Russo et al., 2008), as well as bidder behavior in terms of bid timing and amount (Borle et al., 2006; Park and Bradlow, 2005); another stream of research focuses on the price evolution during an online auction. Related to this are studies on price dynamics

(Wang et al., 2008a,b; Bapna et al., 2008b; Dass and Reddy, 2008; Reddy and Dass, 2006; Jank and Shmueli, 2006; Hyde et al., 2008; Jank et al., 2008b, 2009) and the development of novel models for dynamically forecasting auction prices (Wang et al., 2008a; Jank and Zhang, 2009a,b; Zhang et al., 2010; Jank and Shmueli, 2010; Jank et al., 2006; Dass et al., 2009). Further topics of research are quantifying economic value such as consumer surplus in eBay (Bapna et al., 2008a), and more recently, online auction data are also being used for studying bidder and seller relationships in the form of networks (Yao and Mela, 2007; Dass and Reddy, 2008; Jank and Yahav, 2010), or competition between products, between auction formats, and even between auction platforms (Haruvy et al., 2008; Hyde et al., 2006; Jank and Shmueli, 2007; Haruvy and Popkowski Leszczyc, 2009). All this illustrates that empirical research of online auctions is thriving.

1.1 ONLINE AUCTIONS AND ELECTRONIC COMMERCE

Online auctions are part of a broader trend of doing business online, often referred to as *electronic commerce*, or *eCommerce*. eCommerce is often associated with any form of transaction originating on the Web. eCommerce has had a huge impact on the way we live today compared to a decade ago: It has transformed the economy, eliminated borders, opened doors to innovations that were unthinkable just a few years ago, and created new ways in which consumers and businesses interact. Although many predicted the death of eCommerce with the "burst of the Internet bubble" in the late 1990s, eCommerce is thriving more than ever. eCommerce transactions include buying, selling, or investing online. Examples are shopping at online retailers such as Amazon.com or participating in online auctions such as eBay.com; buying or selling used items through websites such as Craigslist.com; using Internet advertising (e.g., sponsored ads by Google, Yahoo!, and Microsoft); reserving and purchasing tickets online (e.g., for travel or movies); posting and downloading music, video, and other online content; postings opinions or ratings about products on websites such as Epinions or Amazon; requesting or providing services via online marketplaces or auctions (e.g., Amazon Mechanical Turk or eLance); and many more.

Empirical eCommerce research covers many topics, ranging from very broad to very specific questions. Examples of rather specific research questions cover topics such as the impact of online used goods markets on sales of CDs and DVDs (Telang and Smith, 2008); the evolution of open source software (Stewart et al., 2006); the optimality of online price dispersion in the software industry (Ghose and Sundararajan, 2006); the efficient allocation of inventory in Internet advertising (Agarwal, 2008); the optimization of advertisers' bidding strategies (Matas and Schamroth, 2008); the entry and exit of Internet firms (Kauffman and Wang, 2008); the geographical impact of online sales leads (Jank and Kannan, 2008); the efficiency and effectiveness of virtual stock markets (Spann and Skiera, 2003; Foutz and Jank, 2009); or the impact of online encyclopedia Wikipedia (Warren et al., 2008).

Broad research questions include issues of privacy and confidentiality of eCommerce transactions (Fienberg, 2006, 2008) and other issues related to mining Internet transactions (Banks and Said, 2006), modeling clickstream data (Goldfarb and Lu,

2006), and understanding time-varying relationships in eCommerce data (Overby and Konsynski, 2008). They also include questions on how online experiences advance our understanding of the offline world (Forman and Goldfarb, 2008); the economic impact of user-generated online content (Ghose, 2008); challenges in collecting, validating, and analyzing large-scale eCommerce data (Bapna et al., 2006) or conducting randomized experiments online (Van der Heijden and Böckenholt, 2008); as well as questions on how to assess the causal effect of marketing interventions (Rubin and Waterman, 2006; Mithas et al., 2006) and the effect of social networks and word of mouth (Hill et al., 2006; Dellarocas and Narayan, 2006).

Internet advertising is another area where empirical research is growing, but currently more so inside of companies and to a lesser extent in academia. Companies such as Google, Yahoo!, and Microsoft study the behavior of online advertisers using massive data sets of bidding and bidding outcomes to more efficiently allocate inventory (e.g., ad placement) (Agarwal, 2008). Online advertisers and companies that provide services to advertisers also examine bid data. They study relationships between bidding and profit (or other measures of success) for the purpose of optimizing advertisers' bidding strategies (Matas and Schamroth, 2008).

Another active and growing area of empirical research is that of *prediction markets*, also known as "information markets," "idea markets," or "betting exchanges." Prediction markets are mechanisms used to aggregate the *wisdom of crowds* (Surowiecki, 2005) from online communities to forecast outcomes of future events and they have seen many interesting applications, from forecasting economic trends to natural disasters to elections to movie box-office sales. While several empirical studies (Spann and Skiera, 2003; Forsythe et al., 1999; Pennock et al., 2001) report on the accuracy of final trading prices to provide forecasts, there exists evidence that prediction markets are not fully efficient, which brings up interesting new statistical challenges (Foutz and Jank, 2009).

There are many similarities between the statistical challenges that arise in the empirical analysis of online auctions and that of eCommerce in general. Next, we discuss some of these challenges in the context of online auctions; for more on the aspect of eCommerce research, see, for example, Jank et al. (2008a) or Jank and Shmueli (2008a).

1.2 ONLINE AUCTIONS AND STATISTICAL CHALLENGES

A key reason for the booming of empirical online auctions research is the availability of data: lots and lots of data! However, while data open the door to investigating new types of research questions, they also bring up new challenges. Some of these challenges are related to data volume, while others reflect the new structure of Web data. Both issues pose serious challenges for the empirical researcher.

In this book, we offer methods for handling and modeling the unique data structure that arises in online auction Web data. One major aspect is the *combination of temporal and cross-sectional* information. Online auctions (e.g., eBay) are a point in case. Online auctions feature two fundamentally different types of data: the bid history and the auction description. The bid history lists the sequence of bids placed over time and as such can be considered a time series. In contrast, the auction description (e.g., product information, information about the seller, and the auction format) does not change over the course of the auction and therefore is cross-sectional information. The analysis of combined temporal and cross-sectional data poses challenges because most statistical methods are geared only toward one type of data. Moreover, while methods for panel data can address some of these challenges, these methods typically assume that events arrive at equally spaced time intervals, which is not at all the case for online auction data. In fact, Web-based temporal data that are user-generated create nonstandard time series, where events are not equally spaced. In that sense, such temporal information is better described as a process. Because of the dynamic nature of the Web environment, many processes exhibit *dynamics* that change over the course of the process. On eBay, for instance, prices speed up early, then slow down later, only to speed up again toward the auction end. Classical statistical methods are not geared toward capturing the change in process dynamics and toward teasing out similarities (and differences) across thousands (or even millions) of online processes.

Another challenge related to the nature of online auction data is capturing *competition between auctions*. Consider again the example of eBay auctions. On any given day, there exist tens of thousands of identical (or similar) products being auctioned that all compete for the same bidders. For instance, during the time of writing, a simple search under the keywords "Apple iPod" reveals over 10,000 available auctions, all of which vie for the attention of the interested bidder. While not all of these 10,000 auctions may sell an identical product, some may be more similar (in terms of product characteristics) than others. Moreover, even among identical products, not all auctions will be equally attractive to the bidder due to differences in sellers' perceived trustworthiness or differences in auction format. For instance, to bidders that seek immediate satisfaction, auctions that are 5 days away from completion may be less attractive than auctions that end in the next 5 minutes. Modeling differences in product similarity and their impact on bidders' choices is challenging (Jank and Shmueli, 2007). Similarly, understanding the effect of misaligned (i.e., different starting times, different ending times, different durations) auctions on bidding decisions is equally challenging (Hyde et al., 2006) and solutions are not readily available in classical statistical tools. For a more general overview of challenges associated with auction competition, see Haruvy et al. (2008).

Another challenge to statistical modeling is the existence of *user networks* and their impact on transaction outcomes. Networks have become an increasingly important component of the online world, particularly in the "new web," Web 2.0, and its network-fostering enterprises such as Facebook, MySpace, and LinkedIn. Networks also exist in other places (although less obviously) and impact transaction outcomes. On eBay, for example, buyers and sellers form networks by repeatedly transacting with one another. This raises the question about the mobility and characteristics of networks across different marketplaces and their impact on the outcome of eCommerce transactions. Answers to these questions are not obvious and require new methodological tools to characterize networks and capture their impact on the online marketplace.

1.3 A STATISTICAL APPROACH TO ONLINE AUCTION RESEARCH

In this book, we provide empirical methods for tackling the challenges described above. As with many books, we present both a description of the problem and potential solutions. It is important to remember that our main focus is statistical. That is, we discuss methods for collecting, exploring, and modeling online auction data. Our models are aimed at capturing empirical phenomena in the data, at gaining insights about bidders' and sellers' behavior, and at forecasting the outcome of online auctions. Our approach is pragmatic and data-driven in that we incorporate domain knowledge and auction theory in a less formalized fashion compared to typical exposés in the auction literature. We make extensive use of nonparametric methods and data-driven algorithms to avoid making overly restrictive assumptions (many of which are violated in the online auction context) and to allow for the necessary flexibility in this highly dynamic environment. The online setting creates new opportunities for observing human behavior and economic relationships "in action," and our goal is to provide tools that support the exploration, quantification, and modeling of such relationships.

We note that our work has been inspired by the early research of Lucking-Reiley et al. (2000) who, to the best of our knowledge, were the first to conduct empirical research in the context of online auctions. The fact that it took almost 9 years from the first version of their 1999 working paper until its publication in 2007 (Lucking-Reiley et al., 2007) shows the hesitation with which some of this empirical research was greeted in the community. We believe though that some of this hesitation has subsided by now.

1.4 THE STRUCTURE OF THIS BOOK

The order of the chapters in this book follows the chronology of empirical data analysis: from data collection, through data exploration, to modeling and forecasting.

We start in Chapter 2 by discussing different ways for obtaining online auction data. In addition to the standard methods of data purchasing or collaborating with Internet businesses, we describe the currently most popular method of data collection: Web crawling and Web services. These two technologies generate large amounts of rich, high-quality online auction data. We also discuss Web data collection from a statistical sampling point of view, noting the various issues that arise in drawing data samples from a website, and how the resulting samples relate to the population of interest.

Chapter 3 continues with the most important step in data analysis: data exploration. While the availability of huge amounts of data often tempts the researcher to directly jump into sophisticated models and methods, one of the main messages of this book is that it is of extreme importance to first understand one's data, and to explore the data for patterns and anomalies. Chapter 3 presents an array of data exploration methods and tools that support the special structures that arise in online auction data. One such structure is the unevenly spacing of time series (i.e., the bid histories) and their combination with cross-sectional information (i.e., auction details). Because many of the models presented in the subsequent chapters make use of an auction's price evolution, we describe plots for displaying and exploring curves of the price and its dynamics. We also discuss curve clustering, which allows the researcher to segment auctions by their different price dynamics.

Another important facet is the concurrent nature of online auctions and their competition with other auctions. We present methods for visualizing the degree of auction concurrency as well as its context (e.g., collection period and data volume). We also discuss unusual data structures that can often be found in online auctions: semicontinuous data. These data are continuous but contain several "too-frequent" values. We describe where and how such semicontinuous data arise and propose methods for presenting and exploring them in Chapter 3.

The chapter continues with another prominent feature of online auction data: data hierarchies. Hierarchies arise due to the structure of online auction websites, where listings are often organized in the form categories, subcategories, and subsubcategories. This organization plays an important role in how bidders locate information and, ultimately, in how listings compete with one another.

Chapter 3 concludes with a discussion of exploratory tools for interactive visualization that allow the researcher to "dive" into the data and make multidimensional exploration easier and more powerful.

Chapter 4 discusses different statistical models for capturing relationships in auction data. We open with a more formal exposition of the price curve representation, which estimates the price process (or price evolution) during an ongoing auction. The price process captures much of the activity of individual bidders and also captures interactions among bidders, such as bidders competing with one another or changes in a bidder's bidding strategies as a result of the strategies of other bidders. Moreover, the price process allows us to measure all of this change in a very parsimonious matter—via the price dynamics. Chapter 4 hence starts out by discussing alternatives for capturing price dynamics and then continues to propose different models for price dynamics. In that context, we propose *functional regression models* that allow the researcher to link price dynamics with covariate information (such as information about the seller, the bidders, or the product). We then extend the discussion to *functional differential equation models* that capture the effect of the process itself in addition to covariate information.

We then discuss statistical models for auction competition. By competition, we mean many auctions that sell similar (i.e., substitute) products and hence vie for the same bidders. Modeling competition is complicated because it requires the definition of "similar items." We borrow ideas from spatial models to capture the similarity (or dissimilarity) of products in the associated feature space. But competition may be more complex. In fact, competition also arises from temporal concurrency: Auctions that are listed only a few minutes or hours apart from one another may show stronger competition compared to auctions that end on different days. Modeling temporal relationships is challenging since the auction arrival process is extremely uneven and hence requires a new definition of the traditional "time lag."

Chapter 4 continues with discussing models for bidder arrivals and bid arrivals in online auctions. Modeling the arrival of bids is not straightforward because online auctions are typically much longer compared to their brick-and-mortar counterparts and hence they experience periods of little to no activity, followed by "bursts" of bidding. In fact, online auctions often experience "deadline effects" in that many bids are placed immediately before the auction closes. These different effects make the process deviate from standard stochastic models. We describe a family of stochastic models that adequately capture the empirically observed bid arrival process. We then tie these models to bidder arrival and bid placement strategies. Modeling the arrival of bidders (rather than bids) is even more challenging because while bids are observed, the entry (or exit) of bidders is unobservable.

Chapter 4 concludes with a discussion of auction networks. Networks have become omnipresent in our everyday lives, not the least because of the advent of social networking sites such as MySpace or Facebook. While auction networks is a rather new and unexplored concept, one can observe that links between certain pairs of buyers and sellers are stronger than others. In Chapter 4, we discuss some approaches for exploring such bidder–seller networks.

Finally, in Chapter 5 we discuss forecasting methods. We separated "forecasting" from "modeling" (in Chapter 4) because the process of developing a model (or a method) that can predict the future is typically different from retroactively building a model that can describe or explain an observed relationship.

Within the forecasting context, we consider three types of models, each adding an additional layer of information and complexity. First, we consider forecasting models that only use the information from within a given ongoing auction to forecast its final price. In other words, the first—and most basic—model only uses information that is available from within the auction to predict the outcome of that auction. The second model builds upon the first model and considers additional information about other simultaneous auctions. However, the information on outside auctions is not modeled explicitly. The last—and most powerful—model explicitly measures the effect of competing auctions and uses it to achieve better forecasts.

We conclude Chapter 5 by discussing useful applications of auction forecasting such as automated bidding decision rule systems that rely on auction forecasters.

1.5 DATA AND CODE AVAILABILITY

In the spirit of publicly (and freely) available information (and having experienced the tremendous value of rich data for conducting innovative research firsthand), we make many of the data sets described in the book available at `http://www.ModelingOnlineAuctions.com`. The website also includes computer code used for generating some of the results in this book. Readers are encouraged to use these resources and to contribute further data and code related to online auctions research.

2

OBTAINING ONLINE AUCTION DATA

2.1 COLLECTING DATA FROM THE WEB

Where do researchers get online auction data? In addition to traditional channels such as obtaining data directly from the company via purchase or working relationships, the Internet offers several new avenues for data collection. In particular, the availability of online auction data is much wider and easier compared to ordinary "offline" auction data, which has contributed to the large and growing research literature on online auctions. Because transactions take place online in these marketplaces, and because of the need to attract as many sellers and buyers, information on ongoing auctions is usually made publicly available by the website. Moreover, due to the need of buyers and sellers to study the market to determine and update their strategies, online auction websites often also make publicly available data on historical auctions, thereby providing access to large archival data sets. Different websites vary in the length of available history and the type of information made available for an auction. For example, eBay (`www.eBay.com`) makes publicly available the data on all ongoing and recently closed auctions, and for each auction the data include the entire bid history (time stamp and bid amount) except for the highest bid, as well as information about the seller, the auctioned item, and the auction format. In contrast, SaffronArt (`www.saffronart.com`), which auctions contemporary Indian art, provides past-auction information about the winning price, the artwork details, and the initial estimate of closed auctions, but the bid history is available only

Modeling Online Auctions, by Wolfgang Jank and Galit Shmueli

during the live auction. On both eBay and SaffronArt websites, historical data can be accessed only after logging in.

When an online auction site makes data publicly available, we can use either manual or automated data collection techniques. Manual collection requires identifying each webpage of interest, and then copying the information from each page into a data file. Note that for a single auction there might be multiple relevant pages (e.g., one with the bid history, another with the item details, and another with the detailed seller feedback information). Early research in online auctions was often based on manually extracted data. However, this manual process is tedious, time consuming, and error prone. A popular alternative among eCommerce researchers today is to use an automated collection system, usually called a Web agent or Web crawler. A Web agent is a computer program, written by the researcher, that automatically collects information from webpages. Web agents mimic the operations that are done manually, but they do it in a more methodical and in a much faster way. Web agents can yield very large data sets within short periods of time.

Another automated mechanism for obtaining online auction data is using Web services offered by the auction website. A growing number of eCommerce websites offer users the option to download data directly from their databases. This is done via two computer programs that"talk" to each other: the researcher's program requests information from the auction website's server using an agreed-upon format.

Before describing in further detail the more advanced technologies for obtaining online auction data, we note that traditional methods for acquiring data still exist. We therefore start by describing traditional methods and then move on to more popular Web data collection techniques.

2.1.1 Traditional Data Acquisition

One option is to obtain data via a relationship with a company (e.g., via a consulting relationship). A second option is purchasing data directly from the auction house. eBay, for instance, sells data through third-party companies such as AERS (`www.researchadvanced.com`), Terapeak (`www.terapeak.com`), and HammerTap (`www.hammertap.com`). Such data can be richer than its publicly available data, but the cost and the type of bundling, which are geared toward businesses, can often be a limitation for academic researchers with no data acquisition funds. In addition, not every online auction website offers their data for sale.

A third, but currently still an unusual acquisition option, is direct download of data from an auction website. Prosper (`www.prosper.com`) is an online auction platform for loans, connecting individual borrowers with individual lenders (known as peer-to-peer lending). It offers a "Data Export" option where they "provide a complete snapshot of all listings, bids, user profiles, groups, and loans ever created on Prosper. This snapshot is updated nightly. [They] provide this data in a raw XML format. Additionally, [they] also provide the tools to import the XML data into the database of your choice" (see `www.prosper.com/tools`).

Finally, another traditional option is to use data that are shared by other researchers. Accompanying this book will be several data sets of online auctions that we have collected and used in various studies. We have also greatly benefited from similar sharing of data by other colleagues. Online auction data do not pose confidentiality concerns, and therefore we hope that researchers will be inclined to share their collected data.

In the following, we describe two technologies that are the more widely used methods for obtaining online auction data: Web crawling and Web services.

2.1.2 Web Crawling

A Web crawler is a program or automated script that browses the Internet in a methodical, automated manner. Web crawling (also known as "screen scraping") is a popular technique among online auction researchers for obtaining data. The researcher writes a set of computer programs that collect data from the relevant webpages and store them locally. Current popular programming languages for Web crawling are PHP, Perl, and Windows scripting languages. In general, any computer language that can access the Web can be used for Web crawling. In all cases, basic knowledge of HyperText Markup Language (HTML), which is the basic and predominant markup language in which webpages are written, is required. There are three general steps involved in collecting data via Web crawling:

1. *Identifying the Webpages of Interest (Creating the List of URLs):* This is done by browsing the auction website and identifying the structure of the URLs (the web address) of interest. For example, to search for all auctions for a particular book, use the website's "search" function to obtain a results page that contains all the items of interest. The URL of this search result can then be the starting point for the Web crawler. Note that this page will include hyperlinks to the individual pages of each book auction.
2. *Collecting and Saving Webpages Locally:* Auction websites are highly dynamic. New bids arrive constantly, new auctions are being posted, new closed auctions are constantly added to the historical archives, while older auctions are no longer publicly available. Even other components, such as policies regarding bidding or selling rules, or how ratings are given, can and do change. Because of this dynamic nature, it is important to save all pages of interest to one's local computer. The same information might no longer exist later on, or might be repackaged in a different format that would require new programming. The information might even change during the data extraction phase, such that earlier visited pages would contain different data than later visited pages. The advantages of locally saved pages include not only quicker data extraction but also the possibility of extracting further information from the pages in the future (e.g., for the purpose of further research or for revising a paper), as well as for data sharing.
3. *Parsing the Information from the Webpages into a Database:* This last stage relates to the extraction of the data from the locally saved pages. A program is written that identifies the fields of interest on a page and extracts only the

requested information into a database or some other file format (e.g., comma separated values (CSV) or Excel files). Identifying the location of the information of interest on a webpage requires careful examination of its HTML code, and the specification of character strings that uniquely identify the locations of interest. A more efficient approach uses Regular Expressions (RegExp), which is based on *string matching*, to find and extract data from strings (thereby avoiding the need for loops and instead extracting the information in one sweep). For more information on RegExp, see, for example, Goyvaerts and Levithan (2009) or `www.regexp.info`.

To illustrate these three stages, consider the collection of data from eBay on closed auctions for Amazon.com gift certificates. To identify the webpage that includes the list (and links) for the auctions of interest, we search for the keyword "Amazon" within eBay's "Gift Certificate" category. The partial search results page is shown in Figure 2.1. The web address (URL) of this results

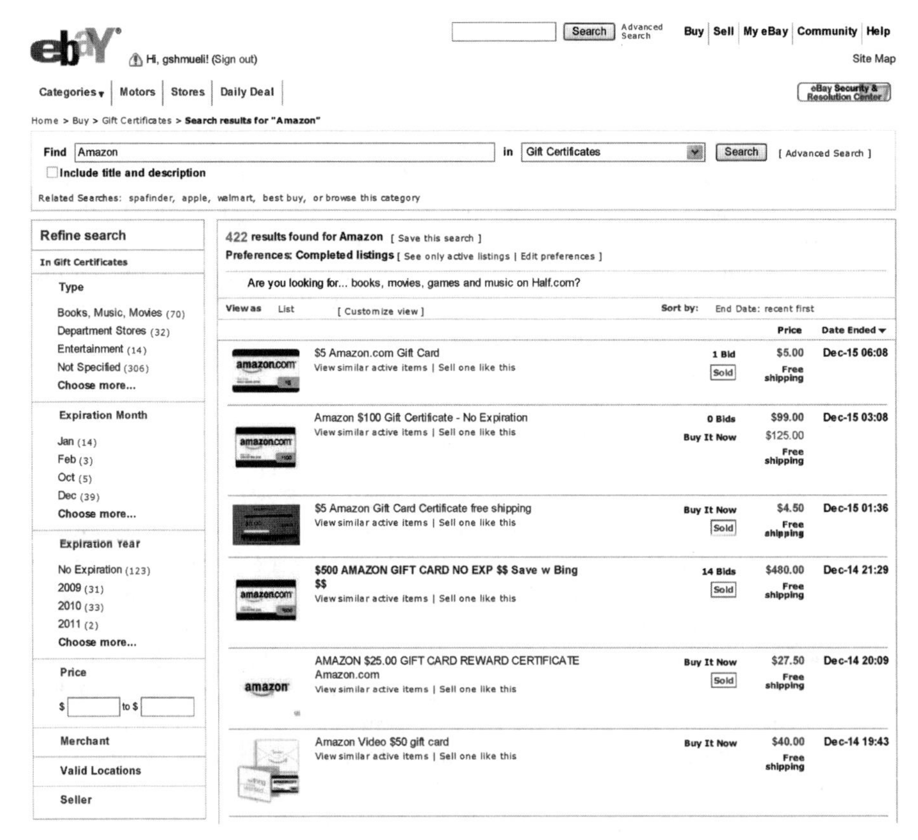

FIGURE 2.1 eBay search results page showing some of the closed auctions for Amazon.com gift certificates (obtained on December 15, 2009).

page is http://completed.shop.ebay.com/Gift-Certificates-/31411/i.html?LH_Complete=1&_nkw=Amazon&_dmpt=US_Gift_Certificates&_pgn=1. The URL includes an indication that these are closed auctions ("completed"), within the Gift Certificate category (in the United States), and that they include the keyword "Amazon." The URL also indicates that this is the first results page, (pgn=1). If you were using a browser, then you could click the "Next" page or a specific page number to reach other results pages. However, noting that the page number is part of the URL (in this example and in many other cases) allows us to easily infer the URLs of the next search pages, which can be used in a faster construction of the pages to be visited by the Web crawler. The next step is therefore to save the search results pages locally. Then, the results pages are used to extract the URLs of the individual auctions. To see how the information can be extracted (or "parsed"), examine the partial HTML code for this same results page,[1] as shown in Figure 2.2. The location of the information on the first three auction is marked, and the URL of each of the corresponding auction pages follows 2–3

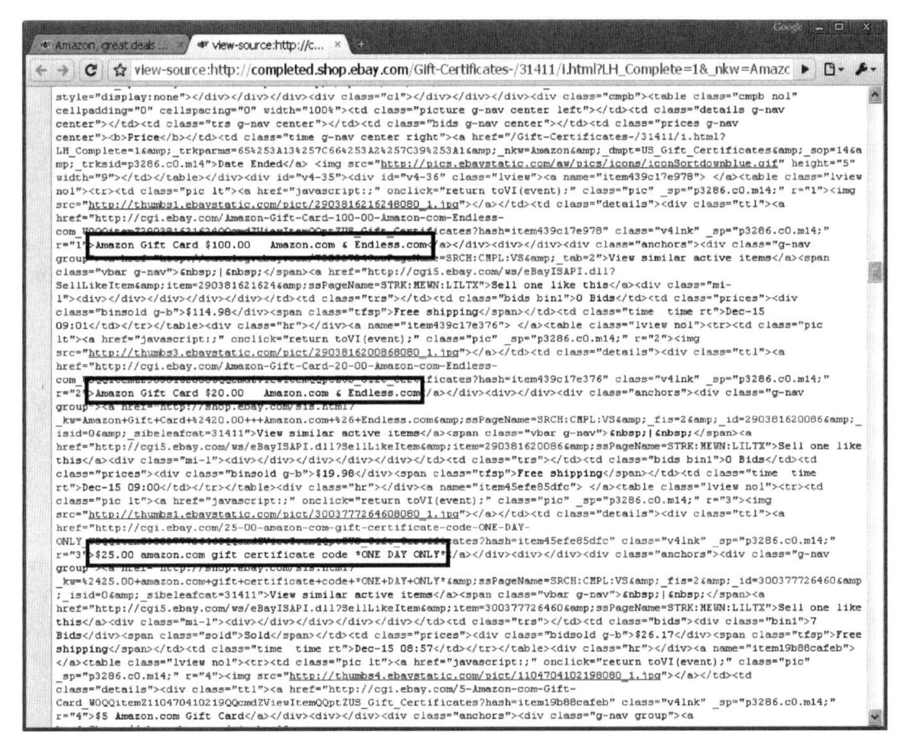

FIGURE 2.2 Part of the HTML code generating the page shown in Figure 2.1. The first three auction titles are marked.

[1] Viewing the HTML source code of a webpage can be done by using the browser's "View Source" or "View Page Source" option.

lines below the marked title. Now a program can be written to extract the individual URLs.

In the next stage, all the individual auction pages are obtained and saved locally. In our example, there are at least two pages of interest for each auction, as shown in Figure 2.3. The first is the general "item description" page with information about the item, the closing time and date, seller information, and so on (top panel), and the second page of related information is the detailed bid history (bottom panel), which is hyperlinked from item description page. Note that the URLs of these two pages (`http://cgi.ebay.com/ws/eBayISAPI.dll?ViewItem&item=300377726460` and `http://offer.ebay.com/ws/eBayISAPI.dll?ViewBids&item=300377726460`, respectively) are also informative, as they include the unique Item ID number. They can be used in the Web crawling script as well.

Finally, after the set of auction pages has been collected and saved locally, a separate program is written to extract the information from these pages. For instance, if we want to collect the time stamps and bid amounts from the bid history page, we must identify the location of these values within the HTML code, and then extract them accordingly. Figure 2.4 shows part of the HTML code for the bid history, where three bid amounts and their corresponding dates and times are marked.

Advantages and Challenges Web crawling is currently the most popular method being used by academic researchers who study online auctions. The biggest advantage of Web crawling is its flexibility and automation. Once the researcher has written the script and has made sure that it works correctly, data extraction is usually quick and the quality of data is high. Web crawling is also still cheaper than purchasing data or using Web services. All these reasons contribute to the popularity of Web crawling in online auction research.

In some cases, Web crawling is the only way to obtain data. For example, if the website only displays data on ongoing auctions (and does not provide any tools for obtaining the data directly), then Web crawling during ongoing auctions can be used for collection. Bapna et al. (2006) describe a Web crawler that obtains information throughout the live auction: "to track the progress of the auction and the bidding strategies the same hypertext page has to be captured at frequent intervals until the auction ends."

The challenges in Web crawling are technological and legal/ethical. From a technical point of view, Web crawling requires some knowledge of programming in a Web scripting language, and data quality relies on robustness of the website and the researcher's programming skills. Extracting data from different websites, and sometimes even from different areas within a particular auction website, requires different programs. Also, changes in the website structure or access require updating the code accordingly. In terms of access, although many websites make their data publicly available, it does not mean that the data are freely available for download. When an auction site (or any website in general) does not want to allow Web crawling, it can proceed in several ways. One way is by implementing technological barriers that make the crawling difficult (e.g., by requiring a user login or information that is normally sent by the user's browser ("cookies"), blocking IP addresses that send

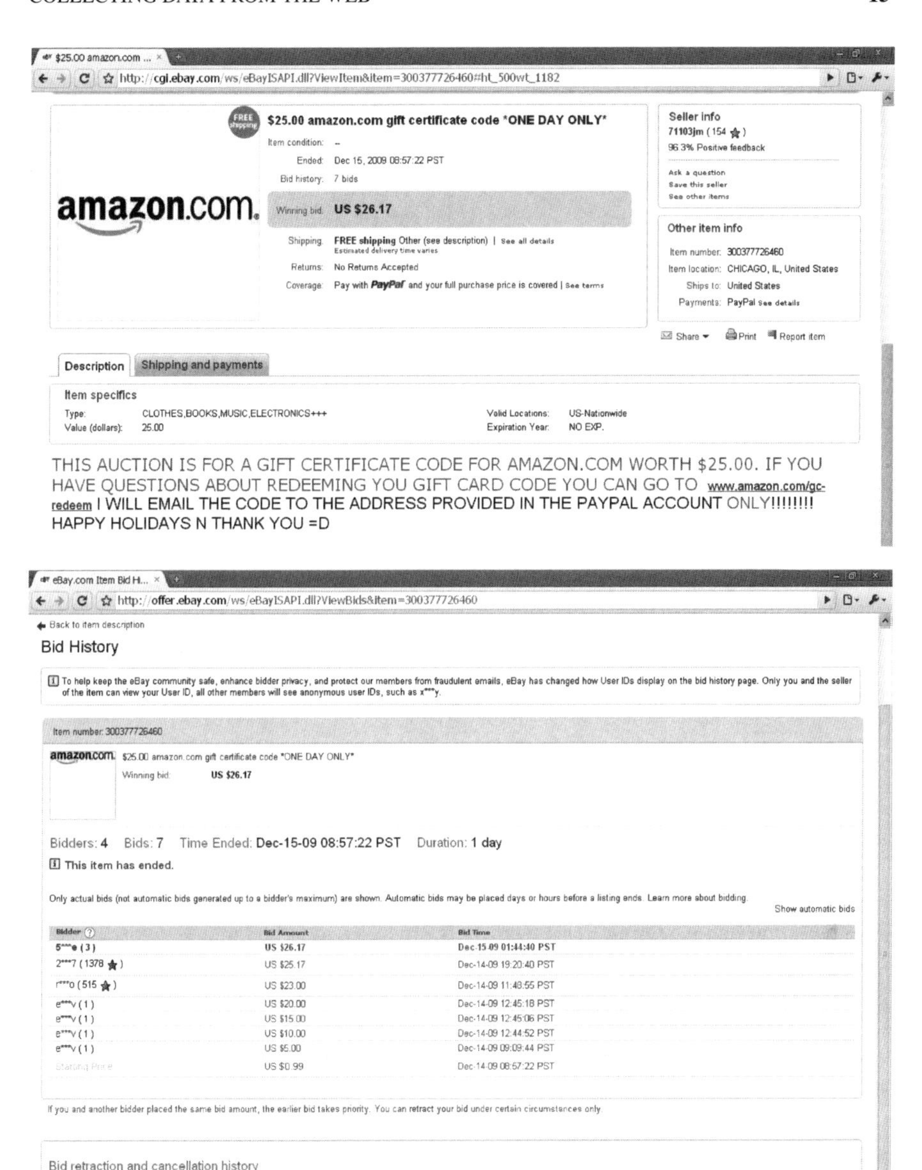

FIGURE 2.3 Item description page (top) and bid history page (bottom) for item #3 in Figure 2.1.

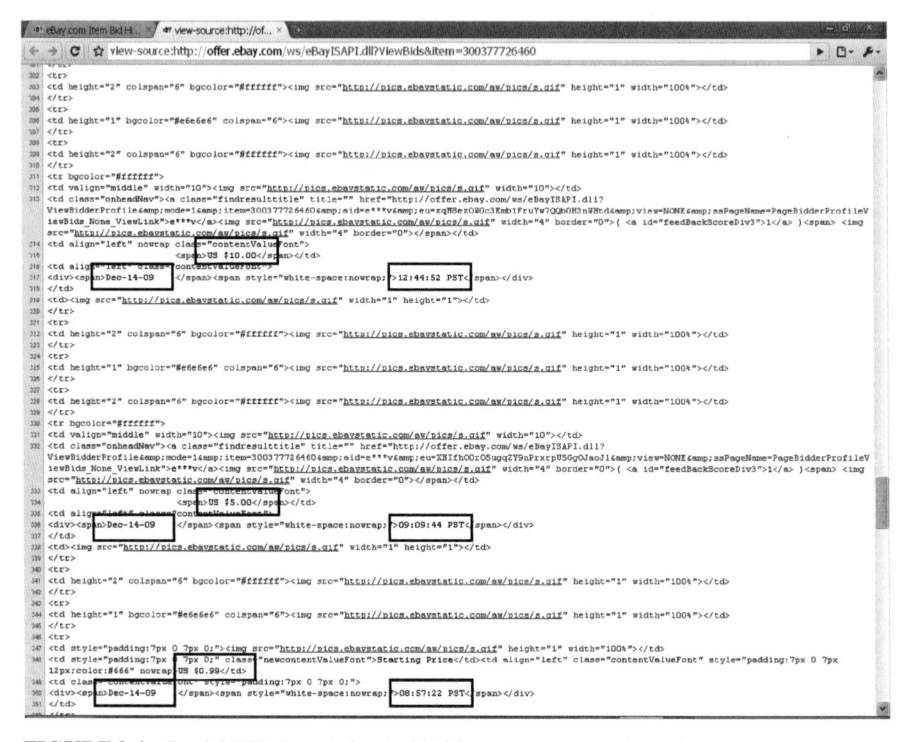

FIGURE 2.4 Partial HTML code for the bid history page shown in the bottom panel of Figure 2.3. Marked are bid amounts, times, and dates for three of the bids.

many requests, or posing a question that only a human can solve (CAPTCHA)). Another approach is stating that crawling is not allowed. The latter is done by placing a file called *robots.txt* on the main page of the website (e.g., on eBay the URL is `http://eBay.com/robots.txt`) and/or by mentioning restrictions in the website's "Terms and Conditions" page. There are several reasons why a website would restrict data crawling. One reason is to avoid overloading their servers, thereby potentially slowing down the service to the intended users. Web crawling therefore uses the website's resources, which are intended to be for sellers and buyers. Restricting Web crawling is also intended against competitor websites or parties, who might use the data for purposes of competition. Although it is unlikely that Web crawling restrictions are aimed at academic researchers, there are ethical and legal issues to consider. Allen et al. (2006) describe the various issues related to Web crawling for academic purposes and suggest guidelines for acceptable use. Additional suggestions are given in Bapna et al. (2006).

2.1.3 Web Services

A growing number of websites are now providing archival data to interested users by allowing querying their database. Among the online auction websites, eBay has been the largest site to offer data through such querying. Prosper.com also offers auction data via Web services. Other sites such as New Zealand's Sella (`www.sella.co.nz`) have announced that they will soon launch such services. The technology, called Web services or API (Application Programming Interface), is a software system designed to support interoperable machine-to-machine interaction over a network. In a typical Web service, the researcher's machine, called the "client," queries the website's database, by interacting with the "server" machine. Most Web services (including eBay's) use a language called XML, which is a markup language for documents containing structured information.

The intention of this section is not to provide a technical explanation of Web services, but instead to describe the use of this technology for collecting online auction data, and to discuss its advantages and weaknesses for researchers.

Advantages and Challenges The main advantage of using a Web service for obtaining auction data (or data in general) is that the data are provided directly by the auction website, hence guaranteeing more reliable and structured data, which are often richer than what is available on the website. The Web service provides a more structured platform for transferring data, thereby guaranteeing a better match between the data requested and that provided, thereby eliminating the need for parsing (as is the case in Web crawling). In addition, it offers a more robust environment, which does not require modifying one's software program when the website format changes. Finally, obtaining data through an API is legal (and often requires the user to register at the website in order to obtain a "key"). Figure 2.5 shows a screenshot of eBay's "Shopping API." It provides data by different types of queries (e.g., by seller ID or

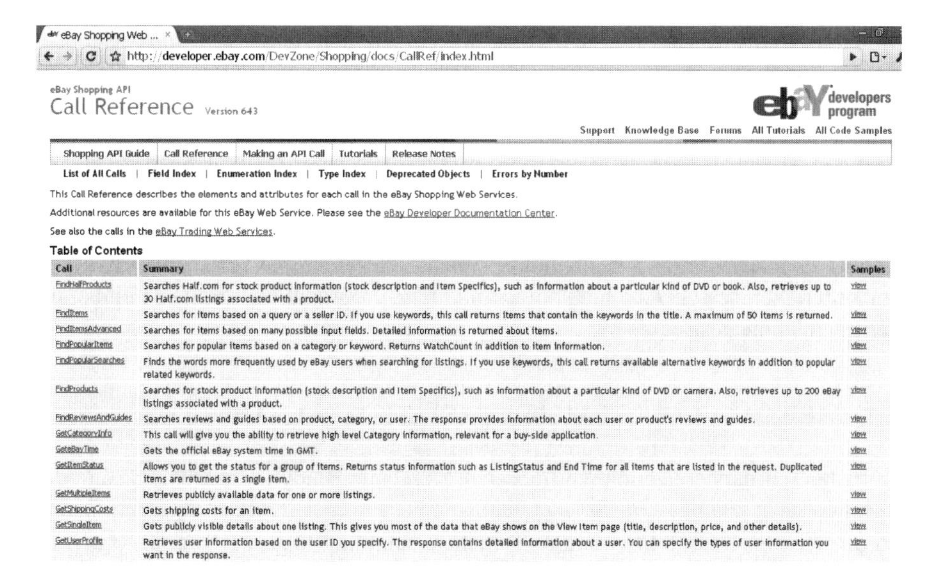
eBay Shopping Web ...

http://developer.ebay.com/DevZone/Shopping/docs/CallRef/index.html

eBay Shopping API

Call Reference Version 643

developers program

Support Knowledge Base Forums All Tutorials All Code Samples

Shopping API Guide | Call Reference | Making an API Call | Tutorials | Release Notes

List of All Calls | Field Index | Enumeration Index | Type Index | Deprecated Objects | Errors by Number

This Call Reference describes the elements and attributes for each call in the eBay Shopping Web Services.

Additional resources are available for this eBay Web Service. Please see the eBay Developer Documentation Center.

See also the calls in the eBay Trading Web Services.

Table of Contents

Call	Summary	Samples
FindHalfProducts	Searches Half.com for stock product information (stock description and Item Specifics), such as information about a particular kind of DVD or book. Also, retrieves up to 30 Half.com listings associated with a product.	view
FindItems	Searches for items based on a query or a seller ID. If you use keywords, this call returns items that contain the keywords in the title. A maximum of 50 items is returned.	view
FindItemsAdvanced	Searches for items based on many possible input fields. Detailed information is returned about items.	view
FindPopularItems	Searches for popular items based on a category or keyword. Returns WatchCount in addition to item information.	view
FindPopularSearches	Finds the words more frequently used by eBay users when searching for listings. If you use keywords, this call returns available alternative keywords in addition to popular related keywords.	view
FindProducts	Searches for stock product information (stock description and Item Specifics), such as information about a particular kind of DVD or camera. Also, retrieves up to 200 eBay listings associated with a product.	view
FindReviewsAndGuides	Searches reviews and guides based on product, category, or user. The response provides information about each user or product's reviews and guides.	view
GetCategoryInfo	This call will give you the ability to retrieve high level Category information, relevant for a buy-side application.	view
GeteBayTime	Gets the official eBay system time in GMT.	view
GetItemStatus	Allows you to get the status for a group of items. Returns status information such as ListingStatus and End Time for all items that are listed in the request. Duplicated items are returned as a single item.	view
GetMultipleItems	Retrieves publicly available data for one or more listings.	view
GetShippingCosts	Gets shipping costs for an item.	view
GetSingleItem	Gets publicly visible details about one listing. This gives you most of the data that eBay shows on the View Item page (title, description, price, and other details).	view
GetUserProfile	Retrieves user information based on the user ID you specify. The response contains detailed information about a user. You can specify the types of user information you want in the response.	view

FIGURE 2.5 eBay's Shopping API, describing the data available through this Web service.

a keyword) and gives data that are unavailable from the eBay website (e.g., the most popular keywords queried on eBay).

There are several challenges for researchers who opt for using Web services. First, not all websites offer such services. Second, the service often requires payment beyond a limited amount of data. For instance, eBay offers two API options (see `http://developer.ebay.com/products/research`):

1. A "Price Research API," which is free for noncommercial use up to 1000 calls a month, gives 30 days of data on historical pricing by keyword, variable date range, and optional daily statistics.
2. A paid alternative is "Advanced Research API" that is available for license from AERS (`www.researchadvanced.com`).

Another limitation of Web services is that the data that one can get is determined by what the online auction website is providing. For example, eBay's Web service provides the aggregate information about seller ratings, but not the detailed ratings (which are publicly available on eBay.com, and are therefore often obtained via Web crawling). You can also see in Figure 2.5 that the call *GetSingleItem* "Gets publicly visible details about one listing. This gives you *most* of the data that eBay shows on the View Item page."

Finally, consider an example of a recent growing research interest in social networks. Treating an online auction market as a network that connects buyers and sellers (see also Section 4.5 and Jank and Yahav (2010)), the researcher would be interested in collecting data on a complete subnetwork. In other words, starting with a given seller, we would then be interested in all the buyers who purchased from that seller (say within a given year), then for each buyer we would collect the list of sellers whom they bought from, and so on. For this purpose, Web crawling is most adequate (where the crawler determines the next collection step dynamically).

In conclusion, traditional channels as well as the technologies of Web crawling and Web services are the main methods for obtaining online auction data, although among them currently Web crawling is the most popular among academic researchers. Being able to collect data from multiple sources is often useful for data validation as well as for flexibility in the types and amounts of data needed. Because of the wealth of publicly available online auction data as well as other Web data, it is extremely useful for empirical researchers to have some training in programming using contemporary scripting languages.

2.2 WEB DATA COLLECTION AND STATISTICAL SAMPLING

In traditional data collection, an important step is the study design. The design determines what and how much data to collect, the timing and order of collection, and so on. There is a large literature on statistical sampling (e.g., in the context of survey sampling) designed for obtaining data that will support answering questions of interest through statistical analysis.

The methodology of statistical sampling is very relevant to online auction research and more generally to eCommerce empirical research, because of the reliance on Web data collection. Sampling includes defining observational units and target and sampled populations, determining sources of sampling and nonsampling errors, choosing appropriate sampling designs, and adjusting sample estimators to reduce bias and increase precision. Sampling online auction data from auction websites not only shares many characteristics with other types of sampling (e.g., surveys), but also has special features that researchers should be aware of and account for. In the following, we discuss sampling in the context of empirical studies that rely on eCommerce data. Online auction data are one important subset of eCommerce data, and we give examples from a variety of other eCommerce applications.

Drawing a representative sample from the population of interest is a fundamental of nearly every empirical study. The seminal question in sample design is "What population does the sample represent?" (like the title of a paper by Rao (1985)). The ability to generalize conclusions from the sample to the population depends on the relationship between the two. The field of sampling deals with the various issues involved in designing and drawing a sample, as well as making inferences from the sample to the population. Classical statistical methods for cross-sectional data, for example, t-tests and regression models, assume that the sample is a "random sample." However, this assumption should not be overlooked since in many situations it is violated. Two examples of violations are a population that is too small to be considered infinite (and thus the probability of sampling one observation affects the sampling probabilities of the other observations) and dependence between observations with regard to the measure of interest. In both cases, the sample design must take into account these special features, and the sample statistics are adjusted to be representative of the population.

In spite of the large and growing body of research on online auctions, and even more generally on eCommerce, which relies heavily on data sampled from the Web, the issue of sampling has not been addressed aside from the conference paper and presentation by Shmueli et al. (2005). In the great majority of Web content studies that rely on sampling from the Web, the data collection phase focuses on the collection mechanism design and the technical details of how the data will be collected and ignores other important sampling issues such as defining the population of interest, the goal of the study (exploratory, descriptive, or analytical), and the relative importance and representativeness of subpopulations (Henry, 1990, p. 47). What are tolerable error rates? What are the main variables of interest? These, in turn, can affect the sample design directly. For instance, when the goal of the study is exploratory, it is more important to get broad coverage of the population than reducing sampling error of resulting estimates. However, when the goal is analytical, it is more important to assure that the sample will be powerful enough for rejecting null hypotheses.

Another common phenomenon in eCommerce research studies is the inexplicit assumption that the collected data comprise the entire population of interest or, more commonly, representative of the population of interest. Assuming that the sample consists of the entire population of interest usually stems from the thought that "we sampled all the transactions in the last week." However, in most cases the purpose is to generalize the results to a longer period of time and thus the sampled week is a subset of the population of interest. In fact, in most cases the population of interest includes a longer time frame than that in which the sample was drawn. For instance, Bapna et al. (2008a) used a sample of transactions on eBay that was collected over a period of 3 months to estimate consumer surplus for that entire fiscal year. Similarly, in an online shopping context, Ghose et al. (2006) sampled several million sales of new and used books on Amazon.com over 180 days (between September 2002 and March 2003) and 105 days the following year (April–July 2004). The authors used this sample to estimate the rate that used books cannibalized on sales of new books (15%), and used it to infer about this rate in the "population," which is implicitly the period of those years. A third example is a study of online price dispersion by Pan et al. (2003). The authors collected three samples of prices on eight product categories taken in November 2000, November 2001, and February 2003. They used the three samples to represent three eras of eCommerce: the "blooming, shakeout, and restructuring of eBusiness" (Pan et al., 2004), and the eight categories to represent the continuum of the global market (based on the price range that they span). In short, a set of data collected from the Web, no matter how large, is most likely to be partial to the population of interest even if it includes all records within a limited time period. A recent advancement in Web agent design addresses the issue of long-term data collection (Kauffman et al., 2000). Using this technology in conjunction with good presampling definitions and choices can lead to improved samples.

Finally, at the analysis stage, consideration of the sample design is needed for determining if and how estimates should be adjusted to compensate for bias and to reduce sampling error. To the best of our knowledge, such postsampling issues have not been published beyond Shmueli et al. (2005). In the following, we discuss some of the main points, but there is definitely plenty of room for further research on the subject.

2.2.1 Sampling and Nonsampling Errors

The first steps of planning a sampling scheme involve answering the following questions:

1. What are the units of interest? What are the units that we will actually sample?
2. What is the target population that we want to study and what is the actual population from which the sample will be taken?

Answers to these questions are crucial for obtaining correct inferences and are directly related to nonsampling errors. Nonsampling errors result from design issues that are unrelated to the fact that a sample is taken rather than the entire population. There are two types of nonsampling errors: selection bias and measurement bias.

Selection Bias When the target population differs from the population to be sampled (the sampled population), this will result in selection bias. Selection bias includes undercoverage of the target population, nonresponse in surveys, and misspecification of the target population. Such errors are just as likely to occur in Web sampled data, but their format can be different. Here are some examples:

- Server problems and Internet congestion during "web rush hour" can lead to many unrecorded transactions (such as bids placed in the last minute of an online auction).
- A website's policy of data refreshing can lead to discrepancies between the target and sampled population: if the data presented on the website are cashed (e.g., by Google or a price comparison engine), then the sampled population might be outdated.
- When using a website's "search" function to retrieve data of interest, choosing the first set of returned records is likely to produce bias in the sample. Search engines do not randomize the results, but rather display them according to some order (e.g., relevance or recency).

It is therefore important to carefully specify the target and sampled populations and compare them before the sample is collected. Skipping this at an early stage can cause complications later on. An example is the study by Venkatesan et al. (2007) that investigates price dispersion across Internet retailers. The collected data on 22,209 price quotes revealed heterogeneity due to the product condition (which previously has not been accounted for). The researchers then reduced their sample by retaining priccs only for products tagged "new," thereby narrowing down the target population. The eliminated data were not random, however, because a subset of retailers was completely eliminated, thereby changing the sample design.

Finally, we note the option of assessing potential selection bias by comparing data that are obtained from multiple sources. For example, comparing the results of a website's "search" function with data that are purchased from the company or from data obtained via their Web service can help in assessing the undercoverage of Web crawling that would be based on the "search" results. To the best of our knowledge, such a comparison has not been published.

Measurement Bias Measurement bias results from the measurement tool: these include interviewer/experimenter effects, non-truthful responses, errors in data recording, and poorly designed questions in questionnaires. Although these types of errors seem less prevalent in Web data, they do exist if in somewhat different format:

- Web agents that collect data can interfere with the website traffic causing a slowing down or other extraordinary effects.
- Fictitious users or non-truthful postings by users still affect the collected data (e.g., bid shilling, where sellers bid on their own items via some proxy).
- Poorly designed websites can lead users to input wrong or even irrelevant information. If the error goes unrecognized, then the sampled population contains errors. If the errors are recognized, the website might drop those records altogether or record them with missing/wrong values.

- Devices to thwart robots, not allowing collection from some area or limiting the query frequency to, say, five times a day.

Sampling Errors Sampling errors result from taking a sample rather than recording the entire population. Although a good sample can provide an accurate representation of the entire population, there are still inaccuracies resulting from including some of the observations and not others. The goal of statistical methods in this case is to quantify the sampling error and thereby allow inference from a sample to the population of interest. Sampling error consists of two main measures: bias and variability. Optimally we prefer unbiased sample estimators that have very low variability. This means that the estimator is accurate, or "on target," as well as producing precise estimates that do not vary much from sample to sample. To construct such estimators, however, requires knowledge of the underlying sampling mechanism. It is not sufficient that Web data samples tend to be very large and that the global population size is huge. It turns out that for different sampling methods the estimators need to be constructed differently to achieve low bias and low variability.

Returning to the issue of defining the observation unit, determining whether it is a transaction, a product, a user, a retailer, or any other entity is directly related to the sampling design. Depending on the website, the data will usually be organized in a way that is most relevant to potential users. However, many websites have a relative database where data are linked to multiple observation units. For example, the data on eBay are organized by auction. However, it is possible to collect data on a particular set of bidders, or on sellers. On Barnes & Noble's online bookstore (`www.bn.com`), the natural unit is a book within a subject category. However, it is easy to focus on authors as the observation units, and to collect prices on their authored books.

Although there might be large amounts of data on the "natural observation unit" of the website, there might be much scarcer data on other units. Returning to the Barnes & Noble's website example, there might be an abundance of romance contemporary fiction books listed on a certain day. However, if we look at authors who published multiple statistics textbooks, there are likely to be fewer observations. Furthermore, if we are interested in comparing subgroups of this population where a certain subgroup constitutes a very small proportion of the population, then it is likely that this subgroup will not be represented in a random sample.

In the following, we give a short description of the main probability sampling schemes, highlighting issues that are relevant to Web sampling. For sampling in general (with relation to surveys), see Lohr (1999) and Levy and Lemeshow (1999). In general, we note that "costs" (in terms of time, effort, money, and any other resources) associated with Web data collection depend on the data collection method. In Web crawling and free Web services, most often the programming requires a "fixed cost" in terms of effort (as well as software, hardware, and time), and then additional per-item "costs" are negligible. For data obtained from paid Web services as well as data purchased from an online auction website or third-party vendor, the cost structure might be different.

2.2.2 Simple Random Sample

Probability samples, as opposed to nonprobability samples such as convenience samples, are characterized by the feature that each unit in the population has a known, nonzero probability of selection into the sample. The current literature that relies on Web data collection implicitly assumes that the data are a probability "random sample" in the sense of sampling from an infinite population (as indicated by the ordinary types of analyses used, from t-tests to regression models). However, when the population is finite, a simple random sample (SRS) that is drawn without replacement from the target population is somewhat different from an ordinary infinite-population "random sample." An SRS is a sample where each observation has an equal probability of being selected from the population. This is the simplest probability sampling scheme. When the population is "infinite" relative to the sample size, a "random sample" coincides with an SRS.

Although SRSs are generally easy to analyze, in Web data sampling they are not always the right design to use. The main reason is that employing an SRS requires identifying and labeling all the records that could potentially be sampled prior to drawing the sample. In many Web data studies, it is hard or impossible to construct such a list. This issue is described in further detail next.

Constructing the Sampling Frame A sampling frame is the list of all the records from which the sample will be drawn, and is in fact the operational definition of the target population. Examples of common sampling frames are telephone directories, email lists, and household addresses. A listing on the Web is usually labeled using a sequence of digits such as an auction ID, an ISBN, or a seller username. Constructing a sampling frame requires identifying each possible record in the sampled population and its label before carrying out the sampling. In some cases, constructing or obtaining such a list is feasible. An example is using price comparison engines such as PriceGrabber.com and Shopping.com. Pan et al. (2003) compared the completeness of lists of retailers from multiple price comparison websites and found that BizRate.com returned the most complete list (except for a few cases where not all retailers were covered). However, in 2004 we performed a search for a popular DVD player on BizRate.com and found that it excludes a major retailer called Amazon.com! Another example is online comparison travel websites such as Expedia, Travelocity, or Kayak. Although they compare flights from many airlines, some airlines (such as Southwest) choose not to allow the comparison websites to include their information. Hence, relying on comparison sites for constructing an exhaustive sampling frame should be carried out cautiously.

In many cases, and especially when the list is potentially very large, enumerating the listings or obtaining their generating mechanism is not disclosed by the website, and therefore it is impractical or impossible to assemble a sampling frame. If querying the website is possible, in many cases the results will yield only a subset of the list, and most likely not a random one. Yet in other cases, it is impossible to search the website directly by the observation unit (e.g., obtaining a list of all open auction IDs on eBay). Of course, if the company is willing to share its data (or the researcher can purchase it), then this problem goes away. However, companies tend to be wary of sharing such data with academic researchers.

Another reason for an inability to construct a sampling frame in eCommerce studies is that the target and sampled populations are in many cases constantly evolving, and therefore the population size N is not a fixed number at the time of sampling. For example, new auctions are added to eBay every moment, so obtaining a list of the currently opened auctions can be illusive.

If a sampling frame does not exist, then probability sampling will yield biased results with no information on the bias itself. The inability to construct a sampling frame before drawing the sample is also common in the offline world, and there are alternative routes including stratified sampling, cluster sampling, and systematic sampling. We discuss each of these below.

Further SRS Issues Let us assume that the goal is to estimate a population mean (e.g., the average price of a certain camera model auctioned online in a certain week) or a population proportion p (e.g., the proportion of transacted auctions on eBay in the last 30 days). If we take an SRS of size n from a very large ("infinite") population, then the sample mean is unbiased for estimating the population mean and is optimal in the sense of lowest variability. Similarly, in a sample of binary valued observations, the sample proportion of 1's is unbiased for estimating the population proportion of 1's and is optimal in the sense of lowest variability. However, if the population size N is not much larger than the sample size n, then we must apply the *finite population correction* (FPC, given by the ratio $(N - n)/(N - 1)$) to obtain unbiased estimators. This situation occurs naturally in many online auction studies due to the ease of collecting large samples of auctions. It is also common because studies often focus on a small subpopulation of auctions such as a certain product during a given period (see Section 2.2 on Stratified Sampling).

Another deviation of SRS estimators from ordinary estimators occurs when the SRS is substituted with an unequal probability sampling scheme: When we are interested in a particular subpopulation that is rare in the population, it makes more sense to set unequal sampling probabilities to oversample this particular group. An example is the different rates of sampling used by Ghose et al. (2006): they sampled books that sell on Amazon.com at low quantities at 2-hour intervals, and books that sell at high quantity were sampled every 6 hours. In such cases to correct for the estimators' bias might require the use of weighting such that observations are weighted in an inverse proportion to their selection probabilities. For example, if members of some subpopulation are twice as likely to be sampled, then their weights will be halved (Henry, 1990, p. 25).

Systematic Sampling One way to avoid the need for creating a sampling frame before the sampling is carried out is to use systematic sampling. The idea is to choose a random starting point and then to sample every ith record to achieve a required sample size ($i = N/n$). This means that the sampling frame is constructed "on the fly." The main danger in systematic sampling is that the records are ordered in a cyclical pattern that might coincide with the sampling cycle. In a Web data collection environment, including online auctions, such a scheme is less appealing because the cost of sampling the remaining records (between sampled records) is negligible. Furthermore, the list that is compiled by the end of the sample might be temporary if the website recycles labels. Examples are auction IDs in eBay and UPS tracking numbers. Alternative sampling designs that do not require an a priori sampling frame are stratified and cluster sampling, as discussed next.

Stratified Sampling Stratified sampling entails partitioning the population of interest into mutually exclusive subpopulations, and then drawing an SRS from each subpopulation. Stratified sampling is a very popular and useful sampling method in the offline world that can yield more precise estimates compared to SRS estimates. The improvement in precision is achieved by creating suppopulations that are more homogeneous, in terms of the quantity of interest, than the overall population.

Auction studies that aim at estimating a global population parameter often tend to use the hierarchical structure of a website perhaps even without considering stratified sampling directly. For example, Bapna et al. (2008b) sampled auctions from eBay and used category-specific values to compute an overall consumer surplus value; Ghose et al. (2006) collected book sales data from Amazon.com by book categories. In such cases, the stratified sampling can lead to more precise overall estimates than those from an equivalent simple random sample. In studies where the categories or another stratifying factor is of interest in itself for obtaining stratum-level estimates or comparing strata, stratified sampling is natural, and there is statistical literature on methods for sample allocations for different purposes. Examples of studies with a comparative goal are comparing auction data in two nonoverlapping periods in 1999 and 2000 to look for shifts in strategic bidding behavior (Bapna et al., 2006) and comparisons of price dispersion across different categories of auctioned items (Pan et al., 2003; Venkatesan et al., 2007).

Although stratification can be used for improving precision in online auction (and eCommerce) studies and for providing stratum-level estimate, it can also be used to achieve further objectives, as we describe next.

Many procedures for collecting Web data rely on querying the website of interest and recording the records that are returned from the query. In most cases, the algorithm used by the website for sorting the returned results and for determining the number of results returned is not disclosed. For example, when querying eBay for all closed auctions on a certain date, it returns only a subset of those auctions. How this subset is chosen and how the records in this subset differ from those that are not returned is hard to assess. The reason for returning a subset is most likely the huge amounts of auctions that match the query. Using the returned subset could therefore result in selection bias.

To reduce selection bias and obtain a more representative sample, we can use stratified sampling with strata reflecting item categories. Since in this scheme queries are category-specific, such a restricted search greatly reduces the number of matching records, which should yield a complete category-level list (at least for that time point—see Section 2.2.2.5 for related issues). In online auction websites, stratification is natural because of the popularity of hierarchical structures and the relevance of the hierarchy categories to popular research questions. General merchandize auctions will usually have a prominent item category hierarchy (and sometimes extra levels of subcategories). Figure 2.6 shows the hierarchy of two general merchandize online auction websites (eBay and uBid.com). The "search" function also easily allows the user to search for a keyword within a certain category. Auction sites for more specialized merchandize will also tend to use some hierarchy that is related to the auctioned items. For instance, art auction sites such as SaffronArt.com support searching by artist and even have a separate page for each artist with all their auctioned works. Prosper.com allows easy sorting of loan requests by loan category, amount, rate, or term. A potential loaner can also limit the search by the Prosper Rating (a risk rating).

In summary, although website hierarchies are designed for ease of user navigation, they can be exploited for the purpose of stratified sampling.

We note that stratified sampling can be achieved either by planned design or by poststratification. In the former case, sample sizes are determined a priori based on knowledge of the strata sizes and perhaps additional strata information such as variability and sampling costs. In the latter case (poststratification), an SRS (or other) sampling scheme is used, and after the data are collected they are broken down by strata. In both cases, there is statistical literature on how to compute and adjust sample estimates.

Finally, in terms of uses of stratified sampling in online auction research, sometimes stratified sampling takes place because data collection is done differently for different strata. An example is adult-only merchandize auctions on eBay, which require extra password logins and therefore require a different Web crawling script. Alternatively, a researcher might purchase data on some category (e.g., eBay data are typically sold by category or subcategory) and supplement them with API or Web crawling of other categories.

Stratified sampling in online auction data collection poses two main challenges: One challenge is determining the sizes of the subpopulations to determine the size of the samples to draw from each stratum. On some auction websites, subpopulation sizes are hard to determine and in all websites these sizes change continuously as auctions close and new ones open. Since many empirical studies use a sample that is a snapshot in time and use it to make inferences about longer time periods, the actual category sizes are unknown until the end of the period of interest. In such cases, if the variability of category sizes is not expected to change dramatically over time, these "snapshot" sizes can be used as approximations. For example, choosing a single category on eBay will reveal the number of current auctions open within that category at that time. Refreshing the browser will most likely show a new value (albeit not dramatically different).

The second challenge is related to the relationship between strata. A main assumption in stratified sampling is that records belong to distinct subpopulations and that there is no overlap between strata. However, categories and subcategories in many websites tend not to be mutually exclusive. For instance, sellers on eBay can choose to list an item in multiple categories. This duplication is more likely to occur at deeper subcategory levels, where the distinction between products/records is smaller.

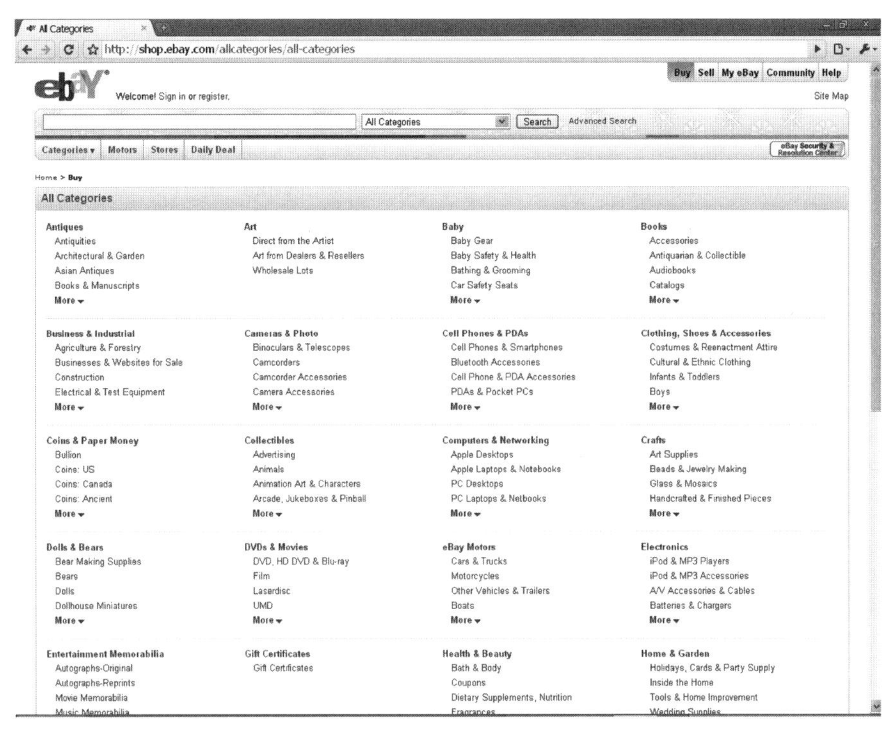

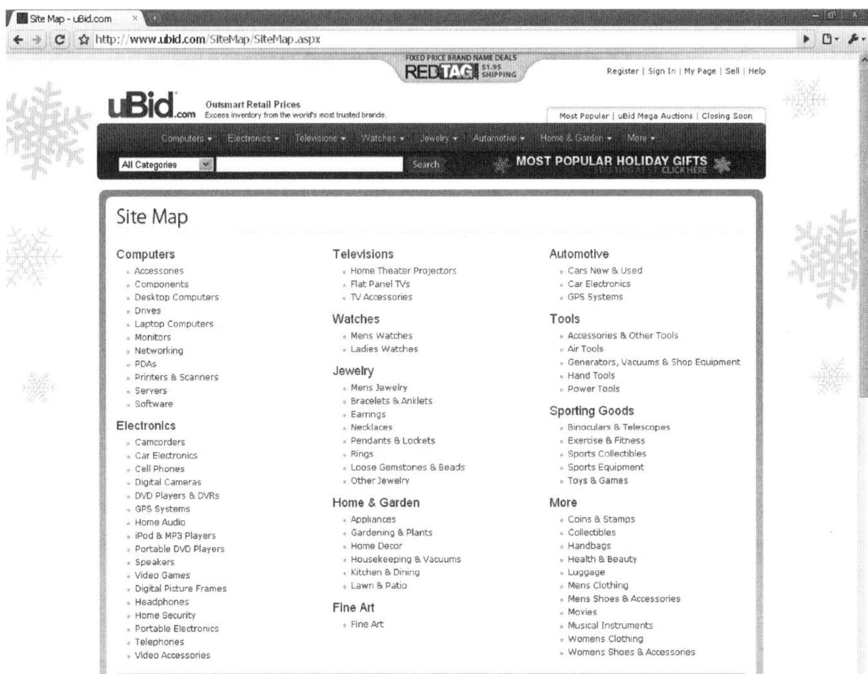

FIGURE 2.6 The most popular hierarchy of general merchandize auction websites is by item category (top: eBay; bottom: uBid).

Therefore, the level of hierarchy used for stratification should be the highest possible that creates strata of "manageable" size (i.e., all records are returned by a query of that level).

Cluster Sampling Cluster sampling is a widely used method aimed at simplifying the sampling and reducing costs. Cluster samples are sometimes referred to as "hierarchical data." The basic idea is to divide the population of interest into subpopulations (clusters) whose members are "close," and in that sense sampling all members is easier, cheaper, and/or faster. A random set of clusters is then selected and all cluster members are drawn into the sample. An example is a survey that randomly selects households, and then surveys each of the household members. Hence, cluster sampling is useful when sampling an additional observation within a cluster requires less resources than sampling an observation in another cluster.

The price of cluster sampling is typically reduced precision of the estimates. In particular, the estimator's precision (i.e., its variance) is proportional to the intracluster correlation. When clusters are of average size M and have intracluster correlation ρ (defined as the correlation between each pair of records within the same cluster, and ranges between $[-\frac{1}{M-1}, 0]$), the variance of a cluster-based estimate is larger than that of an SRS-based estimate by a factor of approximately $1 + (M - 1)\rho$. This implies that the sample size required in cluster sampling is usually much larger than that from an SRS to achieve the same precision. The only exception is when items within a cluster are negatively correlated, which is not typical, and then the cluster sample estimates are more precise than those from an SRS.

The most prominent clustering in auction data (and Web data in general) is temporal. Usually the target population consists of a longer period than the time frame in which the sample is collected. Sometimes multiple snapshots are taken during a time period (panel data). Such a setting can be seen as a cluster sampling design, where the clusters are time points. Then, at each of the selected time points either the entire set of observations is recorded or a sample of them is taken (see Section 2.2.2.6).

Another factor that can lead to clustering in online auctions is concurrency. An SRS assumes that the sampling units (auctions) are independent. However, auctions are often affected by other concurrent auctions that compete with them, supplement them, or otherwise affect them (see more on concurrency in Section 4.3). This happens in eCommerce in general: a set of user ratings for similar movies is likely to be correlated due to the same user rating multiple movies, and the concurrent presentation of multiple movies on the website. From a sampling point of view, we might treat such subsets of concurrent records as clusters.

Another example of a natural clustering in online auctions is sellers. A common research topic is reputation and trust, which is often investigated via seller ratings. When a set of sellers is sampled, and for each seller all his/her transactions are used as records, then we have a clustered sample.

Like systematic sampling, cluster sampling does not require an exhaustive list of the observational units. Instead, it requires only a list of the clusters. The reduction in precision that usually accompanies cluster sampling stems from a high intracluster correlation: records within a cluster tend to be more similar than records across clusters (e.g., education level within the same household). To reduce this effect when the purpose is inference, the number of clusters to be sampled should be large, and the cluster sizes small. In the online auction context, this means that if the goal is to infer about population parameters, one should strive to obtain many nonoverlapping clusters (e.g., time points). The opposite strategy of sampling a few large clusters is useful in two scenarios: when the intracluster correlation is very low, or when the purpose of the study is accurate prediction (Afshartous and de Leeuw, 2005; Shmueli and Koppius, 2008). The point here is that Web sampling actually employs cluster sampling, and therefore the sampling should be actively designed and recognized as such.

One implication is that clusters (e.g., time intervals) should be identified before data collection and then a cluster-scheme sampling should be employed accordingly. The simplest scheme is a one-stage cluster sampling, where a random set of clusters is chosen and all records in those clusters are recorded. This procedure guarantees that each record has an equal chance of being selected (a property called "self-weighting"). In an online auction context, this would mean, for example, specifying a set of time periods where auctions within a period are most likely to interact (e.g., auctions for a specific DVD model that took place between December 1 and 7, 2009).

Multistage Sampling Multistage sampling is a complex form of cluster sampling that does not require a sampling frame. It can further reduce the cost of sampling all the items within a cluster, and is useful when the population is so large that it is easier to sample items by going through a hierarchical selection process. Multistage sampling applies cluster sampling hierarchically with a final random sampling stage. A simple two-stage design first chooses a random set of clusters and then within each cluster takes an SRS. A three-stage design would mean randomly choosing a set of clusters, and then within each cluster randomly drawing a set of subclusters, and finally drawing an SRS within each chosen subcluster. Some eCommerce studies that look at changes over time implicitly employ a two-stage sampling design: SRSs are collected during multiple periods of time, where the target population is a much longer time period. For example, the price dispersion study by Pan et al. (2003) collected data in November 2000, November 2001, and February 2003, and in those time periods the researchers sampled a set of products (which are representative of a larger population of products).

Advanced surveys can have even six or more stages. In addition, they might preselect some clusters instead of drawing a random sample of clusters. Preselection is also usually common in online auction data collection, where a set of categories or items is preselected and they are to represent a wider population of categories or items. As long as the selection at the next stages is random, the bias can be adjusted. Determining the probability of selection in multistage samples and estimating sampling variability is much more complicated. The two popular methods are linearization and resampling, which can be executed through software packages such as SAS.

2.2.3 Further Issues and Conclusions

Publicly available online auction data have opened the door to exciting new research. Yet issues related to the sampling schemes that are employed and could potentially be employed have not received much attention. This gap means that there is room for further research on sampling from the Web as well as potential improvements to data collection designs.

Developing methods for creating sampling frames, evaluating the sizes of populations compared to samples, and identifying clustering factors and sources of homogeneity can help improve the quality of sampling and statistical power. Another challenge to study is the definition of target population, in light of the very dynamic nature of online auction markets and website policies.

Finally, technological skills combined with statistical knowledge could be used for integrating sampling knowledge into Web crawling and querying technologies. Such "smart Web crawlers" can be a powerful contribution to both methodology and practice.

3

EXPLORING ONLINE AUCTION DATA

Once a dataset of online auctions is assembled and cleaned, data usually require an extra preprocessing step before they can be used for any further analysis or even visualization. Depending on the study goal and on the data to be used, the raw auction data might require special transformations or merging with other data on that auction. As mentioned in the previous chapter, data for a single auction can span multiple webpages that contain data on the bid history, item description, detailed seller rating information, and so on. The disparate data also tend to be of different structure: some are cross sectional (e.g., item description) while others are temporal (the time series of bids). Information can also be retrieved on different entities of interest: the auctioned item, the seller, the bidders, the auction format, and so on. This information can also be linked to other information on those entities, thereby generating networks (auctions for similar items, other auctions by the same seller, previous auctions that the bidders participated in, and so forth).

In the following, we discuss several operations that have been used in a variety of online auction studies that support exploration, visualization, and modeling. To motivate and illustrate the usefulness of these operations, we use data visualization. We also adapt and develop new visualizations to support the exploration of the special nature of auction data.

Although most of our discussion relates to data from eBay, which is currently the largest source of data used in online auction research, the principles are general and relevant to data from other online auction sites.

Modeling Online Auctions, by Wolfgang Jank and Galit Shmueli

3.1 BID HISTORIES: BIDS VERSUS "CURRENT PRICE" VALUES

The bid history, which usually contains the bid amounts and time stamps for a single auction, often does not display the bidding information that was observable during the live auction, or it might even not include the full bidding information. Such discrepancies are usually a result of a policy employed by the auction website. However, it is often important to obtain the information that was available to bidders (and/or sellers or auction browsers) during the auction in order to infer about their behavior. The first operation that we discuss is therefore obtaining the "current price" information from the archival bid history.

On Prosper.com (auctions for peer-to-peer lending), none of the bidding information is disclosed during a live auction. Hence, bidders do not know what others have bid and might be placing multiple bids after being notified that they were outbid. Prosper's reasoning for this policy is "this hidden bid auction style guarantees the best rates for everyone involved."

On eBay, most auctions are second-price auctions, where the highest bidder wins the auction and pays the second highest bid (plus an increment). eBay also implements a policy where at any moment during the auction the highest bid is not disclosed, and the second highest bid is shown as the current price. Here too bidders are notified when their placed bid is below the highest (undisclosed) bid. At the conclusion of the auction, as well as in the archival data, the complete sequence of bids and time stamps alongside with the bidder usernames and their feedback scores[1] are given, except for the highest bid that is not disclosed. An example is shown in Figure 3.1. Note that the order in which the bid history page displays the bids is ascending in the bid amounts rather than chronologically. This order makes it appear, at first, that the process of bidding was much more gradual and with higher intensity of bidding than actually occurred. Let us first look at the bid history chronologically by plotting data on a scatterplot that has a temporal x-axis. The plot corresponding to Figure 3.1 is shown in Figure 3.2. We call this a bid history *profile plot*. Looking at the chronologically ordered profile plot reveals the effect of eBay's policy to conceal the highest bid at any point in the auction. We see that after the bid of $175.25 was submitted on January 30, three lower bids followed ($159, $169.55, and $175). The reason for this is that the $175.25 was not displayed during the live auction until it was exceeded. The profile plot also gives information about the intensity of bidding over time. The auction shown in the figure had a very strong start, then a spurt of bids on the second day, and a final spurt at the end of the auction.

To see the *current price* (or *live bid*) that was shown during the live auction, and to study the relationship between the bid amounts and the current price values, we reconstruct the current price values from the bid history by using a function that is based on the principles of the proxy bidding mechanism used by eBay and the

[1] eBay changed its policy at some point such that usernames are not displayed in auctions for high-value items. This change was done to avoid the possibility of fraudulent sellers contacting losing bidders.

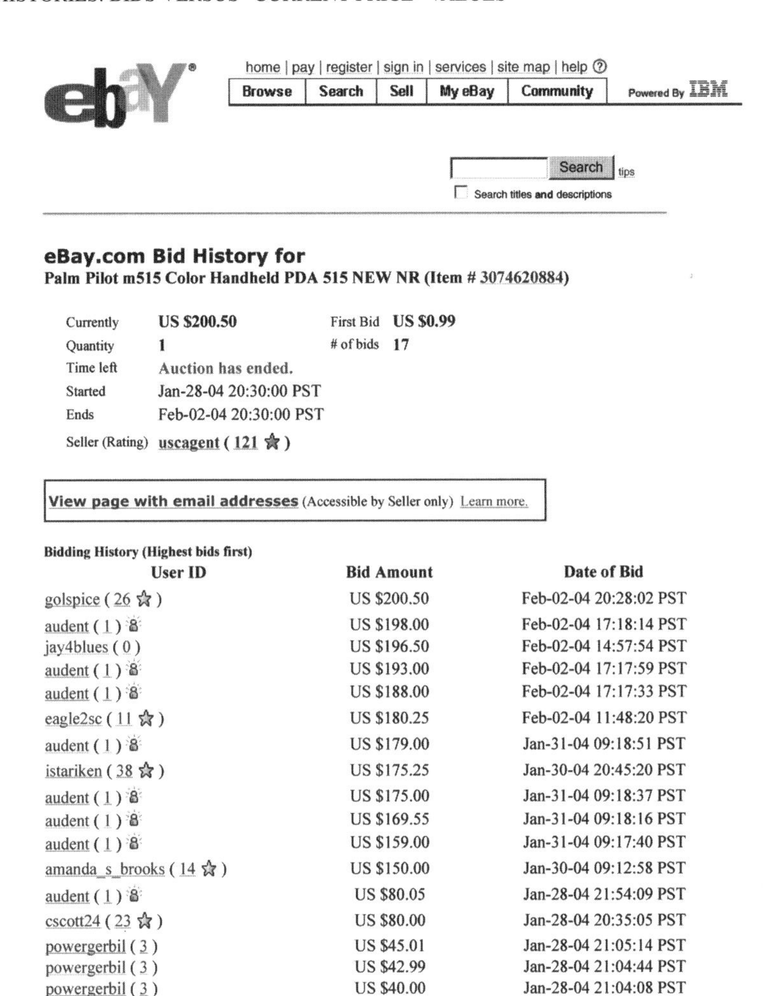

home | pay | register | sign in | services | site map | help

Browse | Search | Sell | My eBay | Community

Powered By IBM

Search tips

Search titles and descriptions

eBay.com Bid History for

Palm Pilot m515 Color Handheld PDA 515 NEW NR (Item # 3074620884)

Currently	**US $200.50**	First Bid	**US $0.99**
Quantity	**1**	# of bids	**17**
Time left	Auction has ended.		
Started	Jan-28-04 20:30:00 PST		
Ends	Feb-02-04 20:30:00 PST		
Seller (Rating)	uscagent (121)		

View page with email addresses (Accessible by Seller only) Learn more.

Bidding History (Highest bids first)

User ID	Bid Amount	Date of Bid
golspice (26)	US $200.50	Feb-02-04 20:28:02 PST
audent (1)	US $198.00	Feb-02-04 17:18:14 PST
jay4blues (0)	US $196.50	Feb-02-04 14:57:54 PST
audent (1)	US $193.00	Feb-02-04 17:17:59 PST
audent (1)	US $188.00	Feb-02-04 17:17:33 PST
eagle2sc (11)	US $180.25	Feb-02-04 11:48:20 PST
audent (1)	US $179.00	Jan-31-04 09:18:51 PST
istariken (38)	US $175.25	Jan-30-04 20:45:20 PST
audent (1)	US $175.00	Jan-31-04 09:18:37 PST
audent (1)	US $169.55	Jan-31-04 09:18:16 PST
audent (1)	US $159.00	Jan-31-04 09:17:40 PST
amanda_s_brooks (14)	US $150.00	Jan-30-04 09:12:58 PST
audent (1)	US $80.05	Jan-28-04 21:54:09 PST
cscott24 (23)	US $80.00	Jan-28-04 20:35:05 PST
powergerbil (3)	US $45.01	Jan-28-04 21:05:14 PST
powergerbil (3)	US $42.99	Jan-28-04 21:04:44 PST
powergerbil (3)	US $40.00	Jan-28-04 21:04:08 PST

FIGURE 3.1 Bid history of a completed eBay auction for a Palm Pilot M515 PDA. See the color version of this figure in Color Plates section.

increment rules that eBay uses (`http://pages.ebay.com/help/buy/bid-increments.html`). Essentially, the current price value at time t is equal to the second highest bid until that time minus the bid increment. The left panel in Figure 3.3 displays both types of values (bids and current prices) and the closing price for the same Palm M515 PDA described in the previous figures. Note that the step function describing the current price values is always below the bid amounts, reflecting eBay's guarantee that the winner will not pay more than an increment above the second highest bid. The plot shows the immediate effect of the \$175.25 bid on the price: it increased the price to an increment above the highest bid until then (\$150), from

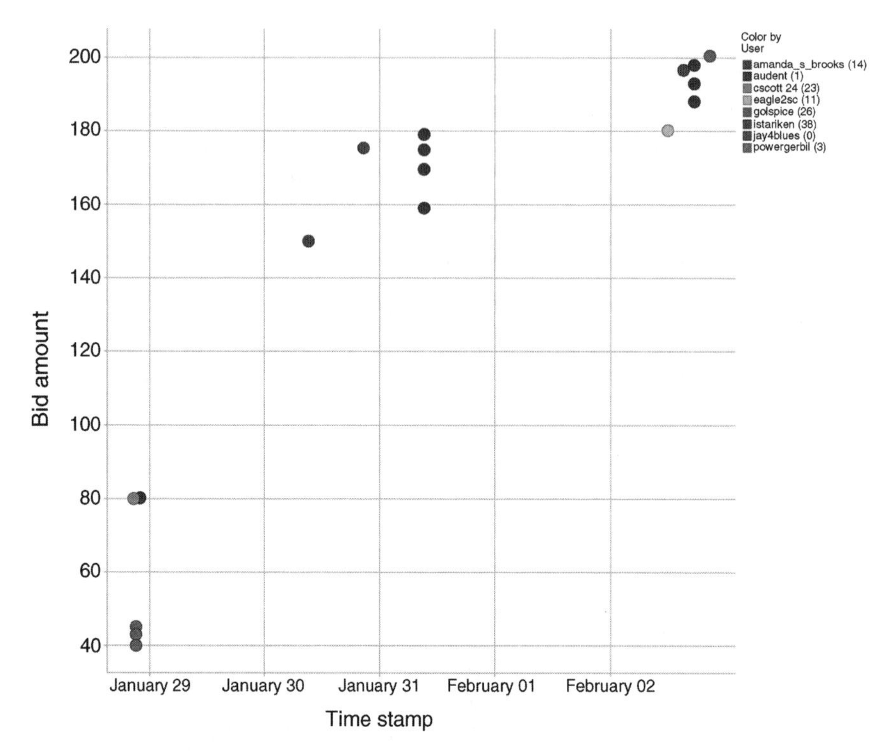

FIGURE 3.2 Profile plot of the bid history shown in Figure 3.1. See the color version of this figure in Color Plates section.

\$81.05 to \$152.5. However, because the bidders participating in the auction saw only the price \$152.5, it explains the arrival of the next three bids of \$159, \$169.55, and \$175, where the first two are lower than the highest bid at that time.

The right panel of Figure 3.3 displays the difference between the current price and the current highest bid. This is useful for studying the ongoing "consumer surplus" in an auction, which is defined (under some conditions) as the difference between the highest bid and the final price. Such a plot shows how fast the current price catches up with the maximum proxy bid. It can be seen, for instance, that the \$150 bid placed on January 30 (following the previous bid of \$80.05) creates a large rectangular area that lasts nearly 12 h until a higher bid (of \$159) is placed. A current limitation of this plot for eBay data is that it cannot display the surplus at the auction end, because currently eBay discloses all bid values except the winner's bid (which is the highest bid).

We note in this respect the study by Bapna et al. (2008a), where a sniping website (used for placing last-moment bids on eBay) owned by the authors was used to obtain full eBay bid histories, *including the highest bid*, for a large sample of auctions. The data were then used for estimating the consumer surplus in online auctions.

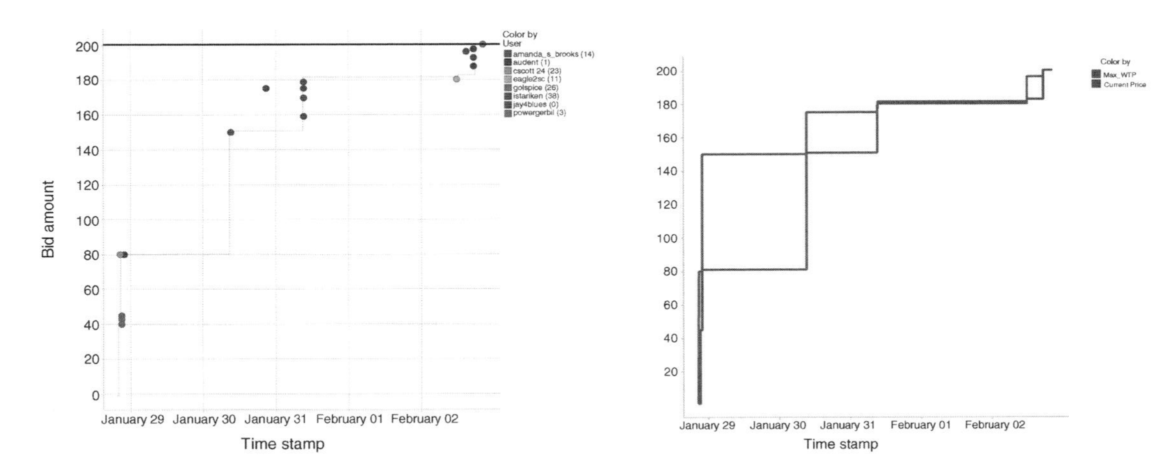

FIGURE 3.3 *Left panel*: profile plot and current price step function (as seen during the live auction). The final price is marked by a horizontal line. *Right panel*: comparing the current price function with the current highest bid. See the color version of this figure in Color Plates section.

3.2 INTEGRATING BID HISTORY DATA WITH CROSS-SECTIONAL AUCTION INFORMATION

In many cases, it is of interest to combine the temporal bid history (or current price function) with other cross-sectional auction information. For example, researchers have been interested in examining bidding strategies. In the following section, we discuss some procedures for exploring the combined information.

3.2.1 Exploring Individual Auctions

Various authors have observed that the number of bidders participating in an eBay auction is usually much smaller than the number of bids placed in that auction (Bajari and Hortacsu, 2003). In other words, a few bidders submit multiple bids within an auction. Bid revision behavior may indicate that the bids (at least all except the highest bid) do not reflect the bidder's truthful willingness-to-pay (WTP) value if indeed such a value exists. To visualize the behavior of different bidders, we must include bidder identifiers in the plot. This can be done via different colors and/or shapes to denote different bidders. Figure 3.3 uses color to differentiate bidders (as can be seen in the legend). In this auction, there were eight bidders, with two placing multiple bids. These two "persistent" bidders placed 9 of the 17 bids. It is interesting to note that one of the persistent bidders placed bids only at the very beginning of the auction, while the other placed series of bids on the third and on the last day of the 5-day auction. Finally, the winning bid came from a single-time bidder. These three types of bidding behaviors have been reported in online auction research and are often classified as *evaluators, participators,* and *opportunists* (Bapna et al., 2003).

Another piece of information that is useful to link to the bidding behavior is bidder feedback (also known as bidder rating). Bajari and Hortacsu (2003) found that experts tend to bid late in the auction relative to nonexperts. Furthermore, Ockenfels and Roth (2002) posited that experienced bidders will tend to place only a single bid during the last minute of the auction. eBay bid histories also include the feedback for users that are typically used as a measure of expertise. If this rating indeed measures expertise, then we would expect to see bids toward the end of the auction coming from bidders with high feedback, and those bids will tend to be single bids. We can use color and/or size on a profile plot to code user rating. In Figure 3.4, we use size for this purpose, with circle size proportional to bidder rating (thereby creating a "bubble chart"). If we disregard multiple bids by the same bidder (where color denotes user), this plot shows when high-rated bidders place bids compared to low-rated bidders. We see that in this auction the two persistent bidders (*powergerbil* and *audent*) have very low ratings, whereas the higher rated (more experienced) bidders tended to place single bids, but throughout the auction.

Here, we showed how bidder-level information can be easily combined with a bid history. Yet bidder IDs and ratings are readily available on the bid history page. A more complicated task is to extract bidder information on other auctions that they participated in. An example is the proportion of a bidder's winnings in all previous auctions that she/he participated in.

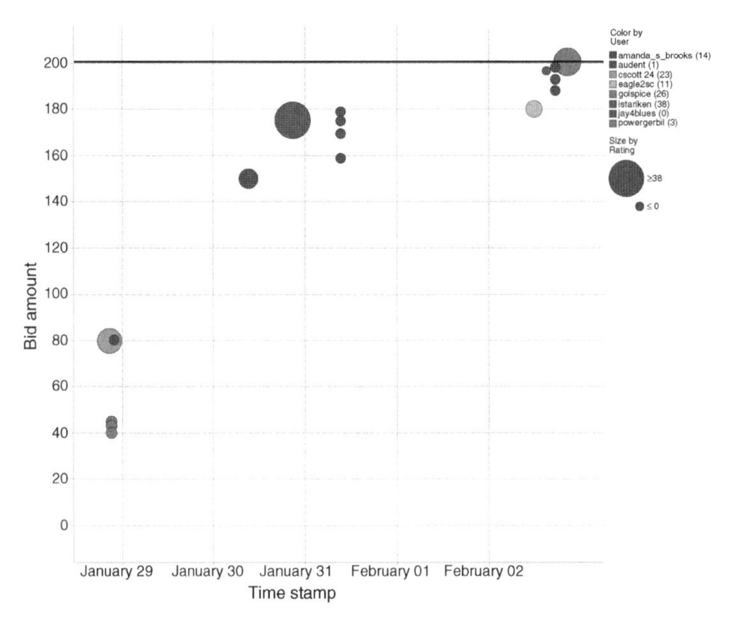

FIGURE 3.4 Profile plot with bidder rating information conveyed as circle size. The final price is marked by a horizontal line. See the color version of this figure in Color Plates section.

In conclusion, the profile plot is easily adaptable to different research questions. With some imagination, many factors of interest can be integrated into it without clutter.

3.2.2 Exploring Multiple Auctions

Examining individual bid histories via profile plots is useful for understanding the bidding mechanism and identifying user behaviors. However, in most cases researchers are interested in examining multiple auctions. Overlaying profile plots of multiple auctions within a single plot conveys some information but loses other, as it is hard to identify which bids belong to which auction. This is illustrated in Figure 3.5, where profile plots of 158 seven-day auctions are overlaid in a single plot. The horizontal lines correspond to the closing prices in each of the auctions. A plot of this type reveals several useful pieces of information about the auctions:

- The intensity of bidding changes over time: There are two dense clusters of bid values at the beginning (days 0 and 1) and especially at the end (day 7), while the middle of the auctions experiences much lower bidding activity.
- The closing prices vary between \$230 and \$280, with \$280 being exceptionally high.
- Many bids were placed above the closing prices of other items. Considering that these were all auctions for the same new Palm M515 PDA, this means that the

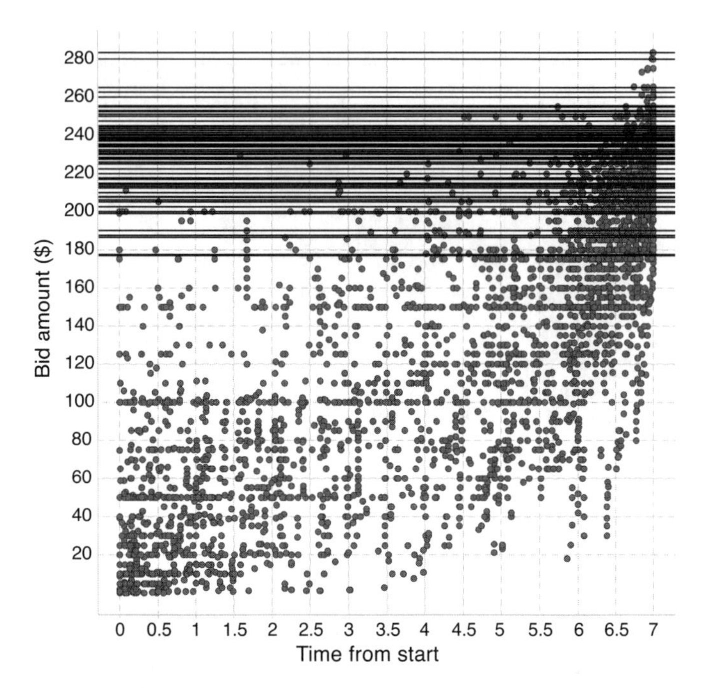

FIGURE 3.5 Profile plot for over 100 seven-day auctions. The final prices are marked by horizontal lines. See the color version of this figure in Color Plates section.

valuation for this item was highly variable: there were many people willing to pay substantially higher prices than others.

The "current auction surplus" plot shown earlier can also be extended to display a set of bid histories by using shading in the area between the bid and current price values to denote the distribution over a set of auctions. For details and examples, see Shmueli and Jank (2005).

When moving to multiple auctions, profile plots encounter several challenges. If the auctions are of different duration, then the profile plot can be used only if the timescales of the different auctions are normalized (e.g., time is presented as a percentage time of the auction duration). From our experience, profile plots are useful for describing a single or several (<30) auctions. Their usefulness is enhanced greatly by plotting auctions that have similar starting prices, that have the same duration, and that take place in a short period of each other. In general, any factor that is known to affect the profile should be used to separate auctions into separate profile plot panels. From a visualization point of view, Aris et al. (2005) explored various representations for sets of unevenly spaced time series of varying duration, using bid histories as an example. Each of their proposed representations has advantages and shortcomings in terms of interaction such as screen resolution, response time for dynamic queries, and learnability. In Section 3.4 and at the beginning of Chapter 3, we describe a price curve representation that solves many of these issues and supports both visualization and modeling.

To learn about the characteristics of auctions for a certain item, bidders and sellers can make use of historical data on closed auctions for similar items. Web tools that are aimed at supporting bidders' efforts, such as Hammertap.com and previously Andale.com (now Vendio.com), supply users with aggregated information on historic auctions on eBay. They provide statistics such as the average selling price and the number of bids for those auctions. In other words, they aggregate bids across time and auctions. From profile plots such as Figure 3.5, it is clear that important information is lost by such aggregation. However, as the number of auctions increases and the number of bids per auction increases, looking at a large number of individual bid profiles (of the bids and/or current prices) might also be overwhelming. The question is how to summarize the entire information on multiple auctions for a certain item without losing valuable information. A solution is, instead of aggregating bid values over the entire auction, to look at bid distributions aggregated over certain time intervals within the auction. The time intervals are chosen according to questions of interest and bidding intensity periods (i.e., smaller intervals during intensive bidding periods). In particular, we use hierarchical aggregation that allows us to look at different temporal breakdowns of bids. Shmueli and Jank (2005) coined the term STAT-zoom for this operation (zooming in or out and summarizing statistically). Figure 3.6 displays the bid distribution at three levels of temporal aggregation: the top panel displays daily aggregation (for each day of 158 seven-day auctions), the middle panel shows hourly

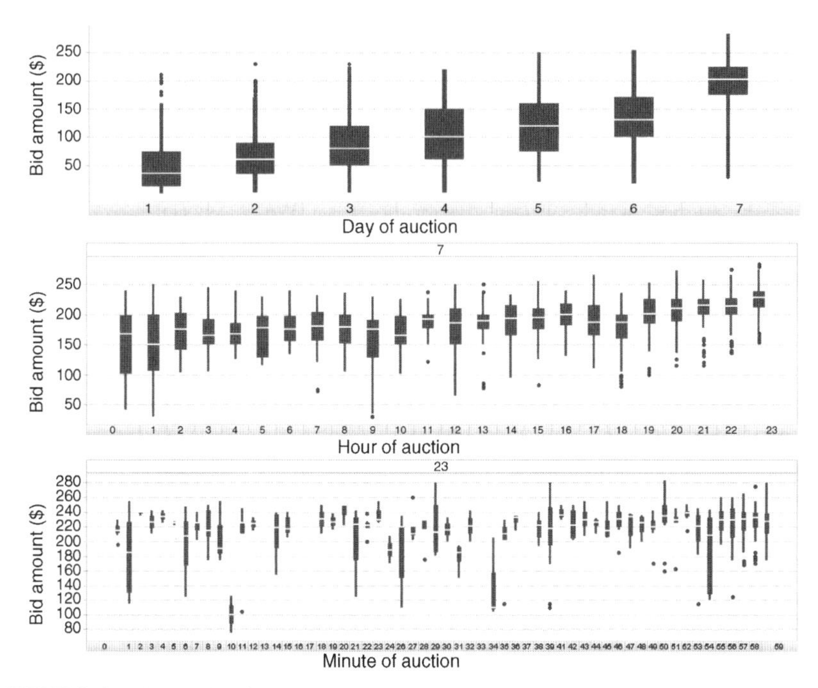

FIGURE 3.6 Bid distributions by varying the temporal hierarchy. Top panel displays daily distributions. Middle panel displays hourly distributions on the last day of the auction. Bottom panel displays minutely distributions for the last hour of the auction. See the color version of this figure in Color Plates section.

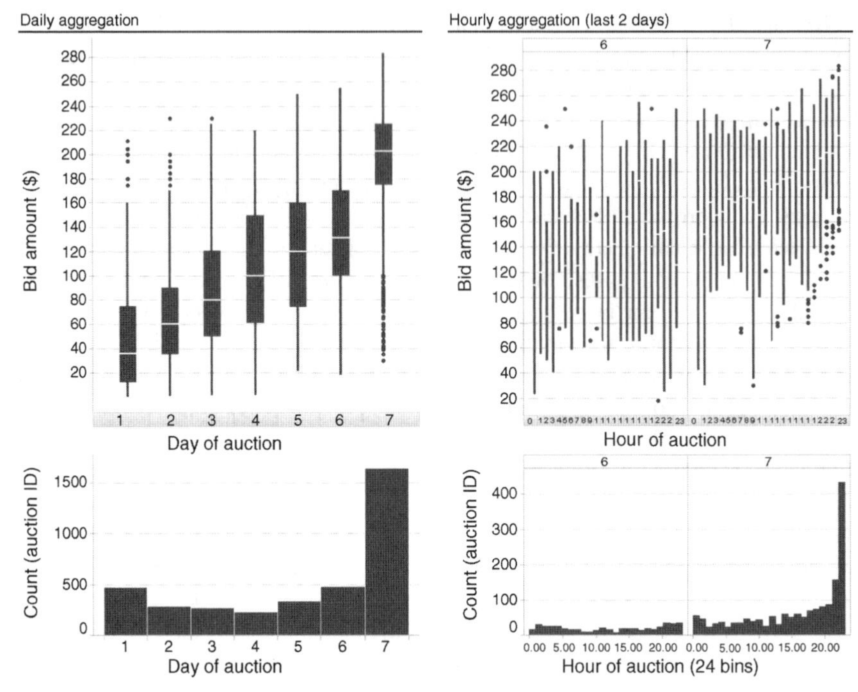

FIGURE 3.7 Bid distribution and intensity by varying the temporal hierarchy. *Left panel*: daily aggregation. *Right panel*: hourly aggregation for last 2 days of auction. See the color version of this figure in Color Plates section.

aggregation for the last day of the auction, and the bottom panel zooms in to the last hour of the auction, aggregating at the minute level. We can also use trellising (i.e., multiple panels) to display multiple days of aggregated hourly data, multiple hours of minutely data, and so on (see an example in the top right panel of Figure 3.7).

To assist the user in determining the bid intensity, which in turn should guide the level of temporal aggregation, we supplement the boxplots with histograms showing the bidding intensity during the aggregated intervals. The boxplots and histogram displays are closely linked, such that the histogram bins match the aggregation level used in the boxplots. The conventional way to display the counts used in side-by-side boxplots is to set each boxplot's width proportional to $\sqrt{n}$, where n is the number of aggregated records in that boxplot (McGill et al., 1978). However, in our context, the conventional method has two disadvantages: First, proportional width is useful more for comparing boxplots (wider ones are based on more bids than narrower ones) than for learning about the actual number of bids, which is of interest here. Second, since the display might include many boxplots, varying-width boxplots would cause more clutter than reveal information.

An example of a combined plot for the 158 seven-day Palm M515 PDA auctions is given in Figure 3.7. The left panel shows daily aggregation and the right panel shows hourly aggregation for the last 2 days of the auction. We can see that the boxplots of bids during days 2–5 are based on approximately the same amounts of bids, whereas the days 1 and 6 have slightly more bids, and day 7 is based on almost four times the amount of bids. Combining the boxplot and intensity information shows that even after controlling for the amount of bids placed on that day, the amount of outliers on day 7 is still extreme and might be indicative of a mixture of two distributions.

3.3 VISUALIZING CONCURRENT AUCTIONS

Concurrency of events is omnipresent in today's commerce, both within the electronic world and even more so if we consider the online and offline domains together. Consumers can gather price information from a variety of sources: store sales circulars, online retailers' websites (e.g., Amazon.com), comparison shopping websites (e.g., Bizrate.comindexwebsite!Bizrate.com), newspaper and online personal ads for used merchandize (e.g., craigslist.com), online auction sites, and so on. The availability of items from multichannel sources gives consumers the power to compare prices and other related information (such as quality, shipping time and cost, trustworthiness of the seller, travel time, and physical inspection). This also means that sources compete with one another and therefore, more than likely, also influence each other. In this section, we focus on concurrency within the online auction channel. Online auctions are different from traditional brick-and-mortar auctions in that they occur simultaneously or within close temporal proximity. Online auctions tend to be longer, ranging over several days rather than minutes. They are also not limited by the geographic barriers of traditional auctions. Sellers and bidders can be located in different parts of the country, or even the world, and still conduct business. The wide and growing popularity of online auctions makes the problem of concurrency an interesting and important topic for research.

Auction theory is mainly concerned with a single auction, and there has not been much theoretical research on concurrent auctions, even in the offline context. One study that does consider simultaneous auctions is by Guerre et al. (2000), who examined the underlying distribution of a bidder's private value in sealed-bid first-price offline auctions. Prior empirical research concerned with the interplay of auctions has focused mainly on sequential offline auctions where auctions occur one after the other but not simultaneously. Sequential auction research comes closer to addressing concurrency and is concerned with price and quality of selection in sequences of auctions. For instance, Allen and Swisher (2000) found that auctions later in an auction sequence obtain higher prices than those in the beginning, probably due to the limited supply at the end of a sequential auction. Deltas (1999) examined sequential ordering across numerous auctions for cattle, where it is common that higher value lots are sold at the beginning of an auction. He found that prices decline throughout an auction, but the rate of decline depended on the size of the lot, such that prices decline faster in small-lot auctions.

While the insight gleaned from these analyses is useful in the traditional auction setting, it is scarcely applicable to the online context. Online auctions present new and challenging questions because of the prevalence of simultaneous events. A great majority of the literature that analyzes online auction data assumes independence across auctions. This assumption is typically made for reasons of simplicity and convenience, while in reality auctions for the same item, competing items, or even related (substitute) items will potentially influence each other, especially if they take place within a close time frame. On eBay, an identical product is often sold in numerous simultaneous auctions. Although each auction contains a "replicate" of the product, the resulting sales, prices, and even the number and level of bids during the ongoing auctions are most likely not independent of each other.

Empirical research has mostly been based on auctions for the same item that have closed during a certain time period (typically within a few months). Examples are coins (Lucking-Reiley et al., 2007; Bajari and Hortacsu, 2003; Kauffman and Wood, 2005) and personal digital assistants (PDAs) (Shmueli and Jank, 2005; Ghani and Simmons, 2004; Jank and Shmueli, 2009). In such situations, it is likely that there is dependence between the auctions because buyers have the option to select which of the competing items to bid on and sellers have the option to decide when to post their item for sale using information on similar previously sold items. Hence, even within the universe of online auctions concurrency is prevalent, and it is challenging to evaluate its magnitude and effects.

There have been only a few attempts at quantifying or addressing concurrency in online auctions. Snir (2006) showed that the expected selling price in S sequential auctions is equivalent to the Sth order statistic. Zeithammer (2006) examined sequential auctions on eBay and found that bidders deflated their bids based on the future expected surplus. Anwar et al. (2006) found that bidders on eBay typically place bids on multiple competitive auctions. Finally, Kauffman and Wood (2005) examined a massive set of concurrent auctions and used the dependence structure to detect collusion (where a seller bids up the price in his/her auction).

In this section and in Section 3.4, we focus on effective visualizations of concurrency (in Section 4.3 we present formal models for concurrency). We are interested in the effect of concurrency not only on the final price of an auction but also on the relationship between the current bid levels and the high bids in simultaneous ongoing auctions. Although our focus is on online auctions, the methods can be used in a wide variety of applications for studying the effect of concurrency on a final result or the evolution that leads to an event.

3.3.1 Auction Calendar

Our first plot is an *auction calendar*. This visualization displays each auction as a horizontal line that extends between its opening and closing times such that the length of the line represents auction duration. The auction calendar can easily display auctions of various durations (e.g., eBay's 3, 5, 7, and 10 days auctions). Figure 3.8 displays an auction calendar for 477 auctions for Palm Pilot M515 PDAs (including the 158 seven-day auctions plotted earlier).

To integrate cross-sectional information into the auction calendar, we can use color for categorical information (e.g., new versus used) and the y-axis for numerical information (final price in Figure 3.8). In Figure 3.8, we added a black line to represent the daily median closing price (computed from the closing prices of all auctions that

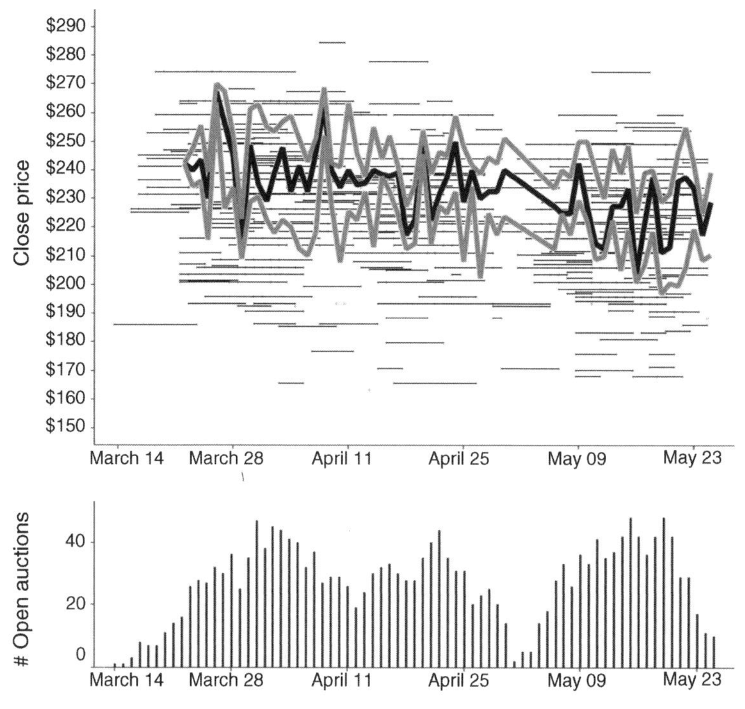

FIGURE 3.8 *Top panel*: Auction calendar for a sample of eBay auctions. Each line corresponds to an auction, extending from its start to end. Vertical axis denotes final price. *Bottom panel*: daily volume of open auctions.

closed on a certain date), and gray lines denote the daily upper and lower quartiles. We also added a lower panel that depicts the daily volume of open/active auctions. The plot reveals the following information:

- The period over which the data were collected extends from mid-March to May of 2003. Around May 1, there is a period of several days with a substantial decrease in auction activity (less than 10 active auctions). Further investigation revealed that this decrease is due to the data collector's spring vacation (hence, unrelated to eBay).
- Other noticeable patterns are a decrease in the median closing price of Palm PDAs from March to May and fluctuations in daily closing prices.

In general, peaks in auction volume and in median closing prices can be related to seasonal factors such as holidays or weekday/weekend. Note that the auction calendar can give a direct sense of temporal effects such as weekday/weekend or seasonal effects without the need to aggregate the data. Such effects are often of interest. For example, Lucking-Reiley et al. (2007) used a bar chart to explore the volume of auction closings by day of week. They found that more auctions tend to close on weekends. In the auction calendar display, color can be used to explore such questions.

In sum, the auction calendar can highlight interesting patterns in the data, reveal information about the data collection process, and serve as a basis for further exploration.

3.4 EXPLORING PRICE EVOLUTION AND PRICE DYNAMICS

For studying concurrency of prices in competing auctions, as well as for modeling and forecasting purposes (see Chapters 4 and 5), we consider the bid history as the observable recording of an underlying continuous price evolution process that takes place during each auction. The first step is therefore to estimate this continuous evolution process from the discrete bid data and represent it in a continuous fashion.

Bids in online auctions are placed at bidder chosen time points. In hard-close auctions, there is typically some bidding activity at the auction start, followed by a period of very little activity, and cumulating in a surge of bidding at the very end of the auction (see Section 4.4). The last-moment bidding is often referred to as "bid sniping" (Bajari and Hortacsu, 2003; Roth and Ockenfels, 2002). The resulting bid histories are therefore time series that are unevenly spaced with very sparse periods and very dense periods. Although we could simply "connect the dots" to obtain the price of the auction at any given time (i.e., the "current price" step function), this would overfit the data, thereby not providing a good representation of the underlying price process.

From a visualization point of view, a smooth curve better displays the temporal nature of a continuous process compared to a series of discrete unevenly spaced bids. Moreover, plots such as profile plots do not scale to multiple auctions in terms of maintaining the auction-specific temporal information. Finally, conceptually a continuous curve is a more appealing representation of a continuous underlying process. Even during the times when no bids are recorded, users are observing the auction (and other auctions) and bidders are contemplating their bidding strategy, thereby affecting the price process.

An alternative to the step function is to represent the price process as a continuous smooth curve. This type of curve representation is prevalent in functional data analysis, an approach where the observations of interest can be any type of continuous representation (Ramsay and Silverman, 2005). In Section 3.4.1, we briefly describe how bid histories are converted into price curves, although we defer the more detailed technicalities to Section 4.1. We focus in this chapter on how price curves can be used for data exploration.

3.4.1 Creating Price Curves from Bid Histories

Obtaining a continuous curve from discrete data is typically achieved through smoothing techniques. After converting the bid amounts from the bid history into current price values, we have a series of time-stamped price values and perhaps a step function. We then fit a smooth curve to the resulting values (or to a grid of points taken from the step function) by using nonparametric smoothers, such as penalized splines or smoothing

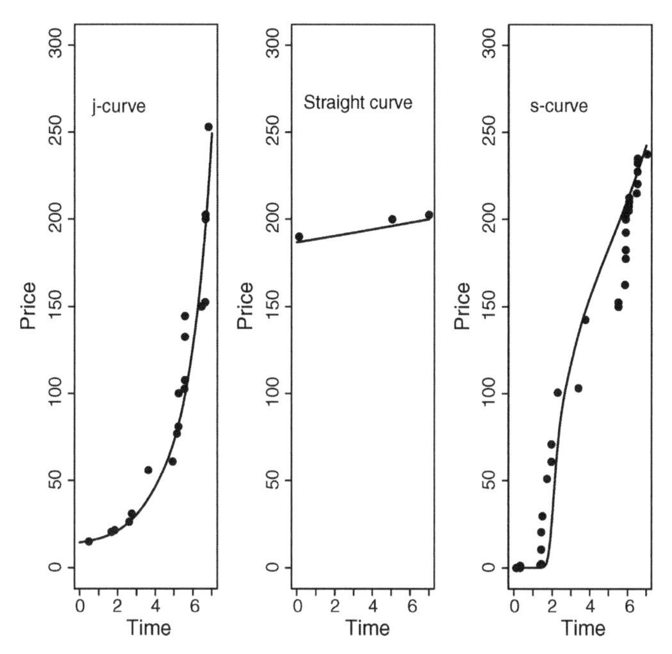

FIGURE 3.9 "Current price" values and fitted price curves for three 7-day auctions. Curves were fitted using monotone splines.

splines, or parametric models (see Chapter 4 for monotone parametric families of curves that are useful in the auction context). Figure 3.9 shows an example of fitting curves to three samples of 7-day auctions. The method used was smoothing splines (for further details see Chapter 4, and for this particular example, see Hyde et al. (2006)).

The advantage of the curve representation is that it captures the complete price evolution in a more compact and easier to visualize way than the raw bid data or step function. Furthermore, an appealing feature of smooth price curves is that we can compute their derivatives to estimate *price dynamics*. The first derivative represents *price velocity* (how fast the price moves during the auction). The second derivative represents *price acceleration* (the rate of velocity change during the auction). Further derivatives can also be computed, but they are harder to interpret in the auction context. An plot showing the price curve, price velocity curve, and price acceleration curve for two single auctions is shown in Figure 3.10. Figure 3.11 shows the three sets of curves (price, velocity, acceleration) for a sample of 187 seven-day Palm M515 eBay auctions. We can see the variability in dynamics even when all auctions were for an identical item, during a short period, and with a known market price.

In our studies of online auction data, we often found that price dynamics are an important manifestation of the concurrency effect and competition, and that price dynamics heterogeneity is often informative. Hence, we opt to explore them as well.

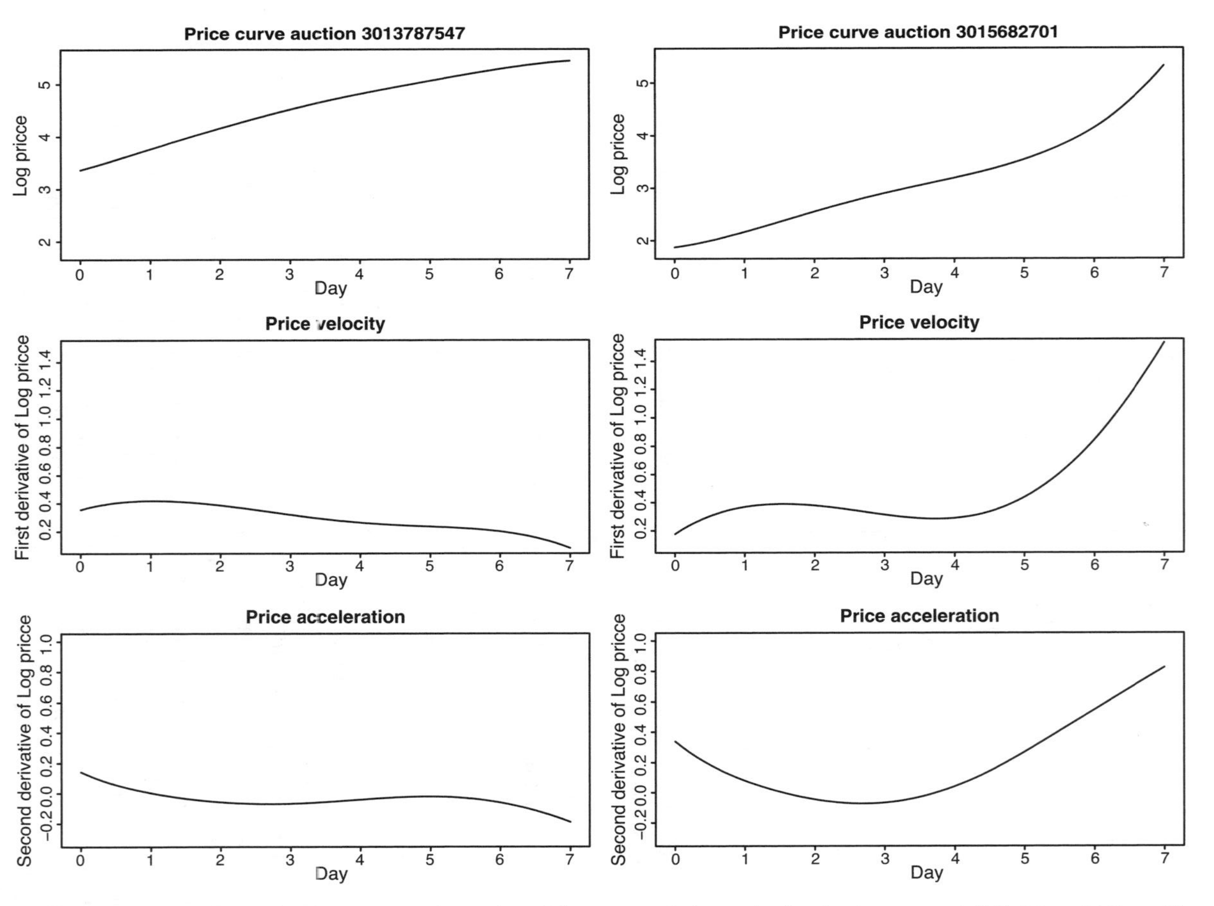

FIGURE 3.10 Price curves and dynamics for two sample auctions. The top panel shows the fitted price curves ($f(t)$). The middle and bottom panels show the price velocity and price acceleration curves, computed as the first and second derivatives ($f'(t)$ and $f''(t)$), respectively.

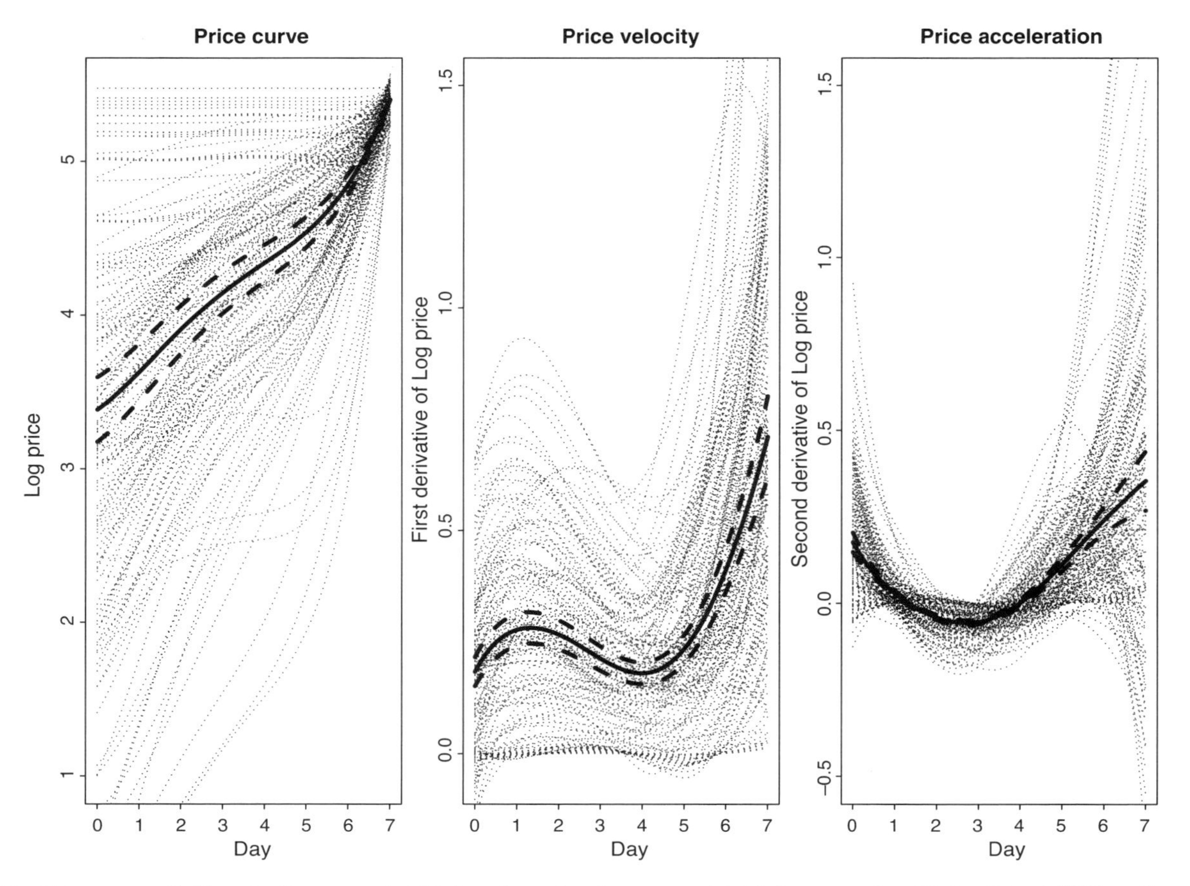

FIGURE 3.11 Price curves (left), velocity curves (middle), and acceleration curves (right) for a sample of 7-day Palm M515 auctions. The heavy solid and dashed curves are the pointwise means and ± 2 standard error curves.

3.4.2 Phase-Plane Plots

A useful visualization for exploring the interplay of dynamics is the phase-plane plot (PPP). In a phase-plane plot, dynamics are plotted one against another. For instance, Figure 3.12 shows a plot of the mean price velocity versus the mean price acceleration. The numbers on the curve indicate the day of the auction. We can see that at the start (day 0) high velocity is accompanied by low, negative acceleration (or deceleration). Acceleration precedes velocity, such that current deceleration translates into lower future velocity and consequently velocity decreases to below 0.5 on day 1. This trend continues until turning positive (between day 4 and day 5), which causes the velocity to pick up toward the auction end. Phase-plane plots are useful for diagnosing whether the interplay of dynamics suggests a system that could be modeled by a suitable differential equation (see Section 4.2).

Another part of data exploration is investigating the distribution of individual variables. Since most parametric models require the response to follow a certain distribution (typically the normal distribution), this step is important for selecting the right model and for assuring the appropriateness of the selected model. One standard tool for investigating the distribution of a numerical variable is the histogram. However, generalizing the idea of a histogram to the functional context is challenging since the input variable is a continuous function. One solution is to plot the distribution of the functional object only at a few select snapshots in time. This can be done by discretizing the object and plotting pointwise histograms (or similar plots such as probability plots) at each time point. Figure 3.13 shows snapshots of the distribution

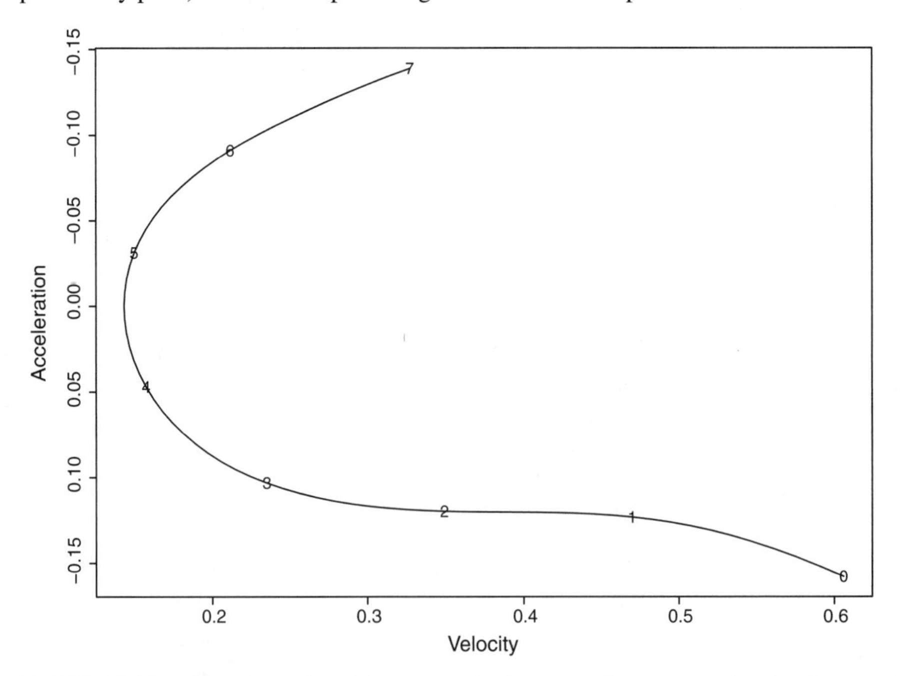

FIGURE 3.12 Phase-plane plot of the mean velocity versus the mean acceleration for a set of auctions. The numbers on the curve indicate the day of the auction.

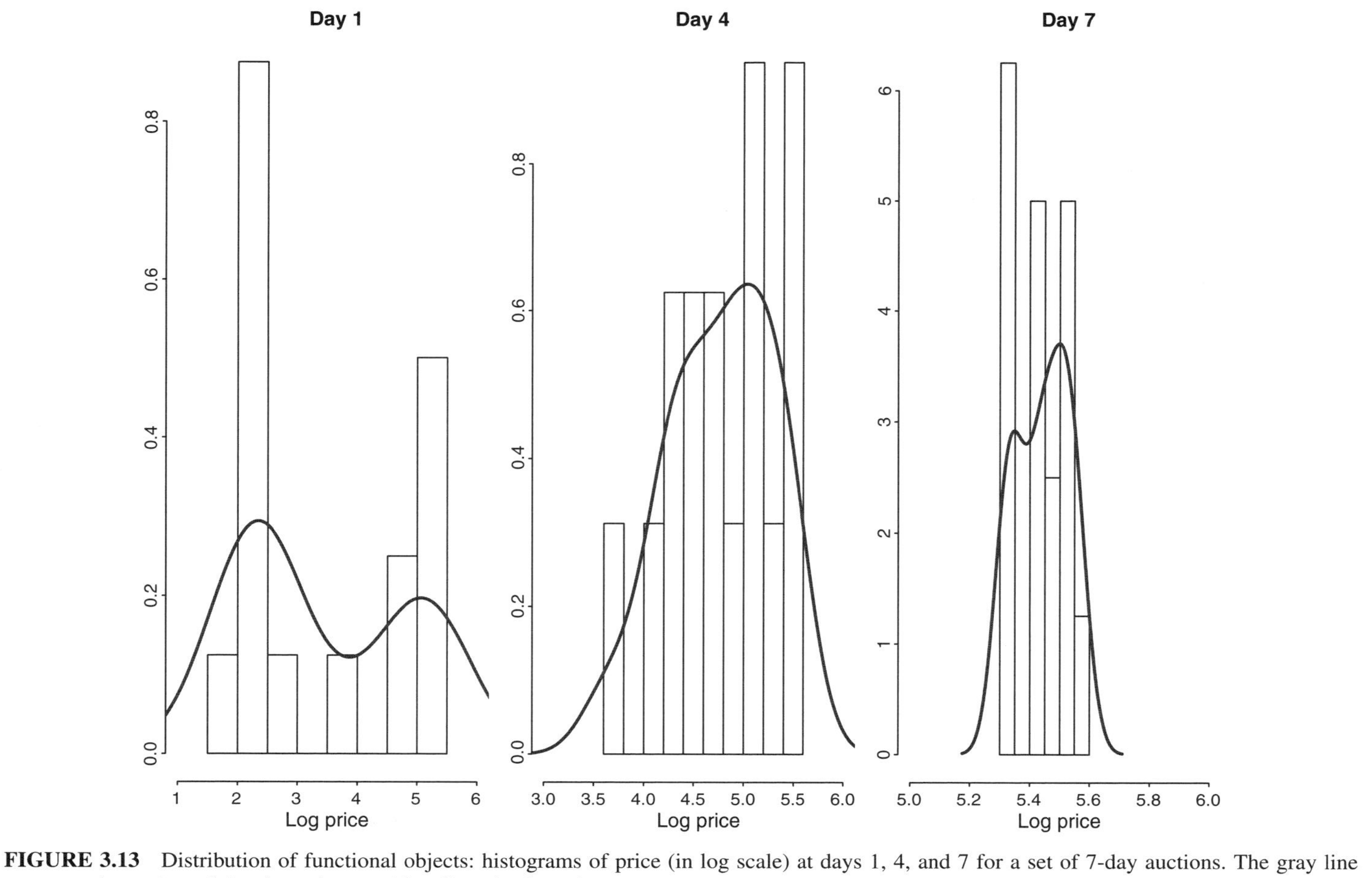

FIGURE 3.13 Distribution of functional objects: histograms of price (in log scale) at days 1, 4, and 7 for a set of 7-day auctions. The gray line corresponds to a kernel density estimate with a Gaussian kernel.

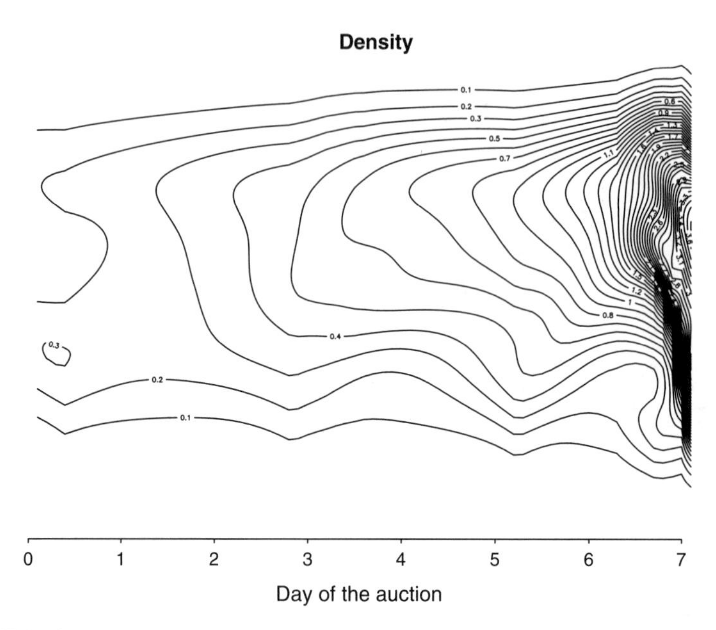

FIGURE 3.14 Contour plot of the price curves density over the 7-day auction. The contour plot is obtained by calculating kernel density estimates as in Figure 3.13 over a fine grid and then interpolating over the 7-day period.

for a set of 7-day eBay price curves at days 1, 4, and 7. These snapshots allow conclusions about the distribution of the entire curve. Note that Figure 3.13 also shows kernel density estimates of the distribution and thus allows conclusions about the evolution of the functional density across days 1, 4, and 7. One can generalize this idea to obtain the density continuously over the entire curve (rather than only at discrete time points). Specifically, Figure 3.14 shows the density estimates evaluated over a fine grid and subsequently interpolated. We can see that at the beginning of the auction the distribution is very flat and starts to peak toward the auction end.

3.4.3 Visualizing Relationships Between Price Curves

In exploratory analysis, after examining individual variables the next typical step is to study relationships between variables. For two numerical variables, this is often accomplished via scatterplots. One way of adapting the ordinary scatterplot to the functional setting is to plot a sequence of pointwise scatterplots. Figure 3.15 shows scatterplots at days 1, 4, and 7 for auction price versus the opening bid (in log scale) for a sample of auctions. We can see that the relationship between the two variables changes over the course of time. While there exists a strong positive effect at the beginning of the auction (left panel), the magnitude of the effect decreases at day 4 (middle panel), and there is barely any effect at all, possibly even a slight negative effect, at the auction end (right panel). This suggests that the relationship between

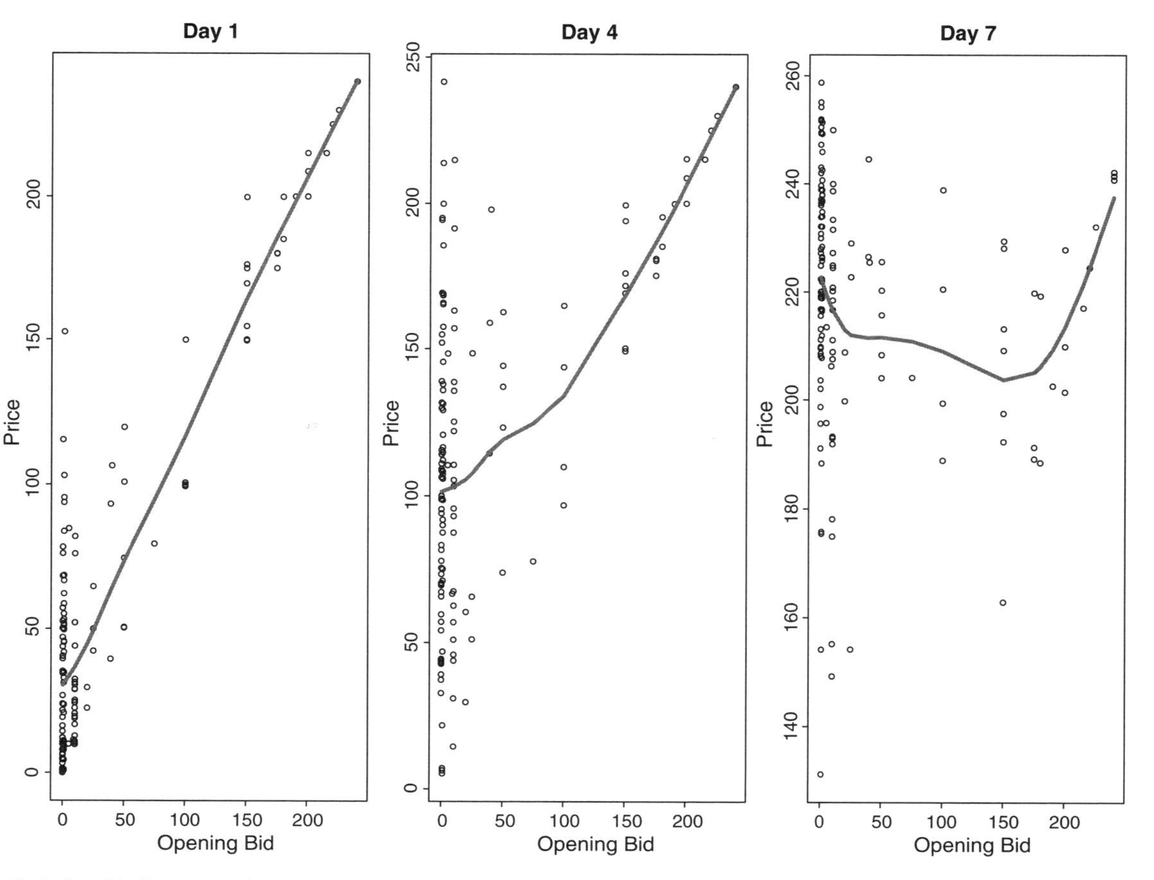

FIGURE 3.15 Relationship between price curves: scatterplots of (log) price versus (log) opening bid at days 1, 4, and 7 of a sample of eBay online auctions. Solid gray line indicates cubic smoothing spline with three degrees of freedom.

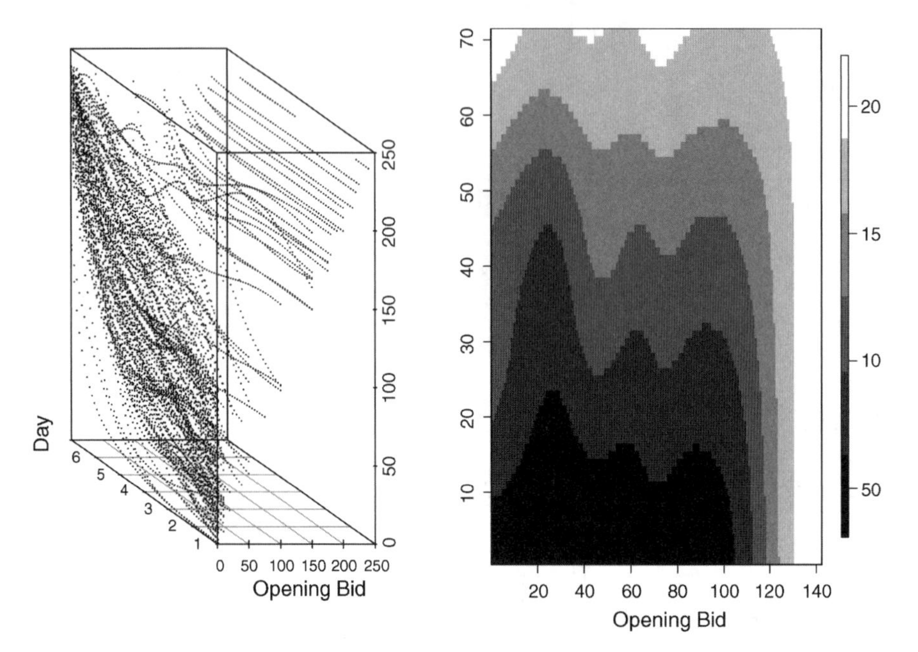

FIGURE 3.16 Relationship between price curves. *Left*: 3D scatterplot of opening bid (x), day of the auction (y), and price (z). *Right*: smoothed price surface, using a Nadaraya–Watson smoother (x = opening bid, y = day of auction.)

the opening bid and the auction price may be modeled well using a time-varying coefficient model. Of course, one aspect that remains undiscovered in this pointwise approach is a possible three-way interaction between opening bid, price, and time. Such an interaction could be detected using a three-dimensional (3D) scatterplot. However, as the left panel in Figure 3.16 illustrates, 3D plots have the disadvantage in that they are often cluttered and not easy to read. We can improve upon the interpretability by smoothing. The right panel in Figure 3.16 shows a smoother image of the price surface using a smoother. Now, the three way relationship between opening bid, price, and time is easier to see.

3.4.4 Rug Plots

Rug plots are a useful display for visualizing a set of auctions by describing their entire price evolution for the purpose of exploring similarities and patterns in price evolutions of concurrent (or partially concurrent) auctions. They are an extension of auction calendars, where instead of representing an auction by a horizontal line it is represented by its price curve or a derivative curve.

Figure 3.17 displays the price curves (top) and price velocity curves (bottom) for a set of eBay auctions. The right panels zoom in to a shorter 1-month period for an additional view. Since the end of an auction is of special interest, the endpoint of each curve is emphasized by a black dot. Because of the rug-like appearance of these

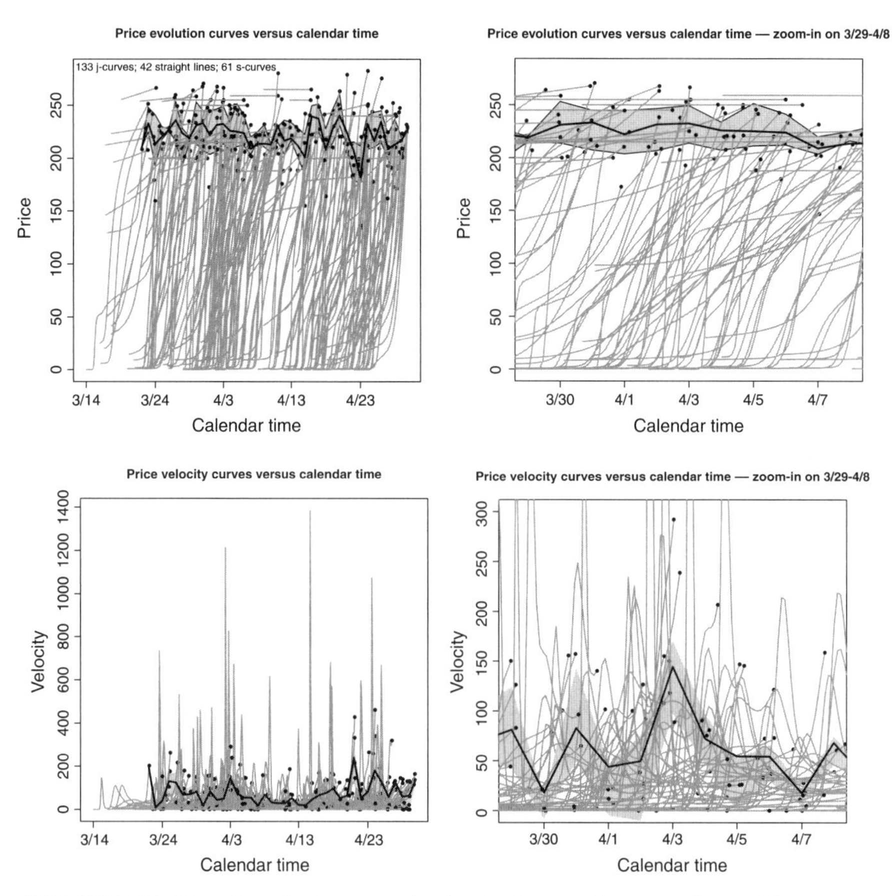

FIGURE 3.17 Rug plots of price curves (top) and price velocity curves (bottom) for a sample of eBay PDA auctions. The right panels display a zoomed-in view during a short period.

displays, they were named *rug plots* (Hyde et al., 2006). Note that price velocity curves are always nonnegative if the price curves are monotone. Zero velocity occurs when the price does not increase, representing a period of bidding inactivity. A small positive velocity reflects a slow price increase at that instant. Similarly, high velocity corresponds to rapid price increases.

To explore the effect of concurrency, we assume that concurrency can affect the shape of a price curve or the shape of the dynamic curve. This is because different curve shapes capture different bidding patterns (e.g., gradual versus bursty bidding or early versus late bidding). Hence, the rug plot can be examined for certain prominent curve types and patterns of temporal proximity between types of curves. Hyde et al. (2006) proposed a taxonomy of three price curve shapes:

"j-shaped" curves. represents auctions with gradual price increases until mid-auction and a price jump toward the auction end (perhaps indicative of sniping).

The corresponding price velocity curve slowly increases and culminates in a peak (left panel of Figure 3.9).

Straight lines. represent auctions with little to no bidding in the middle of the auction. The angle of the line depends on the ratio of the opening and closing prices, and the corresponding velocity curve is constant. When looking at transacted auctions, a straight line price curve represents auctions with a high opening bid (set by the seller), and therefore, very little bidding (middle panel of Figure 3.9).

s-shaped curves. represent auctions where the price increases slowly, then jumps up during mid-auction, and finally slowly increases toward the close. The corresponding velocity curve spikes at the start of the price jump, and then decreases when the price slows down again, resulting in a single hump shape. This can be indicative of jump bidding that occurs during mid-auction, where a single bidder raises the price drastically (right panel of Figure 3.9).

Counts for each of these three curve types within the eBay auction rug plot appear in the top right panel in Figure 3.17.

When searching for auctions for a particular item on an auction website, users are presented with information on all open auctions and sometimes also on historical auctions for that item. A typical user would examine each of the open auctions (their current price, number of bids, and closing time) and perhaps some closed auctions to learn about closing prices. This information is likely to influence how this bidder will bid. In addition, sellers can also use this information to schedule their auction, to set their opening bid, and so on. We therefore expect to see temporal patterns in bidding (and posting) behavior in concurrent auctions. These patterns can manifest themselves as similarities in bidding behavior; for example, when bidders mimic a sniping behavior. Or they can show dissimilarities if, for instance, low-ending auctions lead new sellers to change strategy. An example can be seen in the rug plots (Figure 3.17) that display clustering of curve shapes in certain periods and the lack of such clustering in other periods. One global observation is that the straight line price curves appear to be scattered throughout the data period. Many of these auctions have a high opening bid (sometimes higher than the interquartile range), which is not easily explainable.

Although it is possible to investigate curve types and their temporal patterns in such rug plots, it can become quite tedious when curves are fit nonparametrically, as in this example. Ascertaining types of curves requires visual inspection of each curve, which could be a daunting task for even a moderate data set. A solution that leads to improved rug plot presentation and interpretation was proposed by Hyde et al. (2008). Instead of using nonparametric smoothing methods for deriving price curves, a set of four two-parameter parametric distributions are used, resulting in monotone curves and other desirable properties (see Section 4.1). The following four types of curve taxonomy are also more interpretable in terms of growth models:

Exponential growth. represents auctions where the rate of price increase is proportional to the price level. Similar to "j-shaped" curves, these auctions see gradual price increases until mid-to-late auction and a price jump toward the end.

Logarithmic growth. reflects auctions where early bidding increases the price early in the auction, but because of the existence of a market value (e.g., the price on Amazon.com), price flattens out for the remainder of the auction. This type of price behavior tends to be rare, as most bidders do not wish to reveal their bids early in the auction, but it is observable with inexperienced bidders.

Logistic growth. results in "s-shaped" price curves and reflects the presence of a market value constraint (like population growth).

Reflected (inverse) logistic growth. represents auctions with some early bidding that results in a price increase, followed by little to no bidding in the middle of the auction, and then another price increase as the auction progresses toward its close.

The middle panel in Figure 3.18 displays a rug plot for the same data as in Figure 3.17, and this time based on parametric price curves.

Finally, as in auction calendars, additional auction (or other) information can be integrated into rug plots via color, shading, or multiple panels (trellising). Examples of variables of interest are seller rating, auction duration, and number of bidders. For examples of such plots, see Hyde et al. (2008).

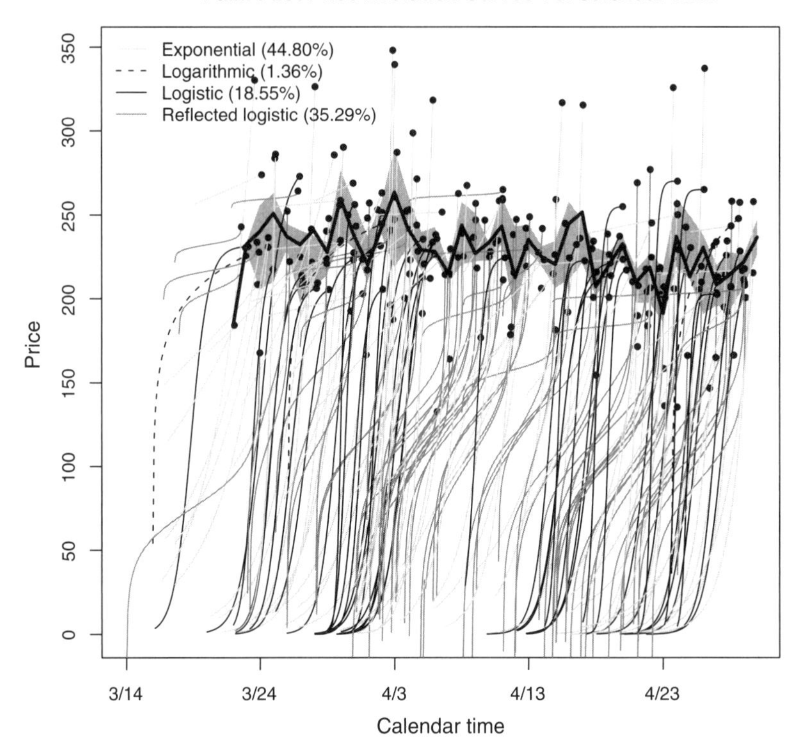

FIGURE 3.18 Rug plots of price curves corresponding to Figure 3.17 (top left), with parametrically fitted price curves. Dots mark the final price of each auction.

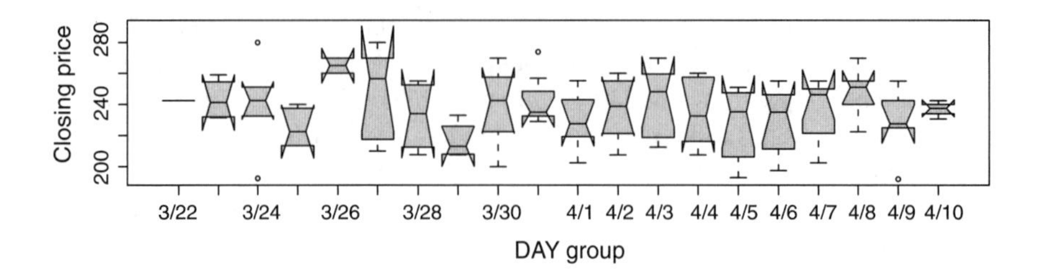

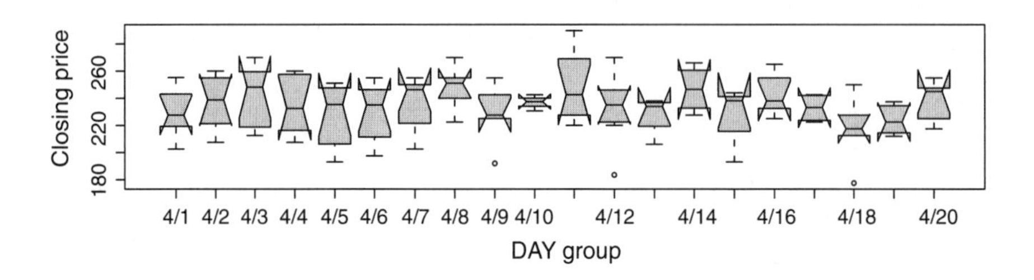

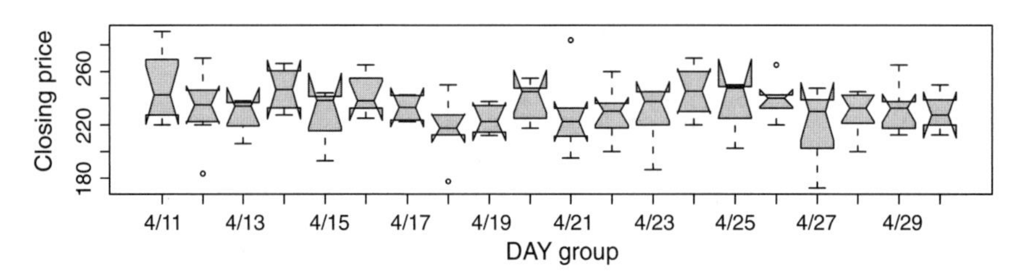

FIGURE 3.19 Notched daily aggregated boxplots of final prices in eBay Palm Pilot auctions. Overlapping notches suggest insignificant differences between adjacent medians.

3.4.5 Time Grouped Boxplots and Moving Statistics

The use of hierarchical boxplots with varying levels of aggregation can be extended for exploring concurrency. In particular, we can plot the final price distribution of groups of concurrent (or partially concurrent) auctions and search for patterns. Hyde et al. (2006) used notched boxplots that are useful for comparing adjacent medians. Figure 3.19 shows such a plot for the Palm data over a few weeks. We can see that although the median daily closing prices are variable, the differences are not statistically significant in most instances.

Another way of visualizing time trends in prices (or any other measure of interest) is by examining moving statistics plots, for example, plotting the moving average or median over time. A moving statistic plot helps in understanding time trends in the data, and unlike side-by-side boxplots, it presents a smoother transition over time. Although moving average plots are very common in many applications, in the auction context there is one caveat: because auction dates (like bids) do not create equal time intervals, moving statistics do not take into account the temporal distance between auctions. This means that moving statistics do not differentiate between auctions that are close in time and those that are far apart. A possible solution is to use weighted statistics where weights reflect temporal distances.

3.5 COMBINING PRICE CURVES WITH AUCTION INFORMATION VIA INTERACTIVE VISUALIZATION

As we discussed earlier in the context of profile plots, combining the temporal information from bid histories with cross-sectional auction information is not straightforward. Moreover, we have shown the limitations of profile plots for exploring more than a small set of auctions. In this section, we describe an interactive visualization approach that supports the exploration of the combined information. In collaboration with colleagues from the Human-Computer Interaction Laboratory of the University of Maryland, we have also developed a visualization tool for exploring online auctions by extending an interactive time series visualization tool (TimeSearcher, `http://www.cs.umd.edu/hcil/timesearcher`) to include attribute data browsing with tabular views and filtering by attribute values and ranges, both tightly coupled to the time series visualization. The tool, sample data, and demo video are available for download at `http://www.cs.umd.edu/hcil/timesearcher/#ts2`.

Figure 3.20 shows the main screen of the visualization tool with an example data set of 34 eBay auctions for magazines. The temporal data are displayed in the left panel, with three series for each auction: "price" (top), "velocity" (middle), and "acceleration" (bottom) that correspond to the price curves and their first and second derivatives. At the bottom of the screen, an overview of the entire time period covered by the auctions is provided to allow users to specify time periods of interest to be displayed in more detail on the left panel. On the right, the attribute panel shows a table of auction attributes. Each row corresponds to an auction, and each column to an attribute, starting with the auction ID number. In this data set, there are 21 attributes. Scrolling provides access to attributes that do not fit into the available space. Users can choose how much screen space is allocated for the different panels by dragging the separators between the panels, enlarging some panels, and reducing others. All three panels are tightly coupled so that an action in one of the panels is immediately reflected in the other panels. Attributes are matched with time series using the auction ID number as a link.

The interactive visualization operations can be divided into time series (bid history) operations and attribute (auction information) operations. The time series operations include zooming temporally; displaying one or more of the panels of price curves, velocity curves, and acceleration curves; filtering curves by dragging TimeBox widgets; and searching for patterns in curves. The attribute operations including sorting auctions by any attribute, highlighting groups of auctions, and obtaining summary statistics of attributes.

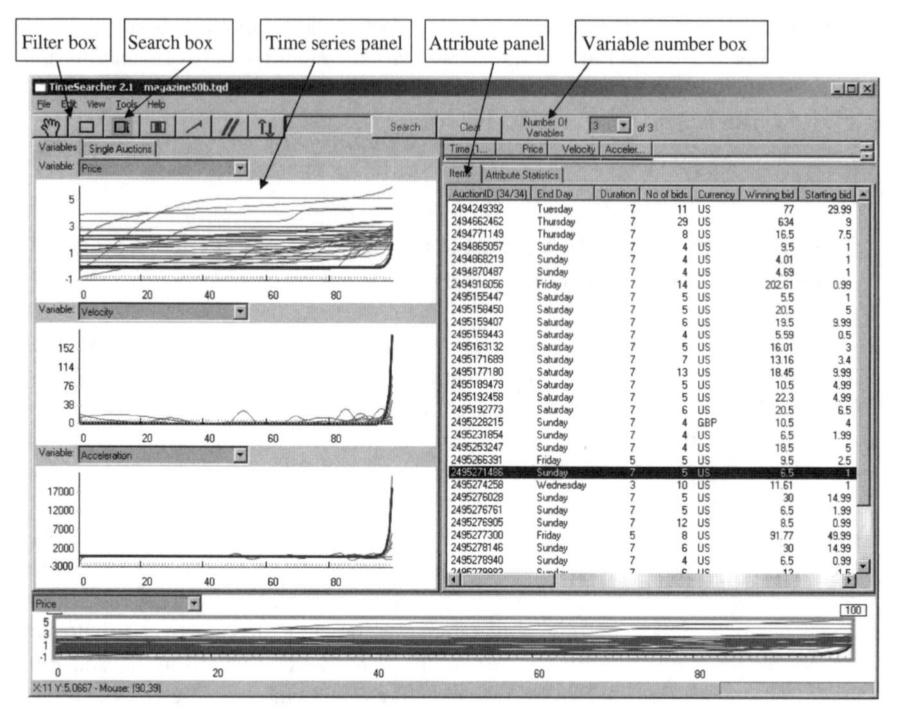

FIGURE 3.20 Main screen of the visualization tool loaded with a sample data set of magazine auctions.

A semistructured process for visual exploration of auction data using the interactive tool was also proposed by Shmueli et al. (2006). The initial steps are aimed at familiarizing users with the data set characteristics. Then, users move to studying more complicated relationships between the auction price, auction dynamics, and auction attributes. These can be hypothesis driven or exploratory.

The data familiarization phase consists of the following steps:

1. Viewing the main screen to learn about the general size, structure, and dimension of the data set.
2. Exploring the time series panels: comparing and contrasting price curves and their derivative by zooming in and out, selecting and deselecting, filtering, and searching for patterns. This gives a sense of the heterogeneity across auctions, the types of prominent patterns, the times when there is activity, and outliers.
3. Studying the attribute panel: examining summary statistics, sorting attributes, and rearranging the order of columns to compare multiple attributes for subsets of auctions.

To illustrate what can be gleaned in the familiarization step, let us consider an example. For the magazines' data set shown in Figure 3.20, the main screen shows the price range for the items. Users can change the log scale of the price curves to $ to get a currency range. They see several auctions with very high end-of-auction dynamics. Zooming in reveals that the majority of auctions actually have relatively slow dynamics. Examining the attributes, users see that most auctions are 7 days long, and all auctions are in U.S. currency, except for one in British pounds. The busiest auction has 29 bids, while the quietest auction has 4 bids. No auctions have a reserve price. One auction is an outlier in terms of ending time, which happens to be the one in British pounds (this explains why the ending time is different; it is because the seller is in a different time zone). Almost half of the auctions have no shipping cost specified. Users also see that there are no missing values for these auctions.

After initial familiarization, users can search visually for patterns and relationships in subsets of auctions. This step involves exploring the relationship between the time series and their attributes. There are two different strategies of doing this: one is an attribute-to-curves exploratory approach, where the attribute data are manipulated and the results observed in the price curves and dynamics. In particular, users compare auction attributes and see how similarities (and differences) manifest themselves in the price curves and associated dynamics. This is an exploratory regression-type operation where the output variables are curves (see *functional regression* in Chapter 5). The second strategy is a curves-to-attribute approach, where the price and dynamics curves are manipulated and the results observed in the attributes. In particular, users start by comparing price/dynamics curves, and then search for common attributes among auctions with similar curves. This is an exploratory curve-clustering approach (see Section 3.7). Clearly, these two interactive techniques are not meant to replace functional regression or curve clustering, but they may prove useful when used in conjunction with analytical methods such as those described in Chapter 5 and Section 3.7. The exploration is completely model free, does not make any distributional assumptions, and cannot estimate sampling error. Thus, unlike formal statistical models, visual exploration cannot assess whether a relationship is specific to the data sample or can be generalized to the population of interest.

The same two approaches can be used for examining hypotheses about patterns and relationships, where the time series and cross-sectional attribute data can assume either the input or the output variable roles. For example, for examining the hypothesis that ending prices are higher for auctions that close on weekends due to the increased throughput, users can look for evidence by examining the closing day attribute. Sorting auctions by this attribute reveals that most of the auctions in the magazine data set ended on weekends and only eight auctions ended on a weekday, supporting the hypothesis. However, to find out whether this belief is well grounded, we compare the price curves of weekday ending auctions to weekend ending auctions and find that the highest closing prices are actually achieved on weekday ending auctions! This, of course, does not prove causality and is not necessarily statistically significant, but it introduces a surprising finding that should be further investigated using analytic models. For further examples of exploratory hypothesis testing, see Shmueli et al. (2006).

An extension of this interactive approach and interface has also been developed for the purpose of forecasting the price of an ongoing auction by Buono et al. (2007) (see Chapter 5 for this forecasting approach).

3.6 EXPLORING HIERARCHICAL INFORMATION

Online auction data are hierarchical by nature, as can be easily seen by navigating a typical online auction website. A general merchandize auction site will usually have item categories, and then one or more levels of subcategories. More specialized auction sites will use a hierarchy relevant to the auctioned items (e.g., artist name for art auctions or loan type for loan auctions). To explore commonalities and differences between auctions while maintaining the hierarchical and clustered information, we can use treemaps. The treemap (Shneiderman, 1992; Bederson et al., 2002) is a space-constrained visualization of hierarchical structures. Treemaps enable users to compare nodes and subtrees even at varying depth in the tree and help them spot patterns and exceptions. Treemaps are interactive and allow dynamic querying. A well-known online application of treemaps is "Map of the Market" by SmartMoney (`http://www.smartmoney.com/map-of-the-market`) that presents the stock market information in an interactive treemap display.

Figure 3.21 displays a treemap for a sample of nearly 11,000 eBay auctions that took place between August 2001 and February 2002. For further information on these data, see Borle et al. (2006). The display is divided into rectangles representing eBay's item categories (e.g., jewelry and watches), then each rectangle is further divided into eBay's subcategories (e.g., premium wristwatches), and finally into brands (e.g., Rolex and Cartier). An individual rectangle in this example is a single brand. We can use color, size, and labels to display three variables of interest. In the figure, we used shading to denote the percentage of negatively rated sellers (seller feedback is determined from buyers' ratings on their previous transactions), where the darker shade denotes a higher percentage. We used rectangle size to represent the average item value (average final price). It is immediately apparent that the highest percent of negatively rated sellers is concentrated in the most expensive items (Rolex and Cartier wristwatches)! This is of special interest because negative seller rating can be an indication of fraud. Considering item value, it is clear why low-rated sellers are concentrated in luxury items (e.g., Rolex watches are sold at approximately $2000).

We can use the treemap to explore other relationships of interest. Figure 3.22 displays the relation between the number of bids in an auction and the opening bid (the starting price set by the seller). The data are again organized by item category, subcategory, and brand. Rectangle size is used here to represent the median opening bid, and color shading is used to represent the median number of bids (darker = more bids). We see that a low opening bid (large rectangle) is associated with many bids (dark). This association is in line with theoretical and empirical findings: a low opening bid attracts many bidders, resulting in many bids.

Treemaps are useful for exploring the many factors that can be measured in online auction data. They can help detect not only relations, but also outliers and unexpected

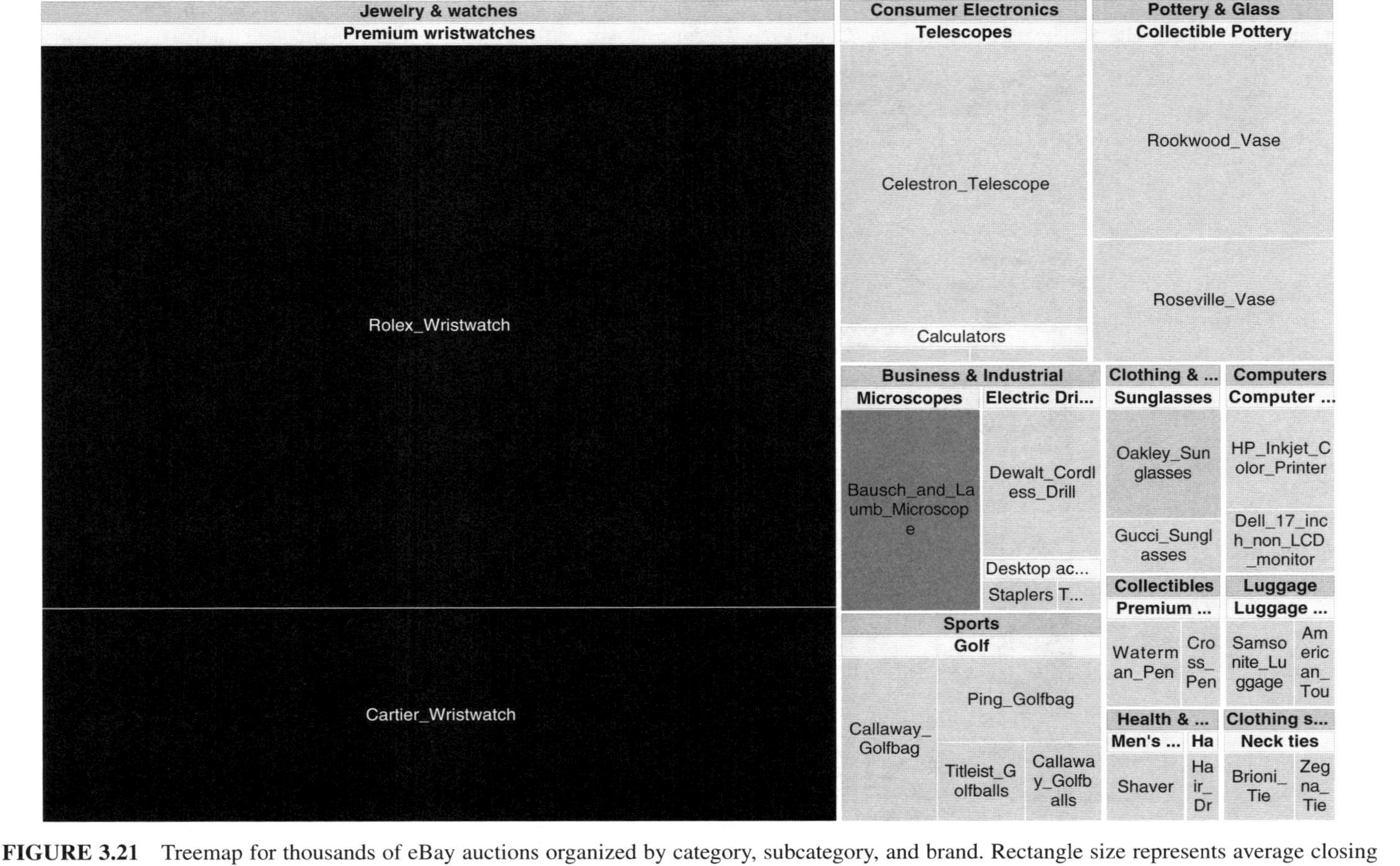

FIGURE 3.21 Treemap for thousands of eBay auctions organized by category, subcategory, and brand. Rectangle size represents average closing price. Shade represents % of sellers with negative feedback (darker=higher %).

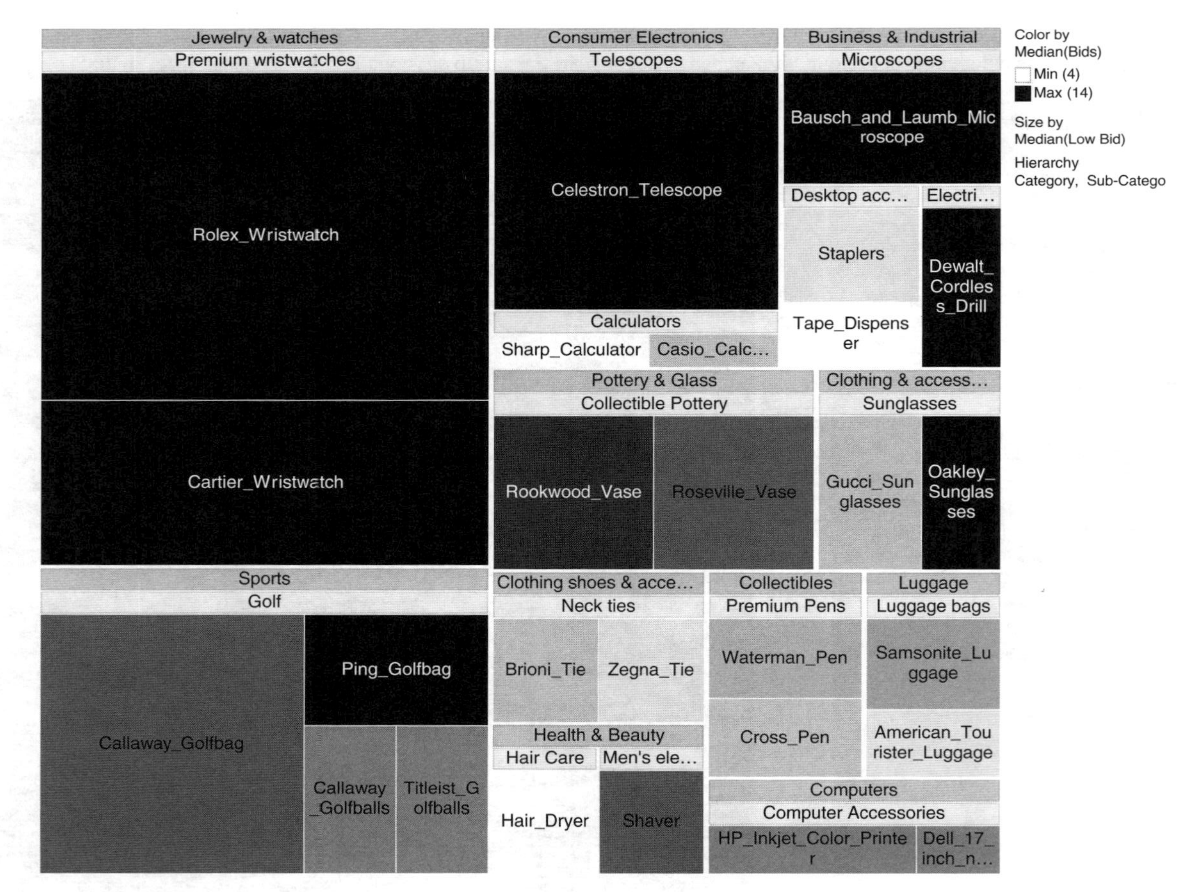

FIGURE 3.22 Treemap for thousands of eBay auctions organized by category, subcategory, and brand. Each rectangle represents a brand. Rectangle size represents median opening bid. Shade represents median number of bids.

patterns. Moreover, they offer a bird's-eye view of large samples of auctions and show an angle that is usually unavailable via other displays.

3.7 EXPLORING PRICE DYNAMICS VIA CURVE CLUSTERING

As we have observed in a few examples and through different visualizations (see in particular Figure 3.11), price dynamics can be quite heterogeneous: while some auctions accelerate toward the end, others decelerate (i.e., the speed of price increase slows down). Similar heterogeneity exists at the beginning of the auction: while in some auctions the price starts out at a very fast pace (but slows down later on), in others it begins very slowly and only later increases its speed. We thus set out to segment price dynamics into more homogeneous subgroups as an exploratory step. Statistically, segmentation is done via clustering. In our case, segmentation is conceptually complicated since we have to cluster *curves* (rather than vector-valued data). We thus first describe a mechanism to perform curve clustering. As in ordinary clustering, performing the actual segmentation is only one component of the entire task; an equally (if not more) important component consists of *interpreting* the resulting clusters. We therefore discuss the interpretation of the curve-clustering algorithm and also compare it with the results of traditional clustering approaches.

3.7.1 Curve-Clustering Principles

As in ordinary cluster analysis, our goal is to segment the data into clusters of homogeneous observations. The clustering provides exploratory information and can support further analyses to understand the characteristics or factors that lead to heterogeneity in the data. Since in our case the data are curves, we use curve clustering to find segments of more homogeneous price profiles.

Curve clustering can be done in several ways. One option is to sample each curve on a finite grid and then to cluster the sampled data gridwise. However, this can lead to unstable estimates (Hastie et al., 1995). A different approach that has been explored in the literature, when curves are obtained by smoothing splines (see Chapter 4), is to cluster the set of curve *coefficients* rather than the functions themselves (Abraham et al., 2003; James and Sugar, 2003). To formalize this approach, let $B = \{\beta_1, \ldots, \beta_N\}$ be the set coefficients pertaining to the N polynomial smoothing splines. Since each of the N curves is based on the same set of knots and the same smoothing parameters, heterogeneity across curves is captured by the heterogeneity across the coefficients. Thus, rather than clustering the original curves, we cluster the set of coefficients B.

Using the above method for measuring distances between pairs of curves, we employ the K-medoids algorithm (with a Manhattan distance) since it is more robust to extreme values than K-means (Cuesta-Albertos et al., 1997). The K-medoids algorithm iteratively minimizes the within-cluster dissimilarity

$$\mathrm{W}_K = \sum_{k=1}^{K} \sum_{j,j' \in I_k} D(\beta_j, \beta_{j'}), \tag{3.1}$$

where $D(\beta_j, \beta_{j'})$ denotes the dissimilarity between coefficients j and j', and I_k denotes the set of indices pertaining to the elements of the kth cluster, $k = 1, \ldots, K$ (Kaufman and Rousseeuw, 1987; Hastie et al., 2001).

We illustrate curve clustering on the set of eBay auctions for new Palm M515 PDAs; see also **?**). Since we are using a nonhierarchical clustering approach, we must first determine the number of clusters, K. To this end, we use several different criteria. The first is the popular method of examining the reduction in within-cluster dissimilarity as a function of the number of clusters. This is shown in Figure 3.23 (left panel), where we see that the within-cluster dissimilarity reduces by about 2 when moving from one to two clusters and also when moving from two to three clusters. However, the reduction is less than 0.5 for a larger number of clusters. The danger with this approach is that it is prone to show kinks even if there is no clustering in the data (Tibshirani et al., 2001; Sugar and James, 2003).

An alternative measure that avoids the within-cluster dissimilarity limitation is based on an information theoretic approach and was introduced by Sugar and James (2003). This nonparametric measure of within-cluster dispersion, d_K, also called the "distortion," is the average Mahalanobis distance per dimension between each observation and its closest cluster center. Rather than using the raw distortions, Sugar

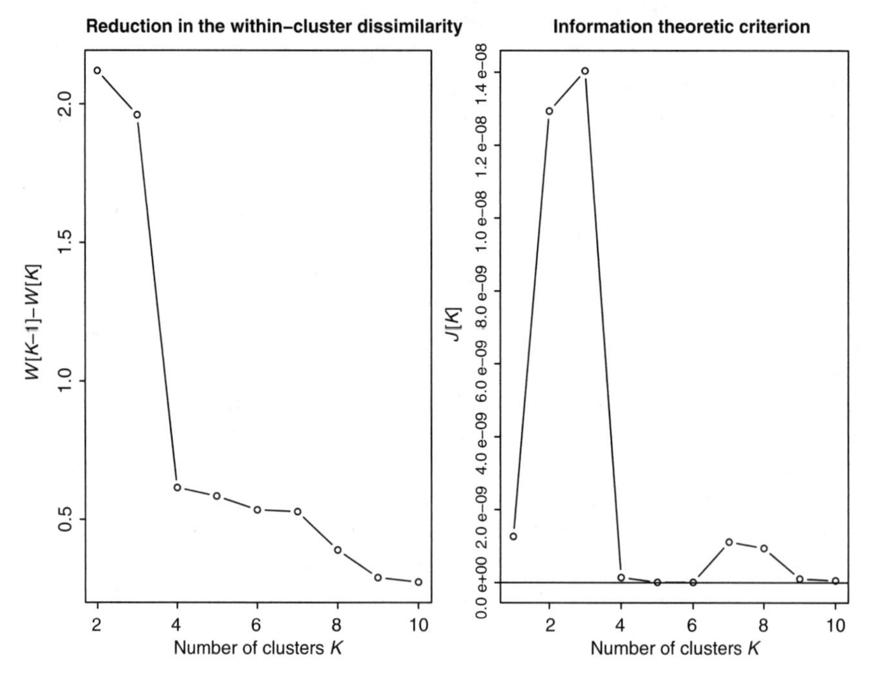

FIGURE 3.23 Choosing the number of clusters: Left panel shows the reduction in the within-cluster dissimilarity as the number of clusters increases. Right panel shows the jump plot of the transformed distortions.

and James (2003) suggest to use the "jump statistic" defined as $J_K = d_K^{-Y} - d_{K-1}^{-Y}$, where $Y = \text{dim}/2$ and dim denotes the dimension of the data. A graph of J_K versus K is expected to peak at the number of clusters K that best describes the data. A jump plot for our data is shown in the right panel of Figure 3.23. The largest jump occurs at $K = 3$, providing additional evidence for three clusters in the data.[2]

3.7.2 Comparing Price Dynamics of Auction Clusters

After determining the number of clusters in the data, we investigate each cluster individually to derive insight into the differences in the price formation process in the different clusters. For our data, we obtain three distinct clusters of sizes 90, 47, and 46 auctions. We start by comparing differences across cluster dynamics, and then supplement the dynamic characterization with differences in static auction features such as opening price, seller reputation, and winner experience.

To compare the price dynamics across clusters, we plot cluster-specific price curves and their derivatives together with 95% confidence interval bounds (Figure 3.24). A comparison of the price curves (top panels) reveals that auctions in clusters 1 and 2, on average, start out at a higher price than those in cluster 3. Also, the average price curves in clusters 2 and 3 increase more steeply at the auction end. Differences in price dynamics are more visible in the price velocities (middle panels) and accelerations (bottom panels): Cluster 1 is marked by high acceleration at the auction start, followed by a long period of stalling/decreasing dynamics that extends almost to the end of day 6. The price then speeds up again toward closing. In contrast, cluster 2 experiences hardly any price dynamics during the first 5 days, but toward the auction end the price speeds up rapidly at an increasing rate. In fact, the maximum acceleration is reached at closing. Since acceleration precedes velocity, a maximum acceleration at closing does not translate into maximum speed. In other words, auctions in cluster 2 close before they reach maximum speed. The picture for cluster 3 is different: Like cluster 1, it experiences some early dynamics, albeit of a larger magnitude. Like cluster 2, the dynamics slow down during mid-auction with a speeding up toward the auction end. However, while auction prices in cluster 2 do not reach their maximum speed before closing, the ones in cluster 3 do! Notice that acceleration in cluster 3 reaches its maximum a day *before* the auction end and then levels off. This means that the price velocity in cluster 3 is "maxed out" when the auction finishes.

3.7.3 Characterizing Auction Clusters Via Phase-Plane Plots

As shown in Section 3.4, PPPs are a popular way of visualizing the interplay of derivatives in a dynamic system. For each cluster, Figure 3.25 shows PPPs of the average

[2] We also clustered a lower dimensional representation of the spline coefficients using principal components and obtained similar results. Further evidence for three clusters was also obtained by performing *functional principal component analysis* (Ramsay and Silverman, 2005).

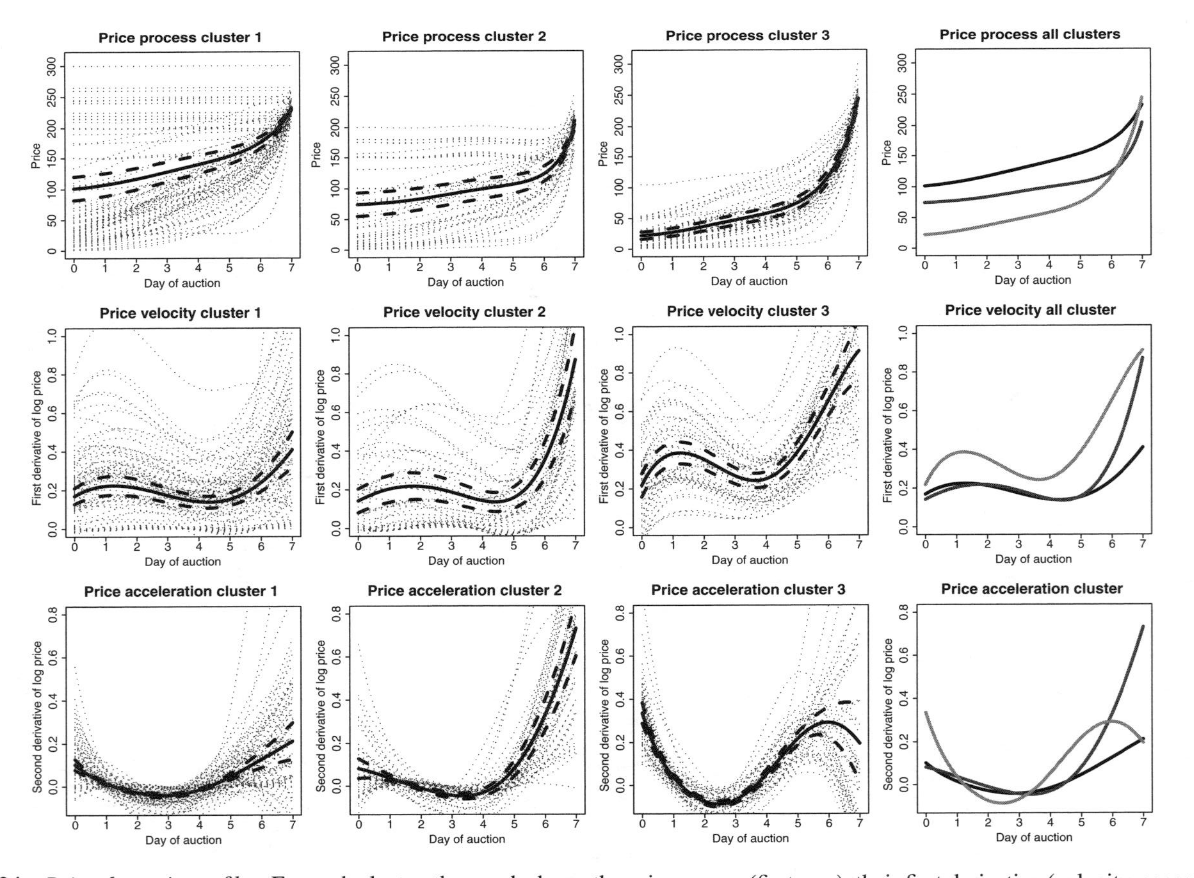

FIGURE 3.24 *Price dynamic profiles*: For each cluster, the graph shows the price curves (first row), their first derivative (velocity, second row), and their second derivative (acceleration, third row). The thick solid curve represents the average curve and the thick dashed curves correspond to 95% confidence bounds. The last column overlays the mean curves from the three clusters.

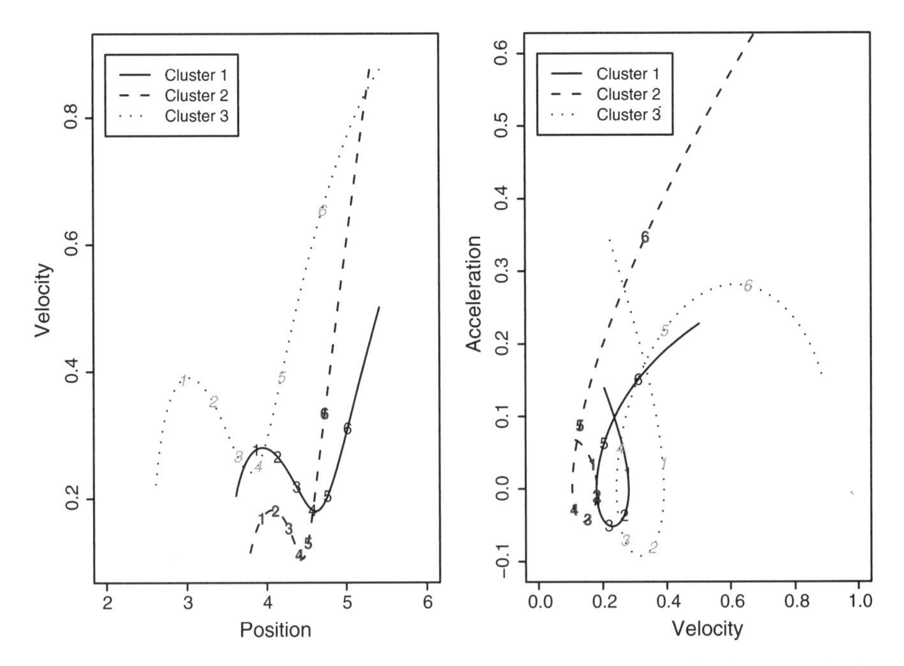

FIGURE 3.25 Phase-plane plots for the three clusters. *Left panel*: price level versus velocity. *Right panel*: velocity versus acceleration. Plots are based on the average dynamics per cluster. Days 1–6 are marked.

price velocity versus the average price position (left panel) and the average price acceleration versus the average price velocity (right panel). Several observations can be made: First, in cluster 2 the price velocity over the last day increases from ~ 0.3 to ~ 0.9. If we recall the identity $\partial \log f(t)/\partial t = \frac{\partial}{\partial t} f(t)/f(t)$, then a (log) price velocity near 0.9 implies that with every single unit increase in t, the price increases by 90%. Second, we see that the size of the curve for cluster 1 is small compared to the other two clusters. By "size" we mean the range in x and y directions. This describes the same phenomenon as shown earlier where auctions in cluster 1 start out high, end high, and have comparatively little change in dynamics. In contrast, price and velocity change dramatically in cluster 3. Third, the right panel shows the relationship between velocity and acceleration. Note that all three PPPs form a "loop" that for each of the three clusters starts and ends at different times of the auction. During this loop, the auction reaches a point of "zero acceleration." At this stationary point, there is nearly no change in velocity. The different clusters leave this point at different phases of the auction. Cluster 3 leaves this point first followed by the other two clusters. This again underscores the different dynamics across the three clusters. Clusters 2 and 3 also differ in another important aspect: While auctions in cluster 3 show increasing velocity but decreasing acceleration at the end (downward curvature of the PPP), both increase for cluster 2.

3.7.4 Comparing Dynamic and Nondynamic Cluster Features

To gain insight into the relationship between price dynamics and other auction information, we compare the three clusters with respect to some key auction features. First, we compare the opening and closing prices of the three clusters that are conceptually important. Although the price curves approximate the actual price during the auction relatively well, the actual price at the auction start and closing have special economic significance, containing additional information about the price formation. For the data in this example, we indeed find that the three clusters differ with respect to the opening and closing prices.

We investigate additional auction characteristics that are relevant to the price formation: the day of the week that the auction closed, the seller's rating (a proxy for reputation), the winner's rating (a proxy for experience), the number of bids, and the number of distinct bidders that participated in the auction (both proxies for competition). Table 3.1 gives summary statistics by cluster for each of the numerical variables. Figure 3.26 compares the closing days of the three clusters, and a χ^2 test (p-value = 0.02) further confirms that cluster and day of week are statistically dependent.

Table 3.1 and Figure 3.26 suggest several possible strategies for bidding on an auction. Take for instance auctions in cluster 2. These auctions are characterized by high opening bids, Tuesday closing days, and low final prices. Alternatively, auctions closing on Saturday or Sunday tend to originate from cluster 1 with high final prices.

In light of these results, we characterize three types of auctions based on the three clusters and their dynamics:

Steady Auctions (Cluster 1). Auctions in this cluster have moderately high opening prices and low competition, yet they achieve high closing prices. A plausible explanation for this steadiness is the highly experienced sellers. A highly rated seller in the online environment is usually regarded as being trustworthy. Such sellers have performed many transactions and therefore gained trust and experience. They also make better design choices, such as scheduling the auction closing on a weekend (when there are likely to be more users logged on). Winners in these auctions are also relatively experienced, indicating that experienced bidders are willing to pay a premium for buying from trustworthy sellers.

TABLE 3.1 Average and Standard Deviation (in Parentheses) by Cluster (on Log Scale)

Clus#	Obid	Price	SelRate	WinRate	Nbids	Nbidders	ClusSize
1	6.98	234.86	40.70	34.04	10.79	6.19	90
	(1.43)	(1.01)	(1.13)	(1.19)	(1.13)	(1.10)	
2	6.18	216.30	26.43	32.13	17.16	8.27	47
	(1.66)	(1.02)	(1.20)	(1.28)	(1.10)	(1.10)	
3	0.17	233.04	20.52	22.41	24.87	12.53	46
	(1.63)	(1.01)	(1.15)	(1.29)	(1.06)	(1.05)	

Opening bid (Obid), closing price (Price), seller rating (SelRate), winner rating (WinRate), and number of bids and bidders (Nbids, Nbidders). ClusSize denotes the cluster size.

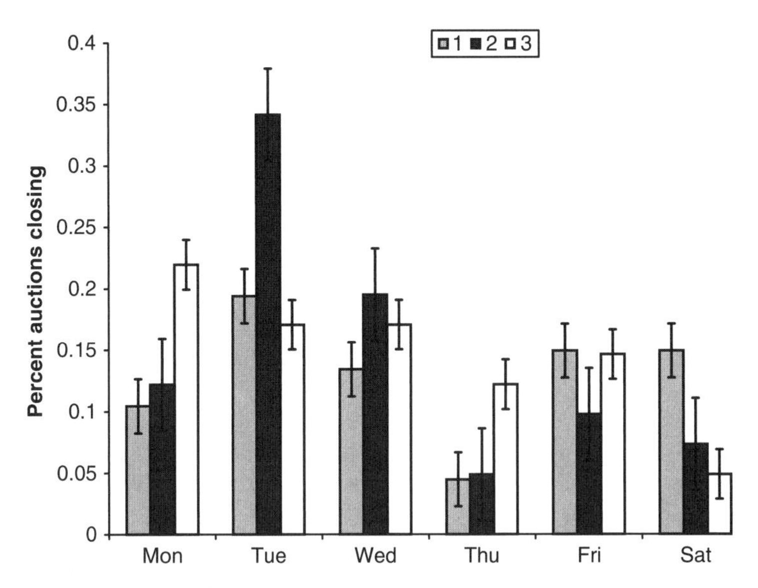

FIGURE 3.26 Percent of auctions that close on each day of the week, by cluster. Bars denote standard errors.

Low-Energy Auctions with Late Dynamics (Cluster 2). Auctions in this cluster have the highest average opening prices and the lowest closing price. In between, the price hardly moves until the last day, when suddenly it accelerates and remains in high acceleration when the auction closes. It appears that the high opening price deters potential bidders from participating in the auction, and in turn the low competition level leads to lower prices. Or perhaps there are other reasons that make these auctions unattractive (e.g., they tend to end on a weekday).

Bazaar Auctions (Cluster 3). Auctions in this cluster are the most dynamic ones: price increase moves quickly with acceleration peaking early on and then again before the close. These auctions have, on average, the lowest opening price, the highest competition, and high closing prices. The low opening price appears to be the only attractor of these auctions: sellers have low ratings and auctions close mostly on weekdays. We call them "bazaar" auctions because the marketplace seems to work efficiently in reaching the final price through dynamic competition.

A Comparison with "Traditional" Clustering To illustrate the usefulness of clustering the price curves and dynamics over ordinary clustering of the attribute data, we continue the data illustration and show the amount of information that is *lost* by not taking a functional approach and instead performing a traditional cluster analysis.

The price curves describe the process of bidding *between* the start and the end of the auction. In contrast, a traditional approach ignores the price curves and focuses only on the beginning and the end of this process, that is, on the opening bid and the final price only. We can think of this as a *static* approach since it ignores all the dynamic price information.

TABLE 3.2 Cluster-Specific Summaries (Similar to Table 3.1) for Clustering on the Opening Bid and the Final Price Only

Clus#	Obid	Price	SelRate	WinRate	Nbids	Nbidders	ClusSize
1	20.24	201.16	46.74	35.59	12.73	6.83	93
	(1.35)	(1.01)	(1.18)	(1.29)	(1.15)	(1.12)	
2	15.14	240.53	40.86	41.45	12.22	6.52	45
	(1.4)	(1.01)	(1.20)	(1.25)	(1.17)	(1.14)	
3	0.01	238.18	11.23	13.38	26.78	13.94	45
	(1.00)	(1.01)	(1.01)	(1.20)	(1.03)	(1.02)	

Table 3.2 shows the results of such a static clustering approach. Taking the same approach (including using the same K-medoids algorithm), we obtain three clusters. Although cluster sizes are similar to those in Table 3.1 (90:47:46 versus 93:45:45), the three clusters are in fact very different from those in Table 3.1. The ordinary clustering segments auctions into three groups mostly by their opening bid: auctions that use eBay's "default" opening bid of 0.01, auctions with medium opening bids, and auctions with high opening bids. The three clusters are hardly separable by any of the other attributes (that were not included in the clustering in either case). In fact, clusters 1 and 2 are almost identical with respect to seller rating, winner rating, and competition (number of bidders and bids). Furthermore, cluster variances are also much larger. The observation that competition is not captured by this static clustering can be attributed to the fact that competition pertains to what happens *during* an auction and can therefore only be captured by the price curve. In contrast, the opening bid completely ignores competition and the final price only very weakly captures it, and thus the result is a loss of valuable auction information.

Another way of assessing the information loss due to a static clustering approach is by looking at the dynamic price profiles. Figure 3.27 shows the price curves and their dynamics (similar to Figure 3.24) for the static clustering (i.e., using only opening bid and final price). The mean trend in the curves is generally comparable to the curve-clustering result (top right panel), but the intercluster heterogeneity is very large. The effect is especially notable in the velocity and acceleration curves (middle and bottom right panels). We can thus conclude that there exists significant variation in the price dynamics of auctions and that in order for a method (such as clustering) to be able to capture this variation, it is necessary to operate on the dynamic data rather than merely considering static measurements.

3.7.5 Price Dynamics Clustering: Conclusions

The existence of several auction types and the relationship between auction characteristics and price dynamics can be of use to sellers, bidders, the auction house, and, in general, to decision makers in the online marketplace. For instance, the results of our case study indicate that low-rated sellers might be better off setting a low opening

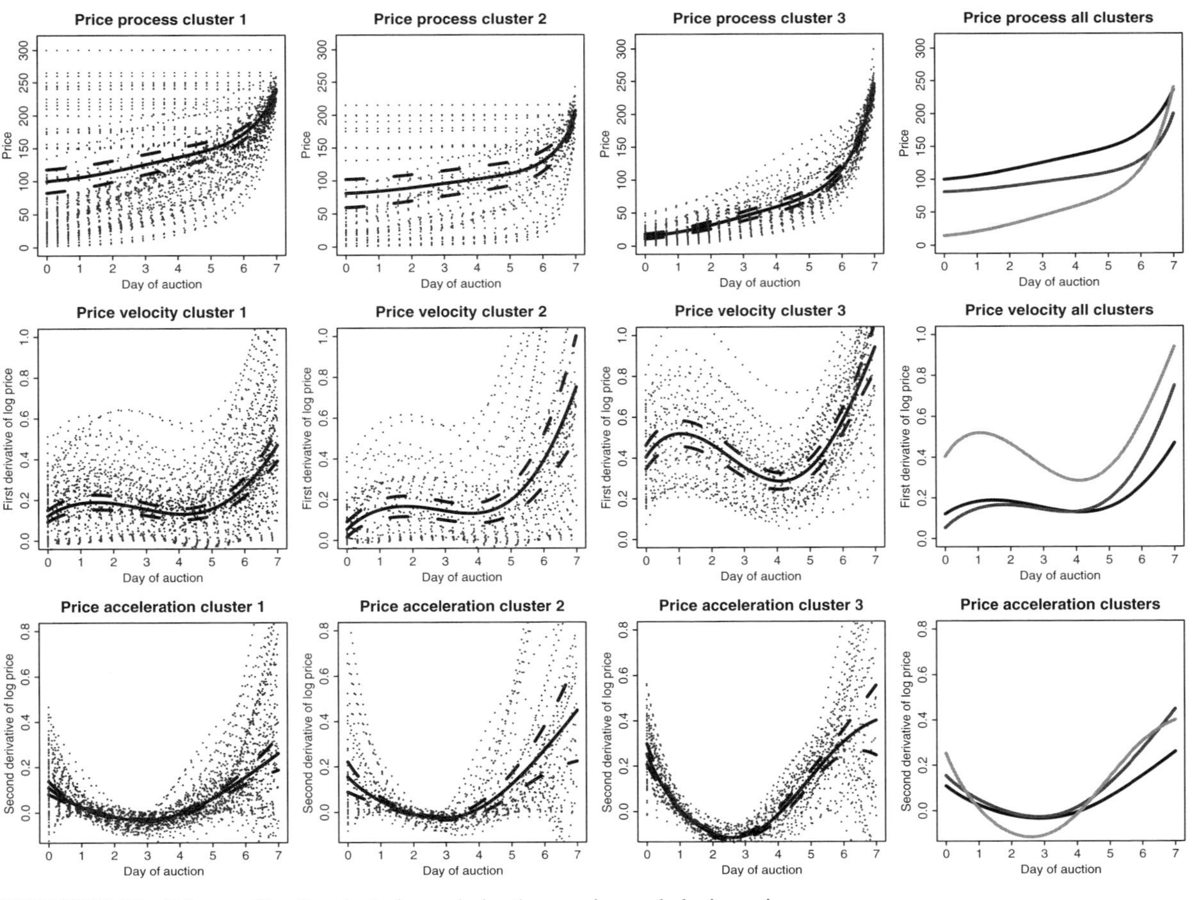

FIGURE 3.27 Price profiles for clustering only by the opening and closing prices.

price, a weekend closing, and other features that will increase competition and result in high price velocity at the end. Furthermore, on eBay, for example, a seller is allowed to change the opening price and a few other settings before the first bid is submitted. We can therefore imagine a system where a seller evaluates, based on the dynamics in the beginning of his/her auction, what type of auction it is and dynamically changes the auction settings to improve dynamics.

An auction house, like eBay, can use information on different types of auction dynamics to decide on varying pricing strategies and to determine optimal bid increments. Pricing strategies and bid increments affect the price dynamics and their optimal choice could avoid the typical lack of price movements during the middle of an auction. Moreover, price dynamics can be used to determine fraud. Fraudulent sellers could be suspected before the end of the transaction if their auction dynamics do not follow typical patterns.

One of the limitations of the current curve-clustering approach is that it is geared only toward auctions of a certain fixed duration (e.g., 7-day auctions). While this can be generalized to auctions of different durations (e.g., Bapna et al., 2008b), one distinct feature of eBay auctions is the fixed ending time. It is yet to be investigated how the same approach can be used for auctions with flexible ending times. Another avenue for future research is to investigate exactly what external factors affect auctions that do show less of a "closed-system behavior" (e.g., those in cluster 2). A first step in that direction is proposed in Jank et al. (2008b), who use novel *functional differential equation tree* methodology to incorporate predictor variables into dynamic models. This approach is discussed in Chapter 4.

3.8 EXPLORING DISTRIBUTIONAL ASSUMPTIONS

In this section, we tackle a special characteristic of auction data that results in *semicontinuous* data for variables such as price and surplus. These variables are often central in online auction research and used as response variables in regression models or in statistical tests such as T-tests. The problem that arises is that the auction mechanism creates data that are neither continuous nor discrete, and therefore standard probability distributions do not provide adequate representations. The need for capturing such data arose in a study in which the goal was to accurately estimate consumer surplus generated in online auctions (Bapna et al., 2008a). Our preliminary investigations showed that surplus, generally defined as the difference between the auction price and the second highest bid, was not captured well by any standard statistical distribution. The reason is related to the fact that surplus data are *user generated* and combined with the online auction mechanism, the resulting data exhibited a special structure. This structure proved to be important: By incorrectly assuming normality (or any other continuous parametric distribution), our results underestimated surplus by several million dollars. Therefore, in an attempt to obtain more accurate estimates, we set out to find a distribution (or, equivalently, a transformation) that best characterizes and captures the data at hand. In the case of online auctions, one observes bids such as "$5" or "$205", and hardly ever bids such as "$5.91" or "$205.63" as would be prescribed by a continuous distribution. As a result, bid data (and hence surplus data) are *semicontinuous* (Shmueli et al., 2008). By semicontinuous data, we mean data that are inherently continuous but get contaminated by an inhomogeneous discretizing mechanism. Such data lose their basic continuous structure and instead are spotted with a set of "too-frequent" values. Semicontinuous data, in general, arise for a variety of reasons that range from human tendencies to enter rounded numbers or to report values that conform to given specifications (e.g., reporting quality levels that conform to specifications), to contamination mechanisms related to the data entry, processing, storage, or any other operation that introduces a discrete element into continuous data. Examples are numerous and span various applications. In accounting, for example, a method for detecting fraud in financial reports is to search for figures that are "too common," such as ending with 99 or being "too round." In quality control, data sometimes are manipulated to achieve or meet certain criteria. For example, Bzik (2005) describes practices of data handling in the semiconductor industry that deteriorate the performance of statistical monitoring. One of these is replacing data with "physical limits": In reporting contamination levels, negative values tend to be replaced with zeros and values above 100% are replaced with 100%. However, negative and >100% values are feasible due to measurement error, flawed calibration, and so on. Another questionable practice is "rounding" actual measurements to values considered ideal in terms of specifications. This results in data that include multiple repetitions of one or more values. We will use the term "too-frequent" to describe such values.

In the aforementioned study, Bapna et al. (2008a) observed that surplus values on eBay had several too-frequent values, most likely due to the discrete set of bid increments that eBay uses and the tendency of users to place integer bids. Figure 3.28

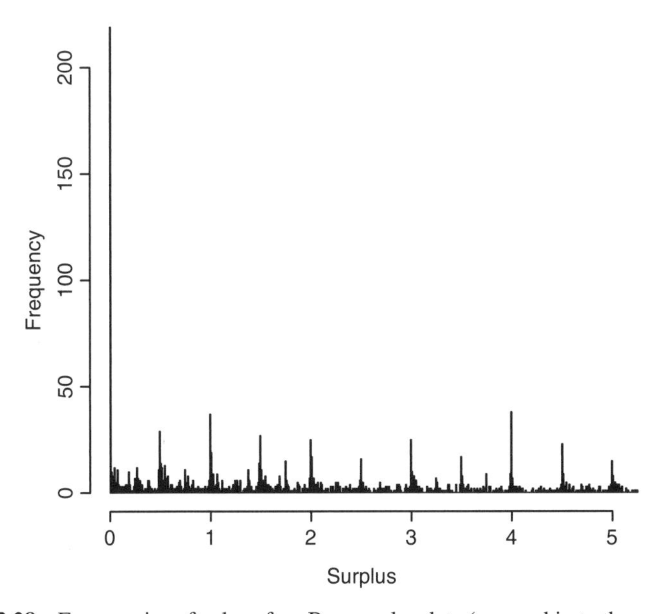

FIGURE 3.28 Frequencies of values for eBay surplus data (zoomed in to the range $0–5).

shows the frequencies of all values observed in the surplus data: the value \$0 occurs 6.35% of the time, and values such as \$0.01, \$0.50, \$1.00, \$1.01, \$1.50, \$2.00, \$2.01, and \$3.00 are much more frequent than their neighboring values.

As we will use the surplus data throughout this section to illustrate and motivate the proposed transformations, we describe a few more details about the mechanism that generates the data. Consumer surplus, used by economists to measure consumer welfare, is the difference between the price paid and the amount that consumers are willing to pay for a good or service. In second-price auctions, such as those found on eBay, the winner is the highest bidder; however, she/he pays a price equal to the second highest bid. Surplus in a second-price auction is defined (under some conditions) as the difference between the price paid and the highest bid. Bapna et al. (2008a) found that although surplus is inherently continuous, observed surplus data are contaminated by too-frequent values, as shown in Figure 3.28.

The main problem with semicontinuous data is that they tend to be unfit for use with many standard statistical methods. Semicontinuous data can appear as if they come from a mixture of discrete and continuous populations. Graphs such as scatterplots and frequency plots may indicate segregated areas or clouds. Such challenges arise in the surplus data described above. Figure 3.29 displays two probability plots for the surplus data: a lognormal fit and a Weibull fit. It is obvious that neither approximates the data well and other distributions fit even worse. The reason is the excessive number of 0 values in the data. Furthermore, using $\log(surplus + 1)$ as the response in a linear regression model yields residual plots that exhibit anomalies that suggest a mixture of populations. Figure 3.30 shows two residual plots exhibiting such behavior; these plots indicate clear violation of the assumptions of a linear regression model.

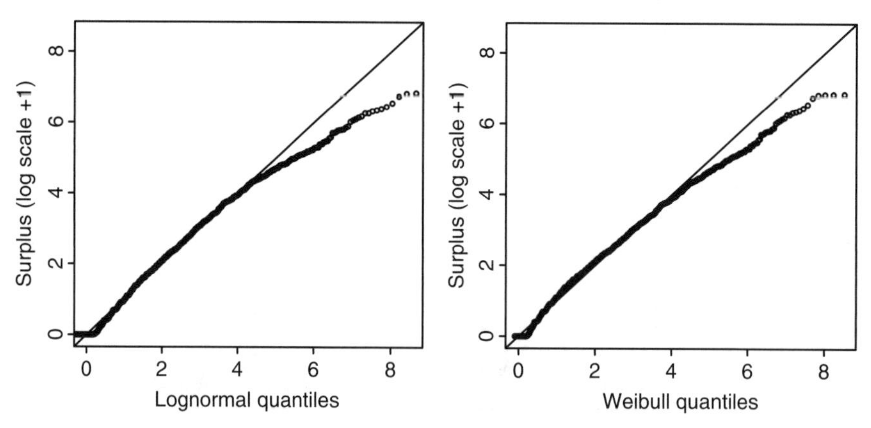

FIGURE 3.29 Probability plots for $\log(surplus + 1)$: lognormal fit (left) and Weibull fit (right).

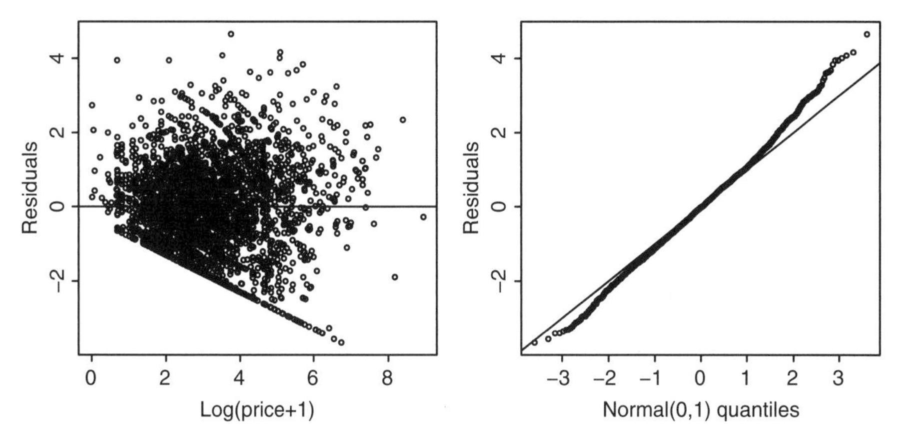

FIGURE 3.30 Residuals from a linear regression model of log($surplus + 1$) on log($price + 1$) (left) and normal probability plot of residuals (right).

One possible solution is to separate the data into continuous and discrete parts, to model each part separately, and then to integrate the models. For example, in a data set that has too many zeros (zero inflated) but otherwise positive continuous values, one could create a classification model for zero/nonzero in the first stage, and then a prediction model for the positive data in the second stage. This approach has two practical limitations: first, partitioning the data leads to a loss of statistical power, and second, it requires the "too frequent" values to be concentrated in a limited area or in some meaningful locations. To solve the first issue, one might argue for a mixture model. Although there exists a plethora of such models for mixed continuous populations or for mixed discrete populations (e.g., "zero-inflated" models, in Lambert (1992)), we have not encountered models for mixtures of continuous and discrete data. Moreover, there is a major conceptual distinction between mixture and semicontinuous data: Unlike mixture data, semicontinuous data are inherently generated from a single process. Therefore, treating them as a mixture of populations is artificial. The ideal solution, of course, would be to find the source of discretization in the data generation mechanism and to eliminate it or account for it. However, in many cases this is impossible, very costly, or very complicated. We therefore strive for a method that "unites" the apparent "subpopulations" so that data can be integrated into a single model and treated with ordinary models for continuous data.

Our proposed solution is a set of two transformations that yield continuous data. We distinguish between two cases: One, where the goal is to obtain data that fit a particular continuous distribution (e.g., a normal distribution, for fitting a linear regression model), and the other where there is no particular parametric distribution in mind, but the data are still required to be continuous. The first approach, when the goal is simply to obtain continuous data, is based on jittering.

As in graphical displays, where jittering is used to better see duplicate observations, our data transformation adds a random perturbation to each too-frequent value, and thereby ironing out the anomalous high frequencies. We call this the *jittering transform.* The second approach, suitable when the data should fit a particular distribution, is based on binning the data in a way that is meaningful with respect to the assumed underlying distribution, and then replacing the too-frequent values with randomly generated observations in that bin. We call this the *local regeneration transform.*

Jittering is used not only in graphical displays, but also in data privacy protection. It is a common method for disguising sensitive continuous data while retaining the distributional properties of the sample that are required for statistical inference. The difference between privacy protection and our application is that unlike in the former, we do not jitter all observations but rather only the too-frequent values.

Local regeneration is related to smoothing and rebinning of histograms. However, there are two fundamental differences. First, the assumption about the data origin is different: In histogram smoothing and binning, the underlying assumption is that extreme peaks and dips in the histogram result from sampling error. Therefore, increasing the sample size should alleviate these phenomena, and in the population such peaks would not appear at all (Good and Gaskins, 1980). In contrast, in semicontinuous data the cause of the peaks is not sampling error but rather a structural distortion created by the data generating mechanism. This means that even very large samples will exhibit the same semicontinuousness. The second difference is the goal: whereas histogram smoothing and binning are mainly used for density estimation (without attempting to modify the data), the purpose of local regeneration is to transform the semicontinuous data into data that fit a particular continuous parametric distribution, similar to the famous Box–Cox transformation.

Both local regeneration and binning attempt to find the best representation of the data in histogram form. The literature on optimal bin sizes focuses on minimizing some function of the difference between the histogram $\hat{f}(x)$ and the real density $f(x)$ (such as the mean integrated squared error, MISE). In practice, $f(x)$ is assumed to be a particular density function (e.g., Scott, 1979), or else it is estimated by using an estimator that is known to have a smaller IMSE (e.g., a kernel estimator). Methods range from rules of thumb to (asymptotically) theoretically optimal bin-width formulas. The methods also vary widely with respect to computational complexity and performance. Simple bin-width formulas such as Sturge's rule and Doane's modifications, which are used in many software packages (e.g., Minitab and S-Plus), have been criticized as leading to oversmoothing (Wand, 1997). On the other hand, methods that have better asymptotic optimality features tend to be computationally intensive and less straightforward to implement in practice. In the semicontinuous data case, we assume that it is impossible to obtain a set of data that are not distorted. Our target function is therefore a parametric distribution.

3.8.1 Identifying Semicontinuous Data

To identify whether a continuous data set suffers from semicontinuity, we examine each of the unique values and see whether there is one that is *too frequent*. One way to check this is to examine a one-way pivot table, with counts for each value in the data, for very large counts. Sorting the counts in descending order can alleviate the search process. However, since such a table loses the ordering of the values, too-frequent values that appear in sparser areas might remain undetected. An alternative that enhances the spotting of too-frequent values while preserving the interval scale of the data is a visualization that is a hybrid between a bar chart and a histogram: the *max-bin histogram* (Shmueli et al., 2008). The max-bin histogram is essentially a histogram with bin widths equal to the smallest unit in the data. It is therefore equivalent to a bar chart with as many nonzero bars as there are unique values in the data, except that its x-axis has a continuous meaning rather than labels. On the max-bin histogram of the raw data, where frequencies are represented by bars, frequencies that are outstandingly high compared to their neighbors indicate values suspected of being "too frequent" for a continuous scheme. Figure 3.28 displays a max-bin histogram for the eBay surplus data—and too-frequent values are visible in the plot. Using plots for continuous data can also assist in identifying semicontinuous data. For instance, probability plots might display "steps" or contain other indications of mixed populations. Scatterplots of the suspected variable versus other variables of interest can also reveal multiple clouds of data. However, depending on the locations and prevalence of the too-frequent values, standard plots might not enhance the detection of such values (e.g., if the too-frequent values are in high-frequency areas of the distribution). The max-bin histogram is therefore a powerful and unique display, ideal for this purpose.

3.8.2 Transforming Semicontinuous Data

Given a set of semicontinuous data, the influence of "bouts" of high-frequency values can be "ironed out" (a term coined by Good and Gaskins (1980)) in one of the following ways: The first approach is to jitter each of the too-frequent values, which means that we add a random perturbation to each such observation. The second approach first defines local neighborhoods by binning the data, and then replaces the too-frequent values with randomly generated observations within their respective neighborhoods. The two methods are similar in several respects. First, they both assume that the too-frequent values are distortions of other close-by values. Second, both methods define a local neighborhood for the too-frequent values, and the transformed data are actually randomly generated observations from within this neighborhood. Third, in both cases, only the too-frequent values are replaced while the other observations remain in their original form. And finally, in both cases the definition of a local neighborhood must be determined. We describe each of the two methods in detail next.

The Jittering Transform Although many statistical methods assume a parametric distribution of the data, there are many nonparametric methods that assume only a continuous nature. This is also true of various graphical displays, which are suitable for many data structures, as long as they are continuous (e.g., scatterplots). After identifying the "too-frequent" values, the transform operates by perturbing each of them by adding random noise. If we denote the ith original observation by X_i, and the transformed observation by $\tilde{X}_i$, then the jittering transformation is

$$\tilde{X}_i = \begin{cases} X_i + \epsilon, & \text{if } X_i \text{ is a too-frequent value,} \\ X_i, & \text{else,} \end{cases} \tag{3.2}$$

where ϵ is a random variable with mean 0 and standard deviation σ_ϵ. The choice of distribution for ϵ depends on the underlying process that the too-frequent values are most likely a result of. If there is no reason to believe that the too-frequent values are a result of an asymmetric processes, then a symmetric distribution (e.g., normal or uniform) is adequate. If there is some information on a directional distortion that leads to the too-frequent values, then that should be considered in the choice of the perturbing distribution. In general, the choice of distribution is similar to that in kernel estimation, where the choice of kernel is based on domain knowledge, trial and error, and robustness.

The second choice is the value of σ_ϵ that should depend on the scale of the data and its practical meaning within the particular application. Specifically, domain knowledge should guide the maximal distance from a too-frequent value that can still be considered reasonable. For example, if the data are measurements of product length in inches, where the target value is 10 in. and a deviation of more than an inch is considered "large", then σ_ϵ should be chosen such that the jittering will be less than 1 in. This, in turn, defines $\sigma_{\max}$, the upper bound on σ_ϵ. A series of increasing values of σ_ϵ is then chosen in the range $(0, \sigma_{\max})$, and the performance of the transformation is evaluated at each of these values to determine the adequate level.

Let 2δ be the width of the jittering neighborhood (symmetric around the too-frequent value). The mapping between δ and σ_ϵ depends on the jittering distribution. If the perturbation distribution for ϵ is normal, then σ_ϵ should be chosen such that $3\sigma_\epsilon = \delta$ since almost 100% of the data should fall within three standard deviations of the mean. If the underlying distribution is Uniform(a, b), then $2\delta = b - a$ and therefore $\sqrt{3}\sigma_\epsilon = \delta$.

Since jittering is performed in a way that is supposed to (inversely) mimic the data mechanism that generated the too-frequent observations, the transformed data should be similar to the unobserved, continuous data in the sense that they are both samples from the same distribution. In fact, since we transform only the few too-frequent values, most of the observations in the original and transformed data are identical.

The Local Regeneration Transform Many standard statistical methods assume a particular parametric distribution that underlies the data. In that case, the jittering transform can still be used, but it might require additional tweaking (in the choice of the perturbation distribution and σ_ϵ) to achieve a satisfactory parametric fit. The reason is that the jittering operation is anchored around the too-frequent values, and the locations of these values are independent of the unobservable, underlying continuous distribution.

A more direct approach is therefore to anchor the transformation not to the too-frequent values, but rather to anchors that depend on the parametric distribution of interest. We achieve this by first binning the observed data into bins that correspond to percentiles of the distribution. In particular, we create k bins that have upper bounds at the $\frac{100}{k}, 2\frac{100}{k}, \ldots, k\frac{100}{k}$ percentiles of the distribution. For example, for $k = 10$ we create 10 bins, each as wide as the corresponding probability between the two distribution deciles. Each bin now defines a local neighborhood in the sense that values within a bin have similar densities. This will create narrow bins in areas of the distribution that have high density, and much wider bins in sparse areas such as tails. The varying neighborhood width differentiates this transformation from the jittering transform, which has constant-sized neighborhoods. This means that the main difference between the two transforms will be in areas of low density that include too-frequent values.

To estimate the percentile of the distribution, we use the original data. If the continuous distribution belongs to the exponential family, we can use the sufficient statistics computed from the original data for estimating the distribution parameters and then compute percentiles. Because too-frequent values are assumed to be close to their unobservable counterparts, summary statistics based on the observed data should be sufficiently accurate for initial estimation. In the second step, each too-frequent value is replaced with a randomly generated observation within its bin. As in the jittering transform, the choice of the generating distribution is guided by domain knowledge about the nature of the mechanism that generates the too-frequent values (symmetry, reasonable distance of a too-frequent value from the unobservable value, etc.).

The local regeneration transform for the ith observation can be written as

$$\tilde{X}_i = \begin{cases} l_j + \epsilon, & \text{if } X_i \text{ is a too-frequent value,} \\ X_i, & \text{else,} \end{cases} \tag{3.3}$$

where l_j is the jth percentile closest to X_i from below (i.e., the lower bound of the bin), $j = 1, 2, \ldots, k$, and ϵ is a random variable with support $[l_j, l_{j+1})$.

The two settings that need to be determined are therefore the parametric distribution and the number of bins (k). For the distribution, one might try transforming the data according to a few popular distributions and choose the distribution that best fits the data (according to measures of goodness of fit). The number of bins can be set in a similar way: trying a few configurations and choosing the one that yields the best fit. In both cases, of course, practical considerations and domain knowledge must be integrated into the decision process.

Goodness of Fit and Deviation Our goal is to find a level of jittering or binning that sufficiently "irons out" the discrete bursts in the data and creates a data set that is continuous and perhaps fits a parametric distribution. In the parametric fitting (using the local regeneration transform), we can measure how well the data fit prespecified distributions at each level of binning. For example, the three most popular distributions used for fitting data and that serve as the basis for many statistical analyses are the normal, lognormal, and Weibull distributions. Evaluating how well data are approximated by each of these distributions, graphical tools such as probability plots, goodness-of-fit measures such as the Anderson–Darling statistic, Kolmogorov-Smirnov statistic, and chi-squared statistic, and the correlation based on a probability plot can be useful.

In the nonparametric case where the goal is to achieve continuous data without a specified parametric distribution, we use the notion of local smoothness. Treating the max-bin histogram as an estimate of a continuous density function, we require frequencies of neighboring bins to be relatively similar. One approach is to look for the presence of an "abnormality" in the max-bin histogram by comparing each frequency to its neighbors. Shekhar et al. (2003) define an abnormal point as one that is extreme relative to its neighbors (rather than relative to the entire data set). They construct the following measure of deviation, α_i:

$$\alpha_i = p_i - E_{j \in N_i}(p_j), \tag{3.4}$$

where $p_i = f_i / \sum_j f_j$ is the observed relative frequency at value i, N_i is the set of neighbors of p_i, and $E_{j \in N_i}(p_j)$ is the average proportion of the neighboring values. They then assume an *a priori* distribution of the data or their histogram, allowing them to determine what is considered an abnormal value for α_i. In our context, the underlying distribution of the data is assumed to be unknown. We therefore use an ad hoc threshold: values i that have α_i larger than three standard deviations of α_i are considered abnormal (assuming that deviations are approximately normal around zero).

An alternative approach is to measure smoothness of the max-bin histogram by looking at every pair of consecutive frequencies $(f_i - f_{i+1})$. Various distance metrics can be devised based on these pairwise deviations, such as the sum of absolute deviations or the sum of squared deviations. We use one such measure here, the "sum of absolute deviations between neighboring frequencies":

$$\text{SADBNF} = \sum_i |f_i - f_{i+1}|. \tag{3.5}$$

A third approach is fitting a nonparametric curve to the max-bin histogram and measuring the smoothness of the resulting curve or the deviation of the bar heights from the curve.

In addition to measuring the fit to a distribution, we also measure deviation from the original data by computing summary statistics at each level of binning or jittering and comparing them to those from the original data.

3.8.3 Performance on Simulated Data

To evaluate the performance of the jittering and local regeneration transformations in practice, we simulate data from known distributions and then contaminate them to resemble observable semicontinuous real-world data. In particular, we choose parameters and contaminations that mimic the surplus data. We apply the transformations to the data and choose the transformation parameters (δ in jittering and k in local regeneration) that best achieve the goal of approximating the underlying (unobserved) distribution or at least obtaining a continuous distribution (which should coincide with the underlying generating distribution).

Data Simulation We simulate "unobservable" data from three continuous distributions: lognormal, Weibull, and normal. We then contaminate each data set in a way that resembles the surplus data. The three data sets are called contaminated lognormal (CLNorm), contaminated Weibull (CWeibull), and contaminated normal (CNorm). The steps for all three simulations are similar except that each simulation starts with a different underlying distribution, different too-frequent values, and different contamination spreads. The initial data, whether lognormal, Weibull, or normal, are thought to be the unobservable data that come about naturally. However, a mechanism contaminates the observed data, thereby introducing a few values with high frequencies. Some of the original characteristics of the data are still present, but the contamination makes it difficult to work with the observed data in their present form. In the following, we describe the details and use the notation for the CLNorm data. Equivalent steps are taken for the CWeibull and CNorm simulated data.

We generate 3000 observations (Y_i, $i = 1, \ldots, 3000$) from a lognormal distribution with parameters $\mu = 0, \sigma = 2$. We choose the too-frequent values $\{s_j\} = \{0, 0.25, 0.50, 0.75, 1.00, 1.50, 2.00, 3.00, 5.00, 10.00\}$ and contaminate 750 of the 3000 observations by replacing values that fall within $\nu = 0.10$ of a too-frequent value by that frequent value (in other words, ν is the width of the contamination neighborhood). The contaminated data are therefore obtained by the following operation:

$$X_i = \begin{cases} s_j, & \text{if } Y_i \in [s_j - \nu, s_j + \nu], \text{ and } i = 1, \ldots, 750, \\ Y_i, & \text{else.} \end{cases} \tag{3.6}$$

We perform similar steps for the Weibull and normal simulations. The underlying distribution for the CWeibull data is given by $Weibull(\gamma = shape = 0.5, \beta = scale = 10)$ with too-frequent values $\{s_j\} = \{0, 0.25, 0.50, 0.75, 1.00,\ 1.50,\ 2.00,\ 3.00, 5.00,\ 10.00\}$ and contamination neighborhood of $\nu = 0.12$. The underlying distribution for the CNorm data is $N(\mu = 4, \sigma = 1)$ with too-frequent values $\{s_j\} = \{2.00, 3.00, 3.50,\ 3.75, 4.00,\ 4.25, 4.50, 5.00, 6.00\}$ and contamination neighborhood of $\nu = 0.05$.

The top two panels of Figures 3.31–3.33 show max-bin histograms of the unobservable and contaminated data. The lognormal and Weibull plots show a zoomed-in region to better see the areas where most of the data are located (i.e., not in the tail(s)). The theoretical density is also drawn on as a solid gray line. We see that the contaminated data have the shape of the original distributions with peaks at the too-frequent values. For the lognormal and Weibull data, only the too-frequent values with high density *overall* are discernable. These are 0, 0.25, 0.50, 0.75, 1.00, 1.50, 2.00, and 3.00 for the CLNorm data and 0, 0.25, 0.50, 0.75, 1.00, and 1.50 for the CWeibull data. For the CNorm data, the too-frequent values 2.00, 3.00, 3.50, 3.75, 4.00, 4.25, and 5.00 stand out; the values 1.80 and 5.12 also stand out; however, this is the result of random noise.

Table 3.3 provides the sample mean, standard deviation, and median for each data set. The summary statistics are very close for the original and contaminated samples. This supports the use of summary statistics from the contaminated data for parameter estimation. These statistics serve as benchmarks for comparing the transformed statistics to make sure that the chosen transformation preserves the main characteristics of the distribution.

Transforming the Contaminated Data: Jittering We start by choosing a range of values for δ that defines the neighborhood for jittering. We choose six values 6σ= 0.01, 0.02, 0.05, 0.08, 0.010, and 0.12 and compare the uniform and normal perturbation distributions for each of these values. Note that we use the *observed* too-frequent values for the transformations to better mimic a realistic implementation.

Max-bin histograms are produced for each δ and perturbation distribution. To be concise, only the best jitter level/distribution (as determined in the next section) is shown for each underlying distribution in the third panel of Figures 3.31–3.33. We see that indeed jittering irons out the too-frequent values and the distribution of the transformed data more closely resembles that of the original data. In addition, for all three distributions, the jittering transformation hardly affects the overall summary statistics, as can be seen in Table 3.3.

Transforming the Contaminated Data: Local Regeneration We start by choosing the levels of binning, the regeneration distribution, and the parametric distributions of interest. For the number of bins, we choose seven values: $k = 4, 5, 8, 10, 16, 20,$ and 25. Each of these values divide 100 easily. For the CLNorm data, we bin the

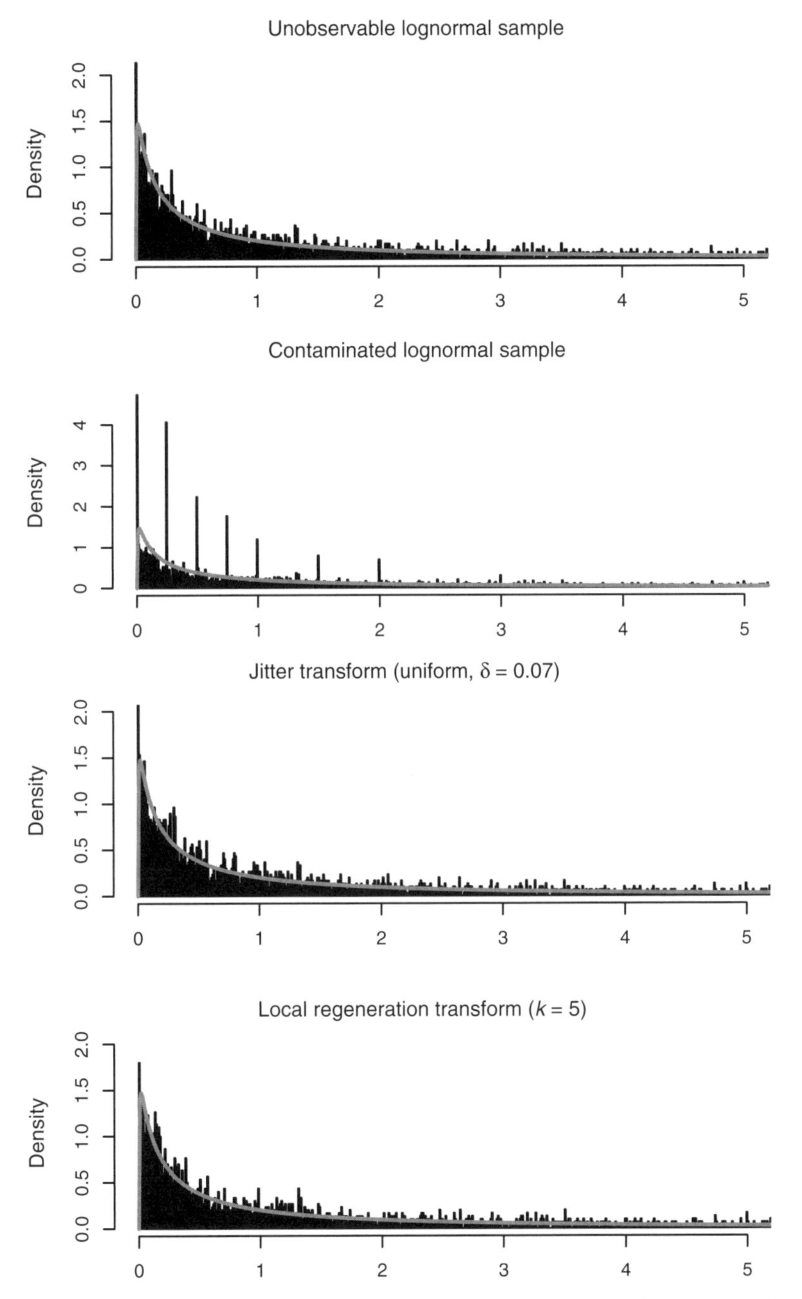

FIGURE 3.31 Histograms of the original, contaminated, jittered (uniform, $\delta = 0.07$), and locally regenerated ($k = 20$) data for the simulated lognormal data. The solid gray line is the lognormal density. Note that the y-axis scale is different for the contaminated data.

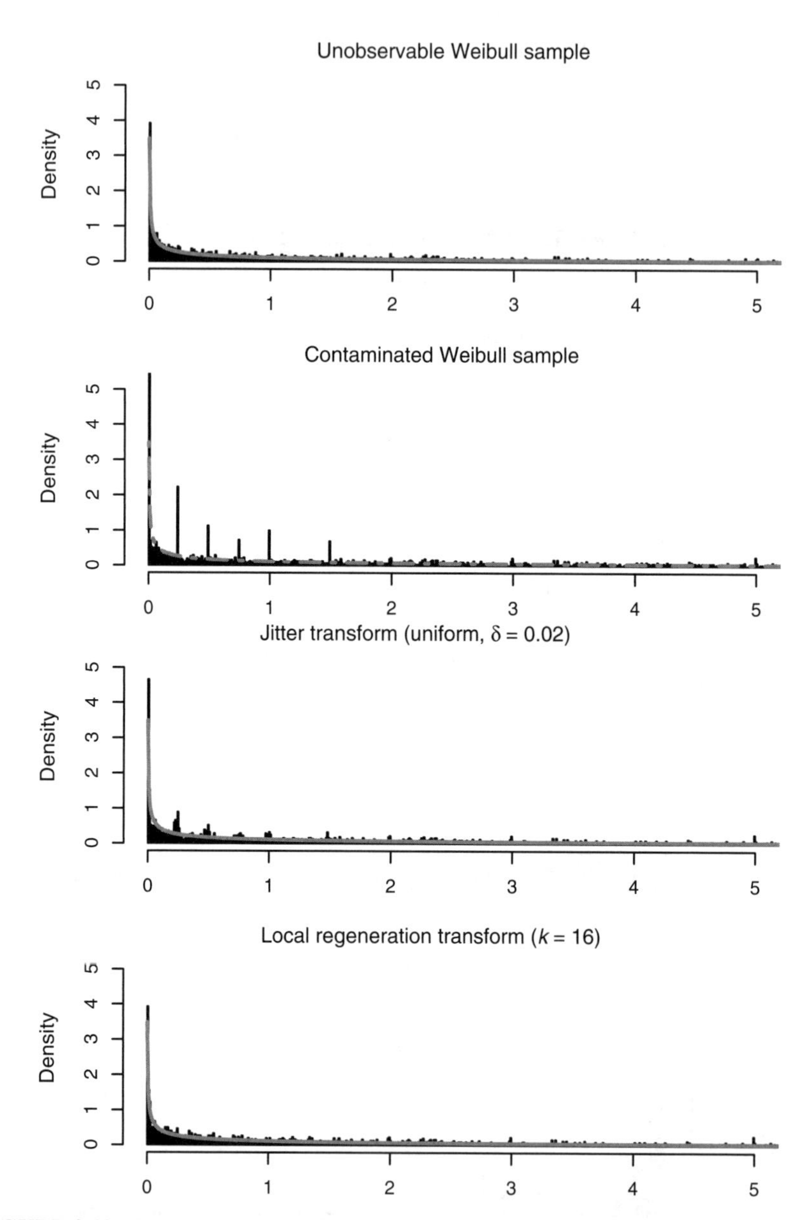

FIGURE 3.32 Histograms of the original, contaminated, jittered (uniform, $\delta = 0.02$), and locally regenerated ($k = 5$) data for the simulated Weibull data. The solid gray line is the Weibull density.

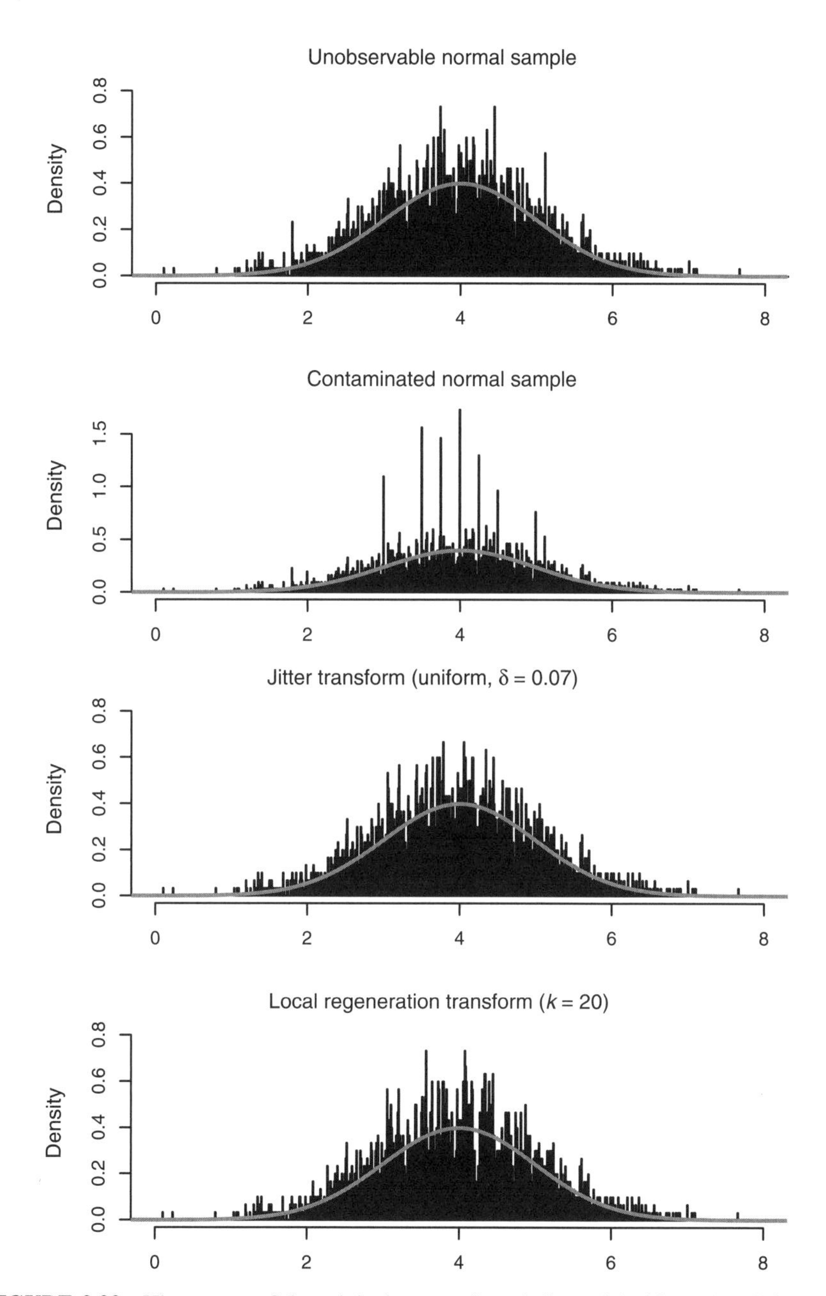

FIGURE 3.33 Histograms of the original, contaminated, jittered (uniform, $\delta = 0.07$), and locally regenerated ($k = 10$) data for the simulated normal data. The solid gray line is the normal density. Note that the y-axis scale is different for the contaminated data.

TABLE 3.3 Summary Statistics for Simulated Lognormal (Top), Weibull (Middle), and Normal (Bottom) Data

Lognormal ($\mu = 0, \sigma = 2$)	Mean (SD)	Median
Original	8.1494 (63.8572)	1.040
Contaminated	8.1482 (63.8574)	1.010
Jittered (uniform, $\delta = 0.07$)	8.1489 (63.8573)	1.030
Local regeneration ($k = 5$)	8.1532 (63.8572)	1.050
Weibull (γ = shape = 0.5, β = scale = 10)	**Mean (SD)**	**Median**
Original	19.7319 (44.5956)	4.465
Contaminated	19.7313 (44.5956)	4.465
Jittered (uniform, $\delta = 0.02$)	19.7316 (44.5955)	4.465
Local regeneration ($k = 16$)	19.7351 (44.5941)	4.465
Normal ($\mu = 4, \sigma = 1$)	**Mean (SD)**	**Median**
Original	3.9759 (1.0227)	3.990
Contaminated	3.9759 (1.0227)	4.000
Jittered (uniform, $\delta = 0.07$)	3.9763 (1.0229)	3.980
Local regeneration ($k = 20$)	3.9783 (1.0208)	3.990

Summaries are for the original data, the contaminated data, jitter-transformed data (with best parameters), and local regeneration-transformed data (with best parameters).

observed too-frequent values corresponding to the percentiles of the $LNorm(\hat{\mu}, \hat{\sigma})$, where $\hat{\mu}$ and $\hat{\sigma}$ are estimated from the CLNorm data, and then we "iron out" the too-frequent values within their corresponding percentile bin. The CWeibull and CNorm data are also "ironed out" using parameter estimates from the contaminated data.

Max-bin histograms are produced for each value of k, but only the best k (as determined in Section 3.8.8) is shown for each underlying distribution in the bottom panel of Figures 3.31–3.33. From the max-bin histograms, we see that the local regeneration transformation yields data that closely resemble the original data. In the Weibull case, it also appears to perform better than the jittering, probably because the too-frequent values are located in low-density areas of the distribution. In such cases, the local regeneration spreads out the too-frequent values over a wider neighborhood (because the percentiles of the distribution are farther away from each other). As with the jittering transform, the summary statistics for the transformed data, for all the three distributions, are extremely close to those of the original data, as shown in Table 3.3.

Choosing the Best Transformation Parameters While visual inspection of the transformation process (and in particular via max-bin histograms) is valuable in determining the parameters of the transformation (δ and the perturbing distribution in jittering, and k in the local regeneration transform), quantitative metrics to assess the data fit can also be helpful, especially for purposes of automation. To determine how well the "ironed out" data fit an underlying distribution, we use three parametric goodness-of-fit measures: the Anderson–Darling (AD), Cramer von Mises (CvM), and Kolmogorov (K) statistics. These are computed for each combination of distribution and δ for the jittering transform, and for each k for the local regeneration transform.

We define a "good" parameter as one that captures the underlying (unobserved) distribution, while altering the data as little as possible. In other words, we seek to obtain a reasonable distributional fit while minimizing the differences between original and transformed too-frequent values.

Jittering: Table 3.4 provides the goodness-of-fit test statistics and corresponding p-values (in parentheses) for the jittering transformation on each of the three simulated data sets. We tried two perturbation distributions (uniform and normal) and six values for δ. Note that when the goal is fitting a parametric distribution, we advocate the local regeneration transform over jittering. However, we do show here the results of the jittering transform for purposes of comparison and also to assess the ability of the jittering transform to recover the original generating distribution.

For simplicity, we discuss only the results using a uniform perturbation (which provided better results than normal perturbation). Recall that we pick the value corresponding to the smallest (and least significant) test statistic. For the lognormal data, the AD statistic suggests $\delta = 0.10$, while the CvM statistic suggests $\delta = 0.07$ (and the K statistic indicates a range of values between $\delta = 0.05$ and $\delta = 0.12$). Looking at the p-values, the least significant statistics are achieved for $\delta = 0.07$ or 0.10. Since our goal is to manipulate the data as little as possible, we choose $\delta = 0.07$. The Weibull data have more straightforward results, where the lowest test statistic for each of the three goodness-of-fit statistics is achieved at $\delta = 0.02$. In contrast, results for the normal data are very fuzzy, with no clear pattern as to which transformation fits best. This result is a manifestation of the robustness of the normality test, which—for our application—is a shortcoming.

Alternatively, we can also evaluate the smoothness of the max-bin histogram using kernel smoothers or deviations between pairs of neighboring frequencies (e.g., SADBNF). See Shmueli et al. (2008) for details.

Local Regeneration: Table 3.5 provides the goodness-of-fit statistics (and p-values, in parentheses) for the local regeneration transformation. In this case, we choose seven values for the number of bins k, ranging between 4 and 25. The lognormal data fit best when $k = 5$ (note that $k = 8$ appears worse than both $k = 5$ and $k = 10$, but this is most likely due to an artifact of these particular data). For the Weibull data, the best fit is achieved with $k = 16$. Finally, for the normal data the picture is again ambiguous, probably due to the strong robustness of these kinds of tests to deviation from normality.

Max-bin histograms of the transformed data (with the best parameters) are plotted in Figures 3.31–3.33, bottom panels. It can be seen that in each of the three cases, the transformed data no longer exhibit too-frequent values, and they appear very similar to the original "unobservable" data.

TABLE 3.4 Goodness-of-Fit Statistics for the Original, Contaminated, and Jittering-Transformed Data

Lognormal ($\mu = 0, \sigma = 2$)	AD	CvM	K	SADBNF
Original	1.07(0.25)	0.15(0.25)	0.02(0.25)	1997
Contaminated	15.66(0.00)	0.33(0.11)	0.04(0.00)	2523
Uniform, $\delta = 0.01$	7.40(0.00)	0.30(0.14)	0.04(0.00)	2224
Uniform, $\delta = 0.02$	4.15(0.01)	0.27(0.17)	0.03(0.00)	2264
Uniform, $\delta = 0.05$	1.52(0.18)	0.18(0.25)	0.02(0.15)	2128
Uniform, $\delta = 0.07$	**1.16(0.25)**	**0.15(0.25)**	**0.02(0.25)**	**2102**
Uniform, $\delta = 0.10$	1.12(0.25)	0.16(0.25)	0.02(0.25)	2131
Uniform, $\delta = 0.12$	1.52(0.18)	0.18(0.25)	0.02(0.18)	2119
Normal, $\delta = 0.01$	13.85(0.00)	0.33(0.12)	0.04(0.00)	2415
Normal, $\delta = 0.02$	8.54(0.00)	0.30(0.14)	0.04(0.00)	2211
Normal, $\delta = 0.05$	3.75(0.01)	0.25(0.20)	0.03(0.02)	2158
Normal, $\delta = 0.07$	2.27(0.07)	0.20(0.25)	0.02(0.08)	2140
Normal, $\delta = 0.10$	1.60(0.15)	0.18(0.25)	0.02(0.19)	2089
Normal, $\delta = 0.12$	1.24(0.25)	0.17(0.25)	0.02(0.25)	2106
Weibull ($\gamma = 0.5, \beta = 10$)	AD	CvM	K	SADBNF
Original	1.18(0.25)	0.17(0.25)	0.02(0.25)	3119
Contaminated	3.92(0.01)	0.23(0.22)	0.03(0.00)	3442
Uniform, $\delta = 0.01$	1.62(0.15)	0.19(0.25)	0.02(0.01)	3259
Uniform, $\delta = 0.02$	**1.44(0.20)**	**0.19(0.25)**	**0.02(0.25)**	**3233**
Uniform, $\delta = 0.05$	2.21(0.08)	0.20(0.25)	0.03(0.02)	3149
Uniform, $\delta = 0.07$	2.70(0.04)	0.22(0.24)	0.03(0.01)	3166
Uniform, $\delta = 0.10$	3.53(0.02)	0.26(0.18)	0.03(0.01)	3165
Uniform, $\delta = 0.12$	4.29(0.01)	0.29(0.14)	0.03(0.00)	3125
Normal, $\delta = 0.01$	2.81(0.03)	0.21(0.25)	0.03(0.02)	3357
Normal, $\delta = 0.02$	1.68(0.14)	0.19(0.25)	0.02(0.11)	3209
Normal, $\delta = 0.05$	1.57(0.16)	0.19(0.25)	0.02(0.09)	3324
Normal, $\delta = 0.07$	1.67(0.14)	0.19(0.25)	0.02(0.05)	3183
Normal, $\delta = 0.10$	2.08(0.09)	0.20(0.25)	0.03(0.02)	3149
Normal, $\delta = 0.12$	2.35(0.06)	0.21(0.25)	0.03(0.01)	3180
Normal ($\mu = 4, \sigma = 1$)	AD	CvM	K	SADBNF
Original	1.50(0.18)	0.26(0.18)	0.02(0.10)	1358
Contaminated	1.52(0.17)	0.27(0.17)	0.02(0.10)	1710
Uniform, $\delta = 0.01$	1.51(0.18)	0.26(0.18)	0.02(0.10)	1442
Uniform, $\delta = 0.02$	1.50(0.18)	0.26(0.18)	0.02(0.10)	1470

TABLE 3.4 ***(Continued)***

Normal ($\mu = 4, \sigma = 1$)	AD	CvM	K	SADBNF
Uniform, $\delta = 0.05$	1.51(0.18)	0.26(0.18)	0.02(0.10)	1462
Uniform, $\delta = 0.07$	1.47(0.19)	0.26(0.19)	0.02(0.10)	1378
Uniform, $\delta = 0.10$	1.50(0.18)	0.26(0.18)	0.02(0.10)	1398
Uniform, $\delta = 0.12$	1.59(0.16)	0.28(0.15)	0.02(0.10)	1422
Normal, $\delta = 0.01$	1.52(0.18)	0.27(0.17)	0.02(0.10)	1592
Normal, $\delta = 0.02$	1.51(0.18)	0.26(0.18)	0.02(0.10)	1408
Normal, $\delta = 0.03$	1.50(0.18)	0.26(0.18)	0.02(0.10)	1410
Normal, $\delta = 0.07$	1.50(0.18)	0.26(0.18)	0.02(0.10)	1406
Normal, $\delta = 0.10$	1.49(0.18)	0.26(0.18)	0.02(0.10)	1428
Normal, $\delta = 0.12$	1.51(0.18)	0.26(0.18)	0.02(0.10)	1420

3.8.4 Transforming Auction Data

We now apply the local regeneration transform to the auction surplus data for the purpose of obtaining data that can be described well by a parametric continuous distribution. Since the data are real (and not simulated), the underlying distribution is unknown. The best data fit is given by a three-parameter Weibull distribution (when using a log scale + 1), as can be seen from the probability plots in Figure 3.29. Without the log transformation, the data do not fit any distribution well.

By inspecting the max-bin histogram in Figure 3.28, too-frequent values appear to be located at {0, 0.01, 0.50, 0.51, 1.00, 1.01, 1.49, 1.50, 1.75, 2.00, 2.01, 2.50, 3.00, 3.50, 4.00, 4.50, 5.00}. These same values also appear in a max-bin histogram applied to the log-transformed data (top panel in Figure 3.34).

The first step is to estimate the three parameters of a Weibull distribution from the observed (contaminated) data. These turn out to be $\gamma = \text{shape} = 1.403793$, $\beta = \text{scale} = 2.129899$, and $\tau = \text{threshold} = -0.10379$. We then apply the local regeneration transformation using k bins from the estimated Weibull distribution, where k = 4, 5, 8, 10, 16, 20, and 25 (Table 3.6). We show the max-bin histograms of the transformed data only for select k values in Figure 3.34. The overlaid gray line corresponds to the three-parameter Weibull density estimate. Note that the y-axis for the original data is much larger than that for the transformed data. This emphasizes the extreme frequency of the zero value. For $k = 20$ and $k = 10$, there are still too many zero and near-zero values compared to a (theoretical) Weibull distribution. However, $k = 5$ appears to "iron out" the data quite well. This can also be seen in the corresponding probability plots in Figure 3.35, where a small "step" (near zero) is visible in all plots except for the data transformed with $k = 5$. Furthermore, the probability plot with $k = 5$ also appears closest to a straight line, even in the right tail. Thus, $k = 5$ appears to be the best choice, which is also supported by the goodness-of-fit statistics in Table 3.6.

TABLE 3.5 Goodness-of-Fit Statistics for the Original, Contaminated, and Local Regenerating Transformed Data

Lognormal ($\mu = 0, \sigma = 2$)	AD	CvM	K
Original	1.08(0.25)	0.15(0.25)	0.02(0.25)
Contaminated	15.66(0.00)	0.33(0.11)	0.04(0.00)
$k = 25$	4.36(0.01)	0.29(0.15)	0.03(0.00)
$k = 20$	3.24(0.02)	0.23(0.23)	0.03(0.00)
$k = 16$	2.83(0.04)	0.23(0.22)	0.03(0.02)
$k = 10$	1.38(0.21)	0.12(0.25)	0.02(0.16)
$k = 8$	1.77(0.12)	0.29(0.14)	0.02(0.06)
$\boldsymbol{k = 5}$	**0.90(0.25)**	**0.10(0.25)**	**0.02(0.25)**
$k = 4$	2.48(0.05)	0.38(0.09)	0.03(0.02)

Weibull ($\gamma = 0.5, \beta = 10$)	AD	CvM	K
Original	1.18(0.25)	0.17(0.25)	0.02(0.25)
Contaminated	3.92(0.01)	0.23(0.22)	0.03(0.00)
$k = 25$	2.03(0.09)	0.19(0.25)	0.02(0.10)
$k = 20$	1.37(0.22)	0.18(0.25)	0.02(0.25)
$\boldsymbol{k = 16}$	**1.46(0.25)**	**0.16(0.25)**	**0.02(0.25)**
$k = 10$	1.55(0.17)	0.19(0.25)	0.02(0.12)
$k = 8$	1.83(0.12)	0.18(0.25)	0.02(0.07)
$k = 5$	2.37(0.06)	0.20(0.25)	0.03(0.03)
$k = 4$	2.33(0.06)	0.18(0.25)	0.02(0.05)

Normal ($\mu = 4, \sigma = 1$)	AD	CvM	K
Original	1.50(0.18)	0.26(0.18)	0.02(0.10)
Contaminated	1.52(0.17)	0.27(0.17)	0.02(0.10)
$k = 25$	1.42(0.20)	0.24(0.21)	0.02(0.10)
$k = 20$	1.35(0.22)	0.24(0.22)	0.02(0.10)
$k = 16$	1.45(0.19)	0.25(0.20)	0.02(0.10)
$k = 10$	1.34(0.23)	0.22(0.24)	0.02(0.10)
$k = 8$	1.41(0.21)	0.25(0.20)	0.02(0.10)
$k = 5$	1.60(0.15)	0.24(0.21)	0.03(0.04)
$k = 4$	1.40(0.21)	0.21(0.24)	0.02(0.10)

To evaluate the deviation between observed and transformed data, we note that approximately 17% of the original data are considered too frequent. Comparing the summary statistics before and after the transformation (Table 3.7) shows that they are remarkably similar for all k values tested. This reassures us that the transformation has not altered the main features of the distribution.

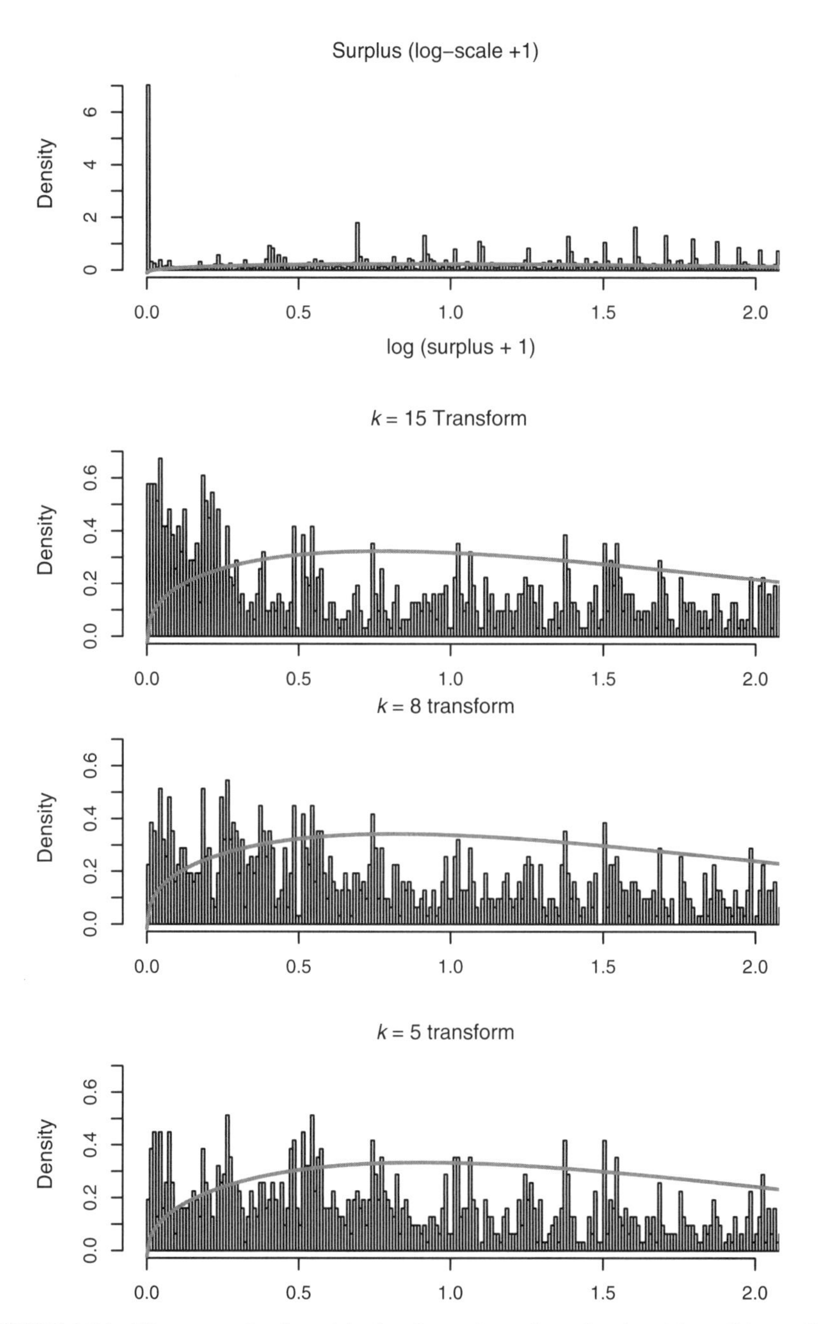

FIGURE 3.34 Histograms for the original and transformed surplus data. The solid gray line on the histogram is the Weibull density.

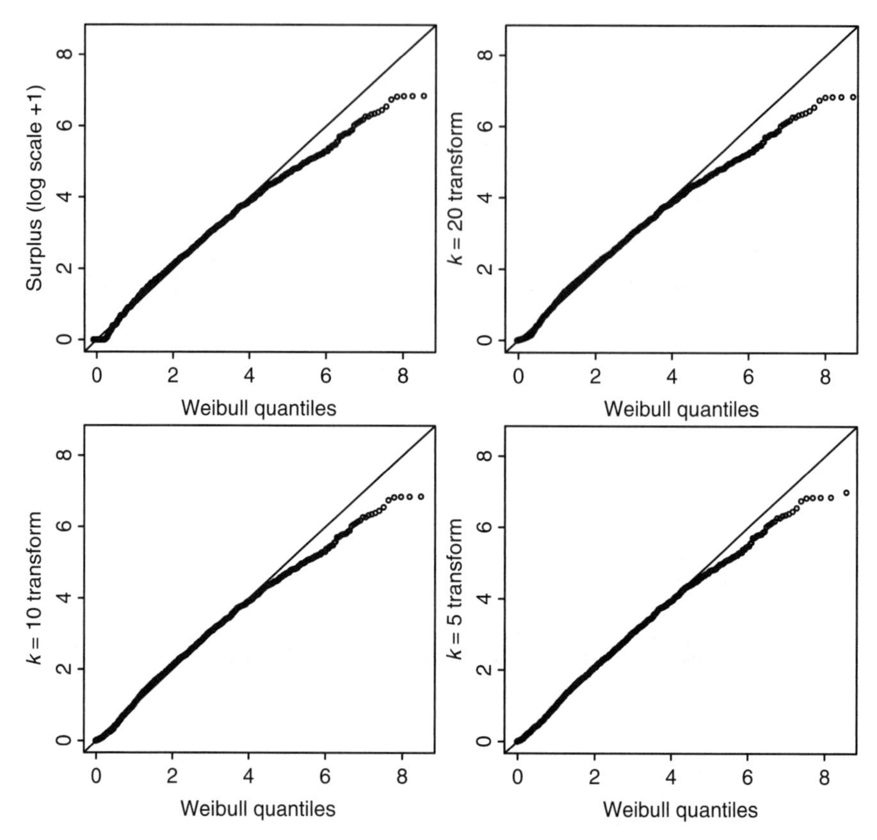

FIGURE 3.35 Weibull probability plots for the original (top left) and transformed ($k = 20$ top right, $k = 10$ bottom left, $k = 5$ bottom right) surplus data.

We can now use the transformed data for further statistical analyses. For example, Figure 3.36 compares the residuals from a linear regression model of surplus on price (both in log form) using the raw data (top panels) versus the transformed data with $k = 5$ (bottom panels). Transforming the raw data removes some of the unwanted patterns from the residuals, thereby making it a more appropriate model for consumer surplus in online auctions.

3.8.5 Conclusions: Data Transformation

Online auction data can have a structure that is not easily captured by standard distributions. One reason is their user-generated nature. We focused on one outcome that is semi-continuous data, that is, data that are too continuous to be considered discrete but spotted with many "too frequent" discrete values. We described two possible solutions for such semicontinuous data—jittering and local regeneration—both of which aim at generating data that are more continuous than the original (observed) data.

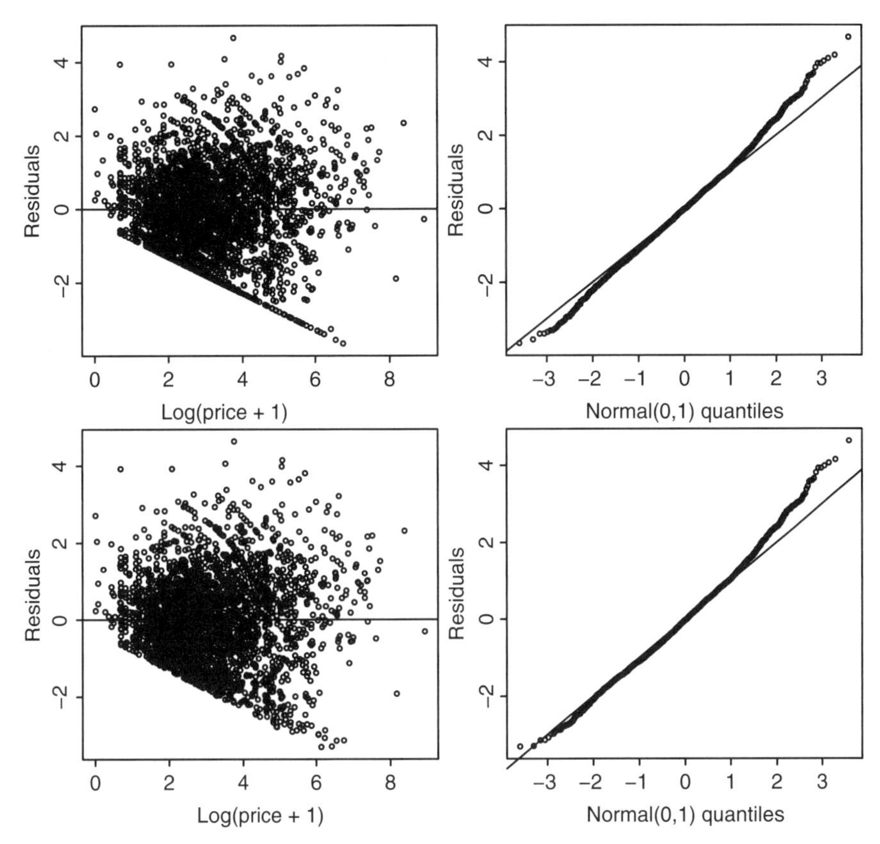

FIGURE 3.36 Before (top) and after (bottom) the transformation. *Left panels*: Residuals from a linear regression model of log(surplus + 1) on log(price + 1). *Right panels*: Normal probability plots for residuals.

TABLE 3.6 Goodness-of-Fit Statistics (and p-Values) for Surplus Original and Transformed Data (log scale + 1) Fit to a Weibull Distribution

	Anderson Darling	Cramer von Mises	Kolmogorov
Surplus	23.4733(0.0010)	2.4524(0.0010)	0.0567(0.0010)
$k = 25$	22.4062(0.0010)	2.7194(0.0010)	0.0589(0.0010)
$k = 20$	20.9232(0.0010)	2.6829(0.0010)	0.0591(0.0010)
$k = 16$	18.3008(0.0010)	2.4052(0.0010)	0.0536(0.0010)
$k = 10$	12.2451(0.0010)	1.8150(0.0010)	0.0474(0.0010)
$k = 8$	9.7820(0.0010)	1.5658(0.0010)	0.0466(0.0010)
$\mathbf{k = 5}$	**5.2922(0.0029)**	**0.8394(0.0062)**	**0.0352(0.0010)**
$k = 4$	6.5514(0.0010)	1.0882(0.0021)	0.0402(0.0010)

TABLE 3.7 Summary Statistics for the Original and Transformed Surplus Data

Weibull Fit	Mean (Standard Deviation)	Median
surplus	19.1882 (64.7689)	4.49
$k = 25$	19.2020 (64.7653)	4.48
$k = 20$	19.1941 (64.7672)	4.40
$k = 16$	19.1957 (64.7668)	4.45
$k = 10$	19.2061 (64.7640)	4.49
$k = 8$	19.2149 (64.7619)	4.52
$k = 5$	19.2228 (64.7589)	4.49
$k = 4$	19.2732 (64.7450)	4.54

One method is directly aimed at fitting a parametric continuous distribution, while the other one is nonparametric and leads to continuous data of an unspecified parametric form. The idea behind both transformations is to replace too-frequent values with values randomly generated within their neighborhood. The difference between the two transformations is with respect to the definition of a neighborhood. While jittering defines a fixed-size neighborhood that is anchored around the too-frequent values, local regeneration uses percentiles of the fitted parametric distribution to define neighborhoods. In the latter case, the size of the neighborhood depends on the shape of the fitted distribution, with wider neighborhoods in tail or other low-density areas. The transformed data from the two transformations therefore defer when too-frequent values are located in low-frequency areas of the data.

The proposed transforms, and in particular the local regeneration transformation, are similar in flavor to the well-known and widely used *Box–Cox transformation*. In both the cases, the goal is to transform data into a form that can be fed into standard statistical methods. Similar to the Box–Cox transformation, the process of finding the best transformation level (λ in the Box–Cox case, δ in jittering, and k in local regeneration) is iterative.

While we illustrate these transformations in the context of univariate data sets, one could equally imagine semicontinuous *multivariate* data. To transform multivariate semicontinuous data, the jittering or local regeneration transformations could be applied univariately to each of the variables, using the correlation (or another dependence measure) for choosing the parameter k or δ. For the case of fitting the data to a specific parametric multivariate distribution (with the multivariate normal distribution as the most popular choice), binning could be formulated in the multivariate space. However, this would also require making certain assumptions about how the contamination mechanism affects the different variables and whether the contamination itself is dependent across the variables.

3.9 EXPLORING ONLINE AUCTIONS: FUTURE RESEARCH DIRECTIONS

The visualizations and exploration tools described here are meant for displaying and studying data that have already been collected and stored. Such historic data are usually used for learning about a variety of different phenomena such as bidding strategies and a seller's trustworthiness. One of the next steps is to observe and process the data in real time. Several of the visualizations that we suggested can be used for real-time visualizations with little or no change: An auction profile can be used for monitoring an ongoing auction as long as the incoming bids are available. In eBay, for example, a bid is only disclosed once it has been exceeded, so that at any point in time we have available the sequence of bids until then, excluding the highest bid.

Because the auction duration is known at the auction start, the horizontal axis in the various visualizations can be set accordingly. An example of a slight modification needed for an auction calendar that is updated in real time is to incorporate right censored auctions.

Methods based on hierarchical aggregation (STAT-zooming) require more significant modification. Finally, real-time data and their availability also call for new visualizations that would directly target their structure and the goals of their analysis.

With respect to implementation of the proposed visualizations, most can be coded by using standard software. We generated all the plots using Matlab, R, and TIBCO Spotfire. The latter was use for all the interactive visualizations. Similarly, the combination of time series and cross-sectional and networked data require adequate database structure.

Another issue that we have discussed in this section is the unusual data structures that arise from online auction mechanisms and the need to tailor and develop adequate statistical and visualization methods for representing, plotting, and exploring such data. There is room for developing further such methods as well as exploring additional unusual data structures that arise in this context. Network structures are one such challenge.

4

MODELING ONLINE AUCTION DATA

In this chapter, we discuss different options for modeling online auction data. We start with "modeling basics" that relate to the bid history. In particular, we present different approaches for representing the price path as a continuous curve (a functional object). Considering the entire price path (as opposed to only the opening and closing prices) has several advantages: First, the path summarizes the entire auction in a very concise fashion. Of course, if we are not interested in the entire auction but only in select bidders, then we can apply the same principles to the sequence of bids placed by individual bidders. Looking at the price path allows us to quantify price dynamics (mentioned in Chapter 3 and discussed in detail in Section 4.2). By "dynamics" we mean the velocity of price and the rate at which this velocity changes. Price dynamics are a novel concept in auction research that also allows us to make more accurate predictions of the outcome of an auction (see Chapter 5). We will also see that dynamics can measure otherwise unobservable bidding features such as competition between bidders, or auction "fever" (see Chapter 5).

In Section 4.3, we discuss several ways of modeling auction competition. By "competing auctions," we refer to a set of auctions that all sell the same (or similar) item during the same period of time and, as a consequence, compete for the same bidder. The resulting interrelationships across auctions are important and need to be carefully incorporated into the modeling process.

In Section 4.4, we describe different ways for modeling the arrival of bids and bidders during an online auction. While arrival processes are often described with

Modeling Online Auctions, by Wolfgang Jank and Galit Shmueli

a Poisson distribution, we will show that a simple Poisson model does not suffice to capture the different stages of an auction.

Finally, in Chapter 4.5, we discuss a novel aspect of online auctions: the networks generated by bidders and sellers. In particular, we motivate the existence of auction networks and propose a novel way via functional data analysis to capture the effects of sellers and bidders on the outcome of an auction.

4.1 MODELING BASICS (REPRESENTING THE PRICE PROCESS)

Figure 4.1 shows an example of the raw bidding information (bid history) as it can be found on many online auction sites (eBay, in this case). The top displays a summary of the cross-sectional auction information—the item for sale, the current bid, the starting bid, the number of bids received, the auction start and end times, and the seller's username with his/her rating. In this section, we focus on the bottom of the page, which shows the detailed information on the history of bids; that is, it shows the sequence of bids with their associated time stamps. (Note that in this case, bids are arranged by magnitude, not by their temporal order.)

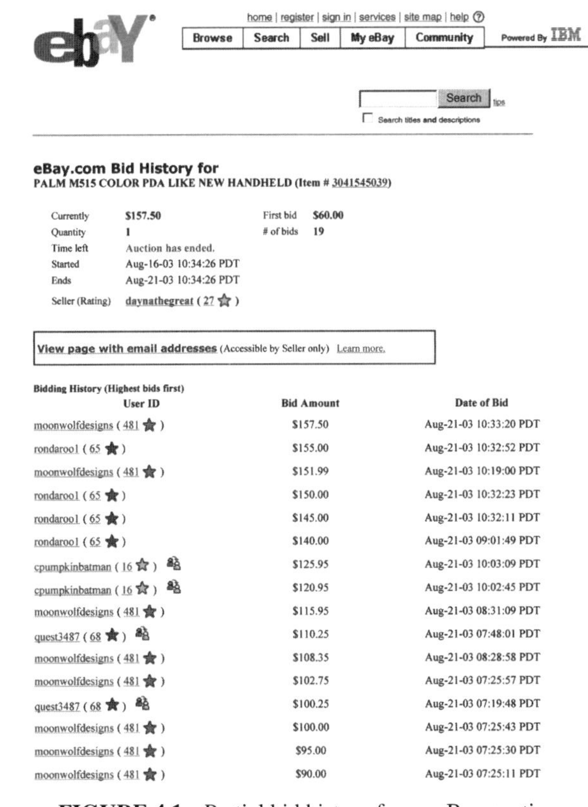

home | register | sign in | services | site map | help

Browse | Search | Sell | My eBay | Community — Powered By IBM

Search tips

Search titles and descriptions

eBay.com Bid History for
PALM M515 COLOR PDA LIKE NEW HANDHELD (Item # 3041545039)

Currently	$157.50	First bid	$60.00
Quantity	1	# of bids	19
Time left	Auction has ended.		
Started	Aug-16-03 10:34:26 PDT		
Ends	Aug-21-03 10:34:26 PDT		
Seller (Rating)	daynathegreat (27)		

View page with email addresses (Accessible by Seller only) Learn more.

Bidding History (Highest bids first)

User ID	Bid Amount	Date of Bid
moonwolfdesigns (481)	$157.50	Aug-21-03 10:33:20 PDT
rondarool (65)	$155.00	Aug-21-03 10:32:52 PDT
moonwolfdesigns (481)	$151.99	Aug-21-03 10:19:00 PDT
rondarool (65)	$150.00	Aug-21-03 10:32:23 PDT
rondarool (65)	$145.00	Aug-21-03 10:32:11 PDT
rondarool (65)	$140.00	Aug-21-03 09:01:49 PDT
cpumpkinbatman (16)	$125.95	Aug-21-03 10:03:09 PDT
cpumpkinbatman (16)	$120.95	Aug-21-03 10:02:45 PDT
moonwolfdesigns (481)	$115.95	Aug-21-03 08:31:09 PDT
quest3487 (68)	$110.25	Aug-21-03 07:48:01 PDT
moonwolfdesigns (481)	$108.35	Aug-21-03 08:28:58 PDT
moonwolfdesigns (481)	$102.75	Aug-21-03 07:25:57 PDT
quest3487 (68)	$100.25	Aug-21-03 07:19:48 PDT
moonwolfdesigns (481)	$100.00	Aug-21-03 07:25:43 PDT
moonwolfdesigns (481)	$95.00	Aug-21-03 07:25:30 PDT
moonwolfdesigns (481)	$90.00	Aug-21-03 07:25:11 PDT

FIGURE 4.1 Partial bid history for an eBay auction.

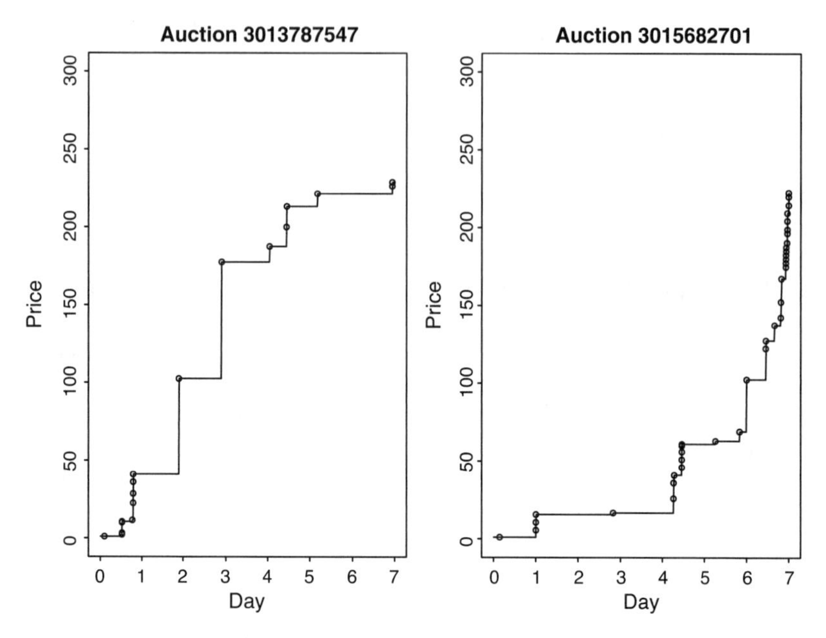

FIGURE 4.2 Live bids (circles) and associated step function of price as seen during the live auction.

A factor that complicates modeling is that, oftentimes, the information displayed after the auction closes is not identical to what is seen during the live auction. In the case of eBay, while bidders see the current price (i.e., the second highest proxy bid plus an increment) during the live auction, Figure 4.1 displays only the proxy bids. To recover the live price, we first must transform the proxy bids (which, in the case of eBay, can be done using a published bid-increment table; see Section 3.1). Figure 4.2 displays the live bid step function for two sample auctions. One important observation in Figure 4.2 is with respect to the *shape* of the step functions: While the price in the left function increases fast early, it levels off toward the end of the auction. This is very different from the right function where price increases are very slow in the beginning, only to culminate in enormous price jumps at the end. From a modeling point of view, finding a single model that can accommodate all possible shapes is challenging. Moreover, we can also see that the distribution of bids is very uneven. In the left price function, most bids arrive early in the auction, but only few arrive toward the end. In fact, the auction middle is very sparse and sees only the occasional bid. In contrast, in the right price function the bid frequency is extremely high , especially toward the end of the auction. Finding models that can handle these very uneven bid arrivals is not straightforward. In fact, many smoothing methods encounter problems when faced with sparse data such as in the middle of the left price function. On the other hand, while too many data points are typically not a problem, there is the risk that the smoother will not represent individual bids sufficiently well.

4.1.1 Penalized Smoothing Splines

One of the (computationally and conceptually) simplest and also most widely used smoothing methods is the penalized smoothing spline (Ruppert et al., 2003; Ramsay and Silverman, 2005). Let $\tau_1, \ldots, \tau_L$ be a set of knots. Then, a polynomial spline of order p is given by

$$f(t) = \beta_0 + \beta_1 t + \beta_2 t^2 + \cdots + \beta_p t^p + \sum_{l=1}^{L} \beta_{pl}(t - \tau_l)_+^p, \tag{4.1}$$

where $u_+ = uI_{[u \geq 0]}$ denotes the positive part of the function u. Define the roughness penalty

$$\text{PEN}_m(t) = \int \{D^m f(t)\}^2 dt, \tag{4.2}$$

where $D^m f$, $m = 1, 2, 3, \ldots$, denotes the mth derivative of the function f. The penalized smoothing spline f minimizes the penalized squared error

$$\text{PENSS}_{\lambda,m} = \int \{y(t) - f(t)\}^2 dt + \lambda \text{PEN}_m(t), \tag{4.3}$$

where $y(t)$ denotes the observed data at time t and the smoothing parameter λ controls the trade-off between data fit and smoothness of the function f. Using $m = 2$ in (4.3) leads to the commonly encountered cubic smoothing spline.

We use the same family of smoothers across all auctions, that is, we use the same spline order, the same set of knots, and the same smoothing parameters. The reason for choosing the same smoothing family is that we do not want to inflict systematic bias due to different smoothers and/or different smoothing parameters. The ultimate goal is that observed differences in the price paths are attributable to differences in the auction, the seller or the bidders *only*, rather than to differences in the way we estimate the price paths.

In the case of penalized splines, our choice of spline order, smoothing parameter, and knots are closely tied to our research goal and the nature of the data. In general, choosing a spline of order p guarantees that the first $p - 2$ derivatives are smooth (Ramsay and Silverman, 2002). Since we are interested in studying at least the first two derivatives of the price curves (to capture price velocity and acceleration), we use splines of order 5. Knot locations are chosen according to expected changepoints in the curves. In our case, the selection of knots is based on the empirical bidding frequencies, and, in particular, accounts for the phenomenon of "sniping" or "last-moment bidding" (Roth and Ockenfels, 2002; Shmueli et al., 2007). To capture the increased bidding activity at the end, we place an increasing number of knots toward the auction end. For instance, for a 7-day auction, we place seven equally spaced knots every 24 h along the first 6 days of the auction, that is, $\tau_l = 0, \ldots, 6$; $l = 1, \ldots, 7$. Then, over the first 18 h of the final day, we place knots over shorter intervals of 6 h each, that is,

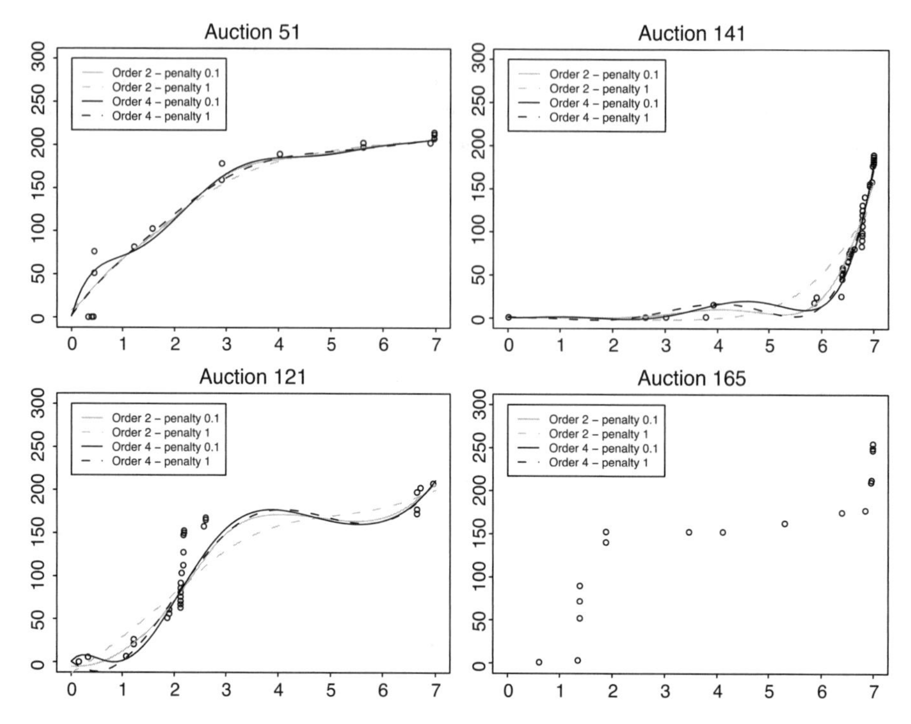

FIGURE 4.3 Smoothing splines for four sample auctions.

$\tau_8 = 6.25$, $\tau_9 = 6.5$, and $\tau_{10} = 6.75$. Finally, we divide the last 6 h of the auction into four intervals of 1.5 h each, letting $\tau_{11} = 6.8125$, $\tau_{12} = 6.8750$, $\tau_{13} = 6.9375$, and $\tau_{14} = 7.0000$.

While the number and the location of the knots have a strong impact on the resulting curve, the choices of the smoothing parameter λ and the spline order p are equally important. Figure 4.3 illustrates this for four sample auctions. Note that for all four auctions, we hold the number and the location of knots constant and only vary λ and p.

We can see that some combinations of λ and p work better than others. However, what works well for one auction may not necessarily work for another auction. For instance, a spline with $\lambda = 1$ and $p = 2$ (gray dashed lines) works well for auction #51, but it fails miserably for auction #165. In general, if we choose a small penalty term (and/or a large spline order), then the resulting spline may fit the data very well, but it may not be monotonically increasing—which, conceptually, does not make any sense for the price process of an ascending auction. On the other hand, if we set the penalty term to be large (To "force" the spline to be almost linear and hence monotone), then the resulting function does not necessarily capture the data very well. This trade-off between data fit and monotonicity is one of the biggest disadvantages of penalized smoothing splines, at least in the context of online auctions. In later sections, we will therefore discuss alternatives to the penalized smoothing spline. Our main focus will be on exploring alternatives that preserve the monotonicity of the price process. We will see though that one typically encounters a trade-off between retaining monotonicity, preserving flexibility of the model, and computational efficiency.

In the following section, we discuss one of the main advantages of representing price paths in a smooth fashion: the ability to estimate an auction's price dynamics.

4.1.2 Price Dynamics

The smooth function $f(t)$ in equation (4.1) estimates the price at any time t during an auction. We refer to $f(t)$ as the price path or sometimes also as the *price evolution*. While $f(t)$ describes the exact *position* of price for any t, it does not reveal how fast the price is *moving*. Attributes that we typically associate with a moving object are its *velocity* (or *speed*) and *acceleration*. Because we use smoothing splines to obtain $f(t)$, velocity and acceleration can be readily computed via the first and second derivatives of $f(t)$, respectively.[1]

Figure 4.4 shows the price curve $f(t)$ with its first and second derivatives for two sample auctions. Although the price path in both auctions is somewhat similar (i.e., monotonically increasing), the price dynamics are quite different: In the first auction (left panel), velocity and acceleration decrease toward the auction end. In contrast, price velocity and price acceleration increase in the second auction (right panel). This suggests that the price formation process can be quite heterogeneous across auctions, even when considering very similar auctions (i.e., the same item, sold during the same time period, in an auction with the same duration, as is the case here). We take this as evidence that price dynamics capture more information than what the price path can convey.

Figure 4.5 shows the price curves and their dynamics for a sample of 183 auctions (which all sell the identical item during the same period of time). There is a clear variation in the price formation process: While some auctions start out at a low price and shoot up toward the end, others maintain a relatively high price throughout the entire auction. Differences in the price formation are easier to detect in the price dynamics plots: The first derivative of the price curve shows that, on average, price increases quickly during the first part of the auction. The velocity then slows down, only to increase sharply again after day 5. A similar picture is seen for the average price acceleration: It is high at the onset of the auction, and then drops below zero ("deceleration") only to sharply increase again toward the end.

Although it is tempting to talk about "typical" dynamics for these data, we notice that there is significant variation among the curves. For instance, not all auctions show increasing price acceleration toward the end. In fact, in some auctions the price decelerates and finishes at large negative acceleration. Similarly, in many auctions there is no increase in the price velocity during the first part of the auction. All this suggests that the price formation process of similar auctioned items is not as

[1] Computation of the derivatives has to be done with care when the data are sparse, as is often the case in online auctions; see also Liu and Müller (2008, 2009).

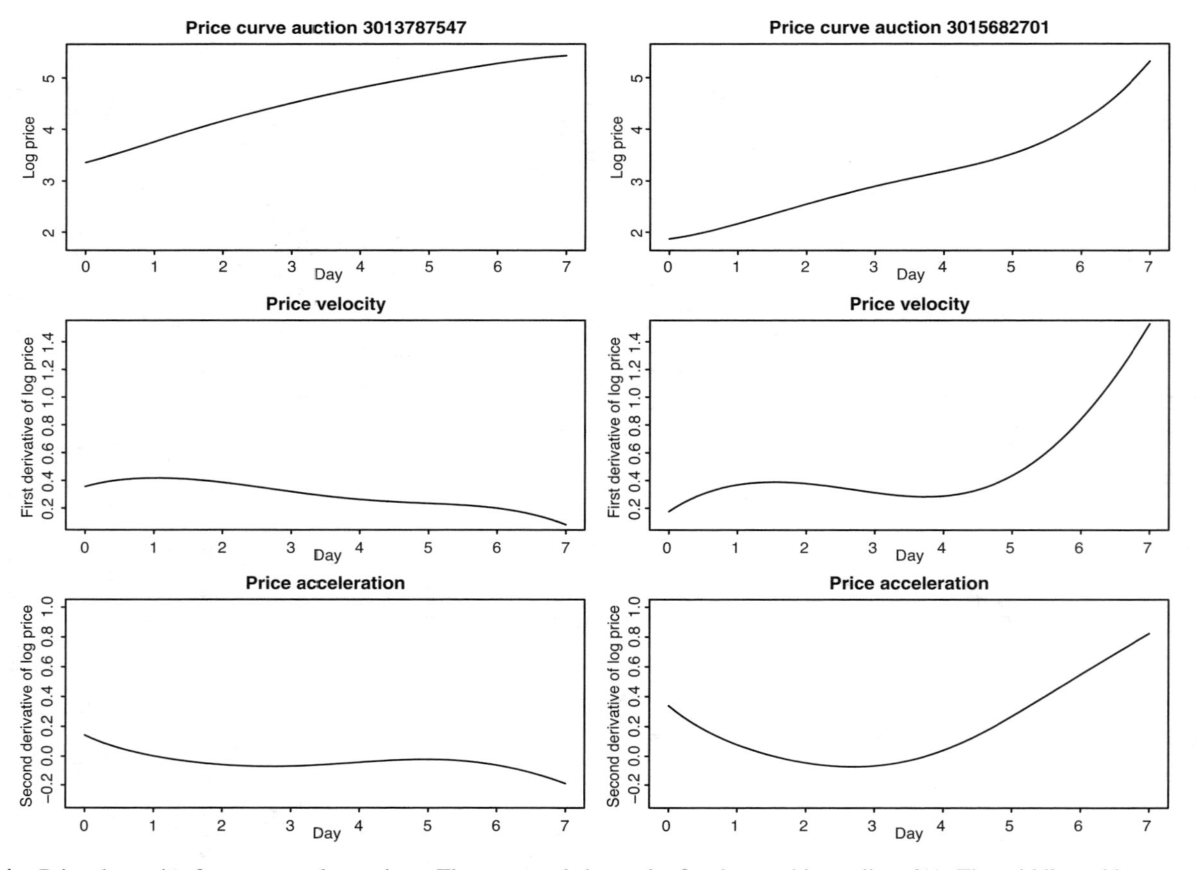

FIGURE 4.4 Price dynamics for two sample auctions. The top panel shows the fitted smoothing spline $f(t)$. The middle and bottom panels show the first and second derivatives $f'(t)$ and $f''(t)$, respectively.

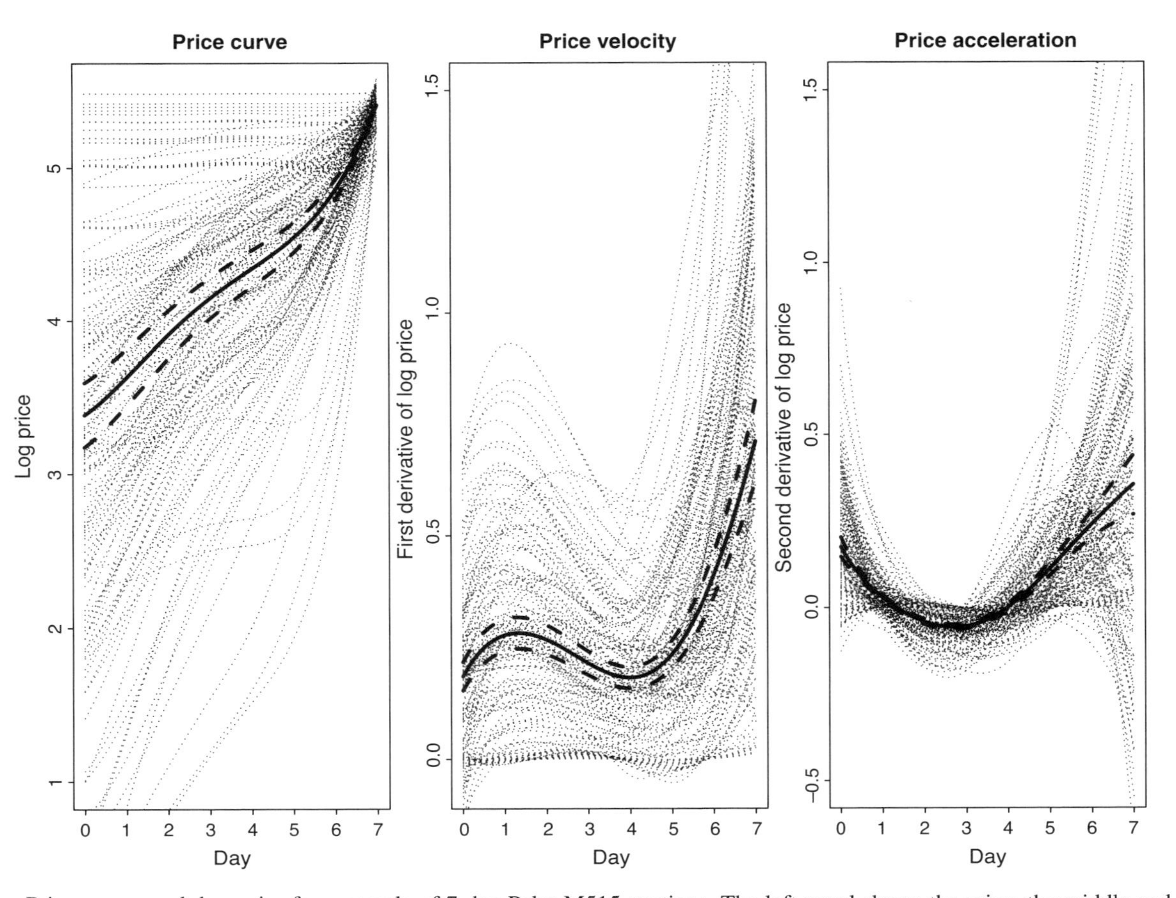

FIGURE 4.5 Price curves and dynamics for a sample of 7-day Palm M515 auctions. The left panel shows the price; the middle and right panels show the velocity (first derivative) and acceleration (second derivative), respectively.

homogeneous as expected (Jank and Shmueli, 2008b; Jank et al., 2008b; Shmueli and Jank, 2008; Wang et al., 2008a,b). In later sections, this observation will lead us to specifically investigate models that can explain differences in the price path and, more importantly, differences in the price dynamics.

4.1.3 Monotone Splines

We have pointed out earlier that using penalized smoothing splines leads to several challenges in the online auction context. First, and most detrimental, is that the created functions are not necessarily monotone nondecreasing—a contradiction to the ascending auction paradigm of, for example, eBay. Second, the functions are often very wiggly (or variable), especially at the ends. This is particularly egregious in the online auction context where the start and the end of the auction are of major importance (e.g., to forecast the outcome of an auction). Finally, there are multiple decisions about smoothing parameters that must be made in advance: the number and position of knots, the polynomial order, and the roughness penalty parameter λ.

Since the bidding process is naturally nondecreasing, Hyde et al. (2006) proposed the use of monotone smoothing splines to represent the price process. The idea behind monotone smoothing (Ramsay, 1998) is that monotone increasing functions have a positive first derivative. The exponential function has this property and can be described by the differential equation $f'(t) = w(t)f(t)$. This means that the rate of change of the function is proportional to its size. Consider the linear differential equation

$$D^2 f(t) = w(t)Df(t). \tag{4.4}$$

Here, $w(t) = \frac{D^2 f(t)}{Df(t)}$ is the ratio of the acceleration and velocity. It is also the derivative of the logarithm of velocity that always exists (because we define velocity to be positive). The differential equation has the following solution:

$$f(t) = \beta_0 + \beta_1 \int_{t_0}^{t} \exp\left(\int_{t_0}^{v} w(v)dv \right) du, \tag{4.5}$$

where t_0 is the lower boundary over which we are smoothing. After some substitutions (see Ramsay and Silverman (2005)), we can write

$$f(t) = \beta_0 + \beta_1 e^{wt} \tag{4.6}$$

and estimate β_0, β_1, and $w(t)$ from the data. Since $w(t)$ has no constraints, it may be defined as a linear combination of K known basis functions (i.e., $w(t) = \sum_k c_k \phi_k(t)$). Examples of basis functions are $\phi_k(t) = t$, representing a linear model or $\phi_k(t) =$

$\log(t)$, which is a nonlinear transformation of the inputs. The penalized least squares criterion is thus

$$\text{PENSSE}_\lambda = \sum_i [y_i - f(t)]^2 + \lambda \int_0^T [w^2(t)]^2 dt. \tag{4.7}$$

While monotone smoothing solves the wiggliness problem of the penalized smoothing splines, some of the same challenges remain and new ones arise. First, monotone smoothing is computationally very intensive. The larger the number of bids, the longer the fitting process. In our experience, monotone splines take almost 100 times longer to fit compared to penalized smoothing splines (Jank et al., 2009). Moreover, as with smoothing splines, the researcher must also predetermine the number and location of knots and the roughness parameter λ to obtain a reasonable data fit.

4.1.4 Four-Member Parametric Growth Family

An alternative to using purely nonparametric smoothing splines or monotone splines is to find a parametric family of models that capture the price paths of auctions. This idea was developed in Hyde et al. (2008). The proposed parametric family is based on four growth models that are able to capture types of price evolution most commonly found in auctions. These include exponential growth, logarithmic growth, logistic growth, and reflected-logistic growth models. The approach is elegant, computationally fast, and parsimonious. It allows automated fitting, and there is no need to specify any parameters in advance. The models are theoretically relevant (i.e., preserve monotonicity) and provide insight into the price process in online auctions. We describe each model in detail next.

Exponential Model: Exponential growth has been used for describing a variety of natural phenomena including the dissemination of information, the spread of disease, and the multiplication of cells in a petri dish. In finance, the exponential equation is used to calculate the value of interest-bearing accounts compounded continuously. The idea behind exponential growth is that the rate of growth is proportional to the function's current size; that is, growth follows the differential equation

$$Y'(t) = rY(t), \tag{4.8}$$

or the equivalent equation

$$Y(t) = Ae^{rt}, \tag{4.9}$$

where t is time and $r > 0$ is the growth constant. Equivalently, exponential decay, when $r < 0$, can model phenomena such as the half-life of an organic event.

From a theoretical standpoint, exponential growth can describe a price process for auctions where there are gradual price increases until mid-to-late auction and a price

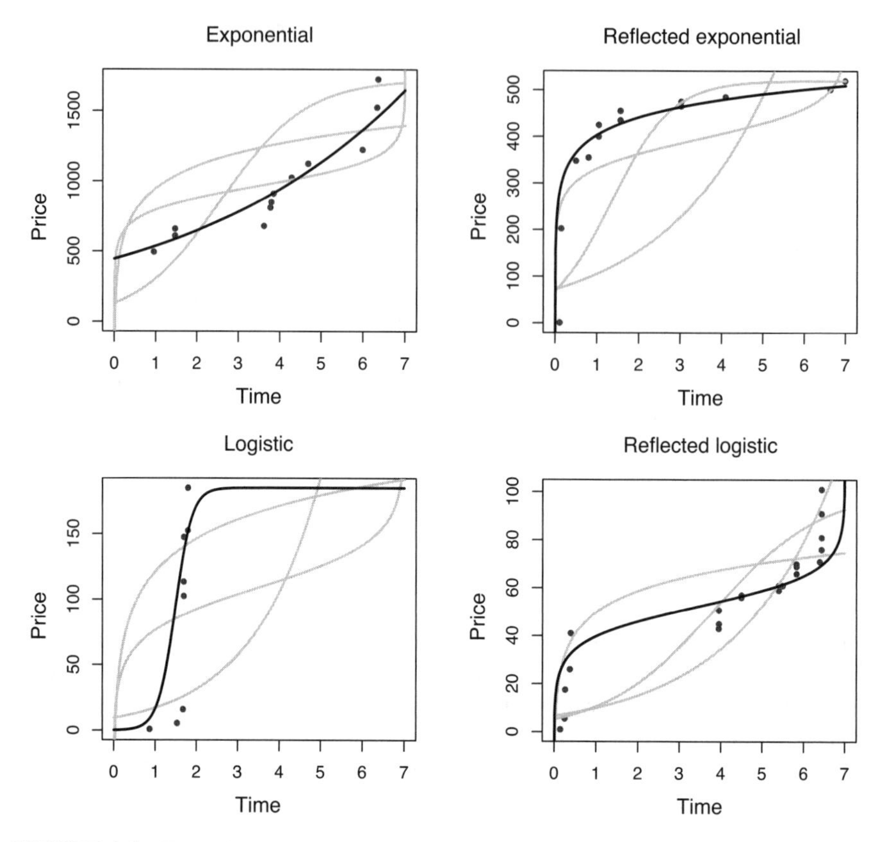

FIGURE 4.6 Example growth curves. The black line shows the model that best fits to the data; the gray lines show the fit of the remaining three growth models.

jump toward the end. This is reminiscent of the "j"-shaped price curves that Hyde et al. (2006) found (see Section 3.4). An example of an auction that is well captured by exponential growth is shown in the top left panel (black line) of Figure 4.6.

Logarithmic Model: The inverse of the exponential function

$$Y(t) = \frac{\ln(t/A)}{r}, \tag{4.10}$$

which Hyde et al. (2006) called logarithmic growth, also approximates price processes well. They chose a form of the logarithmic model that is the mapping of the original exponential model over the line $y = x$. This type of growth occurs when early bidding increases the price early in the auction, but because of the existence of a market value, price flattens out for the remainder of the auction; see also the top right panel in Figure 4.6. This type of price behavior tends to be rare, as most bidders do not wish to reveal their bids early in the auction. However, inexperienced bidders who do not understand the proxy bidding mechanism on eBay have been shown to bid high early

(Bapna et al., 2004). In this model, the velocity starts at its maximum and then decays to little or zero velocity as the auction progresses. The acceleration is always negative, since price increases more slowly throughout the auction and approaches zero at the end of the auction (where very little change in price is occurring).

Logistic Model: While exponential growth often makes sense over a fixed period of time, in many cases growth cannot continue indefinitely. For example, there are only a finite number of people to spread information or a disease; the petri dish can only hold a maximal number of cells. In auctions, it is likely that growth starts out exponentially, then slows as competition increases, and finally reaches a limit or "carrying capacity." In many auctions, the sale item can be purchased in other venues, such as online and brick-and-mortar stores; hence, there is a "market value" that caps price. A typical application of the logistic equation is in population growth. In the beginning, there are seemingly unlimited resources and the population grows increasingly fast. At some point, competition for food, water, land, and other resources slows down the growth; however, the population is still growing. Finally, overcrowding, lack of food, and susceptibility to disease limit the population to some maximal carrying capacity.

The logistic model is given by

$$Y(t) = \frac{L}{1 + Ce^{rt}}, \tag{4.11}$$

and the differential equation is

$$Y'(t) = rY(t)\left(\frac{Y(t)}{L} - 1\right), \tag{4.12}$$

where L is the carrying capacity, t is time, r is the growth rate, and C is a constant. Logistic growth can also be explained in the auction context. It forms a stretched out "s"-shaped curve, discussed by Hyde et al. (2006), where the price increases slowly, then jumps up during mid-auction, and finally levels off toward the end of the auction (see Section 3.4). The closing price is analogous to the carrying capacity L in the logistic growth function. It is necessary for the price to flatten out near the end because the price jumped up to near market value in the middle of the auction, and smart bidders will make sure not to overpay for the item. This price process can been seen in Figure 4.6 (bottom left panel).

Reflected-Logistic Model: Another common price process in online auctions is the inverse of logistic growth, or reflected-logistic growth, given by the function

$$Y(t) = \frac{\ln(\frac{L}{t} - 1) - \ln(C)}{r}. \tag{4.13}$$

This type of growth occurs when there is some early bidding in the auction that results in a price increase, followed by little to no bidding in the middle of the auction, and then another price increase as the auction progresses toward its close. The early price increase is indicative of early bidding by inexperienced bidders (Bapna et al., 2004), and the price spike at the end may be caused by sniping (Bajari and Hortacsu, 2003; Roth and Ockenfels, 2002). An example of reflected-logistic growth is shown in the bottom right panel of Figure 4.6.

Fitting Growth Models: A simple and computationally efficient method for fitting each of the growth models is by linearizing the function and then performing least squares. Since we are especially interested in obtaining an accurate fit at the beginning and end of the auction, it is usually necessary to add two additional points to the data sequence of each auction: the price at the beginning and at the end of the auction. It is not necessary to add these extra points for logistic growth, where the maximum price is already incorporated into the function by defining $L = \max(y_j)$. Furthermore, empirical evidence shows that auctions whose underlying price process is logistic tend to start close to zero, where logistic growth must start.

Fitting Exponential Growth: The exponential growth model from equation (4.9) can be linearized as

$$\ln Y = \ln A + rt. \tag{4.14}$$

This model is fitted directly to the live bids with two additional points in order to constrain the price at the start and the end of the auction. In this case, two parameters, A and r, are estimated. Although we can fix A as the opening price since $Y(t = 0) = Ae^{0r} = A$, we prefer to estimate both parameters for two reasons: first, empirical evidence shows that the two-parameter estimation allows a better fit at the end of the auction, and second, the other three growth models require estimating two parameters. Hence, it is easier to compare model fit.

Fitting Logarithmic Growth: The logarithmic growth model is given in equation (4.10). As with exponential growth, two extra points are added to the live bids to ensure good fit at the start and end of the auction. Note that we cannot reduce this function to one parameter because when $t = 0$, $\ln(0)$ does not exist. Also, we cannot linearize this function. Therefore, rather than using optimization methods for parameter estimation where guesses of the initial value are necessary, we fit $T(y) = Ae^{ry}$ and then linearize as in the exponential growth case. This time least squares minimizes over time instead of price.

Fitting Logistic Growth: The logistic growth model from equation (4.11), where L is the distribution's asymptote, can be linearized as

$$\ln\left(\frac{L}{y} - 1\right) = \ln(C) + rt. \tag{4.15}$$

We know that $\lim_{t\to l} = L$ (since for logistic growth $r < 0$ and $L > 0$). Define $L = \max(\text{price}) + \delta$, where $\delta = 0.01$ is needed so that the left-hand side is defined over all bids y. In this case, there is no need to add the start and closing points to the live bids because defining the asymptote takes care of the fit at the end. Auctions whose underlying price process can be described by logistic growth tend to start out low, so there is also no need to set the start value.

Fitting Reflected-Logistic Growth: The reflected-logistic function is given in equation (4.13). As with logarithmic growth, we cannot linearize this function. We instead fit $T(y) = L/(1 + Ce^{ry})$, where $L = l + \epsilon$ ($\epsilon = 0.00001$), to obtain the coefficients for C and r. We need $\epsilon << \delta$ since the time range is much smaller than the price range. As with logarithmic growth, least squares minimizes over time instead of price. Note that here the extra points are $\left(t = 0.000001,\, y = \min(y_j)\right)$ and $\left(t = l,\, y = \max(y_j)\right)$ so that the left-hand side is defined over all bid times.

Selecting the Best Growth Model: While each of the above four models can be fitted to any set of auction data, we do not know which model fits best. Selecting the most appropriate model is complicated by the fact that we fit some models by applying least squares to price, while for other models we apply least squares to time. The following describes an automated model selection procedure to choose—for each auction—the best among each of the four growth models. The procedure uses a specialized proximity metric that measures the distance between bids and the fitted curve in the two dimensions of time and price. This metric is reminiscent of the Mahalanobis distance . Most model selection criteria measure only the residual distance in y (price, in this case); however, we are interested in capturing the best fit in both price and time because bid times are informative and are random variables (see also Section 4.4). Furthermore, the fit for the logarithmic growth and reflected-logistic growth models are minimized over the x (time) dimension. If we were to choose between models based simply on the price dimension, we would tend to choose the exponential growth and logistic growth models, even though the reflected models may provide a better representation of the price process (as can be determined visually).

Model Selection Metrics: For auction i, let $\{\mathbf{t_i}, \mathbf{y_i}\}$ be the set of live bids (t_{ij}, y_{ij}), where bid y_{ij} is placed at time t_{ij} and the number of bids in the auction is n_i. Define a new set with two additional price points ($n_i^\star = n_i + 2$) that includes the open and the close price of the auction as $\{\mathbf{t_i^\star}, \mathbf{y_i^\star}\} = \{(t = 0,\, y = \min(y_{ij})), \{\mathbf{t_i}, \mathbf{y_i}\}, (t = l,\, y = \max(y_{ij}))\}$, where l is the length of the auction. It is important that we examine the fit at the start and end of the auction because that is where most of the bid activity takes place, they are conceptually important, and also because that is where modeling falls short.

We propose two measures of fit: the weighted sum of squares standardized by the range (WSSER) and the weighted sum of squares standardized by the variance (WSSEV). Both metrics are weighted averages of fit in the y-direction and fit in the x-direction, using weights w_y and w_x, such that $w_y + w_x = 1$. The WSSER for auction i is defined as

$$\text{WSSER}_i = \frac{w_y \sum_{j=1}^{n_i} (y_{ij}^\star - \hat{y}_{ij}^\star)^2}{(\max_j(y_{ij}^\star) - \min_j(\hat{y}_{ij}^\star))^2} + \frac{w_x \sum_{j=1}^{n_i} (x_{ij}^\star - \hat{x}_{ij}^\star)^2}{(\max_j(x_{ij}^\star) - \min_j(\hat{x}_{ij}^\star))^2}. \tag{4.16}$$

Note that the denominator is the squared price range in the y dimension and the squared time range (in our case, the auction length l) in the x dimension. The WSSEV for auction i is defined as

$$\text{WSSEV} = \frac{w_y \sum_{j=1}^{n_i} (y_{ij}^{\star} - \hat{y}_{ij}^{\star})^2}{\text{variance}_j(y_{ij}^{\star})} + \frac{w_x \sum_{j=1}^{n_i} (x_{ij}^{\star} - \hat{x}_{ij}^{\star})^2}{\text{variance}_j(x_{ij}^{\star})}. \tag{4.17}$$

Model Selection Procedure: Using the above metrics, the model selection procedure is now as follows:

1. Select weights, w_y and w_x, representing the importance of fit in the price and time dimensions, respectively.
2. Fit each of the four growth models to the live bids of an auction.
3. Compute model selection metric(s).
4. Choose model with best fit (minimum WSSE).

Typically, we choose equal weights $w_y = w_x = \frac{1}{2}$ since we are equally interested in the price and time dimensions. One may overweight time (large w_x) if capturing bid timing is of special interest. One such case is in studying bid shilling, where sellers may cancel the auction or illegally bid on their own auction if the price has not reached a certain level by a certain time (Kauffman and Wood, 2005). A researcher may overweight price (large w_y) when the focus is on the price level itself (e.g., using this information to make more informed bid decisions.) Note that overweighing price tends to favor exponential and logistic growth models, whereas overweighing time leads to favor logarithmic and reflected-logistic models. Empirical evidence also suggests that both WSSE measures provide very similar results with a slight edge to WSSER.

Figure 4.6 shows the best price curves (using equal weights and WSSER) for four different auctions. Each graph shows the best fit (black line) according to the model selection procedure as well as the fit of the other three models for comparison. We can see that, at least from a visual comparison, the procedure appears to produce very reasonable results.

Pros and Cons: There are several pros and cons to using the four-member parametric growth family. Since the family is entirely parametric, no nuisance parameters require determination. Moreover, since the family is monotonic, it is conceptually appealing for capturing auction price processes. The family is also computationally efficient compared to monotone splines since estimation is based on ordinary least squares. The main disadvantage is that the family is limited to only four basic shapes—exponential, logarithmic, logistic, and reflected logistic—that may be overly simplistic for some auction scenarios. Moreover, because the four models are not nested within a single model, comparing fit (for choosing the best model) is nontrivial. Finally, when fitting the exponential and logistic models, the models minimize error in the bid *amount* space. In contrast, when fitting the two reflected models (logarithmic and reflected-logistic growth), the error minimization is done in the bid *time* space. A comparison is therefore more complicated. In the following section, we discuss another parametric model—the Beta model—that eliminates some of these disadvantages. In fact, it includes the four-member parametric growth family as a special case.

4.1.5 The Beta Model

In light of the above shortcomings, we now discuss a single parametric model that is flexible yet parsimonious for approximating price paths and their dynamics. This model has been introduced by Jank et al. (2009) and is based on the Beta cumulative distribution function (CDF). The Beta distribution is a continuous probability distribution defined on the interval [0, 1] with two shape parameters (α and β) that fully determine the distribution. Its CDF can be written as

$$\mathrm{F}(x, \alpha, \beta) = \frac{\int_0^x u^{\alpha-1}(1-u)^{\beta-1} du}{B(\alpha, \beta)}, \tag{4.18}$$

where $B(\alpha, \beta)$ is the *beta* function ($B(\alpha, \beta) = \int_0^1 u^{\alpha-1}(1-u)^{\beta-1} du$) that serves as a normalization constant in the CDF to ensure that $F(1, \alpha, \beta) = 1$.

The Beta model is very flexible in the types of curves that it can produce. It includes as special cases the four shapes of the four-member growth model family. Figure 4.7 shows the Beta model curves for different values of α and β. The solid line represents the case where price grows rapidly at the beginning of the auction and at the end, but not in the middle, corresponding to logistic growth. The long dashed line represents the situation where rapid growth occurs only at the end, corresponding to exponential growth. The short dashed line shows early rapid growth, corresponding to logarithmic growth. Finally, the dotted dashed line captures a rapid increase in price somewhere in the middle of the auction, corresponding to the reverse logistic growth pattern.

Fitting the Beta Model: Fitting the Beta model to auction data can be done in a way that results in curves that fit well in two dimensions: bid time and bid amount.

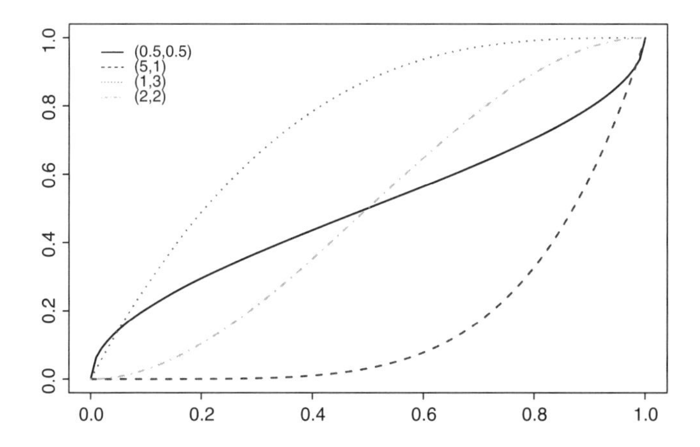

FIGURE 4.7 Beta CDF with different shape parameters (α, β).

As pointed out before, both dimensions are important in the auction context. In particular, a good fit in terms of the bid timing is necessary to accurately capture points of different bidding activities. Periods of vastly different biding activity, such early or last-minute bidding, have been documented well in the auction literature (e.g., Shmueli et al., 2007; Ockenfels and Roth, 2002), and they are important to capture adequately. In terms of bid amounts, a model that adequately captures the bid amounts (i.e., the price at that point of the auction) is necessary for generating accurate forecasts of an auction's final price. Auction price forecasting is of practical interest and different forecasting models have been suggested in the literature (Ghani and Simmons, 2004; Wang et al., 2008a; Jap and Naik, 2008; Jank and Zhang, 2009a; Jank and Shmueli, 2007; Dass et al., 2009; Zhang et al., 2010). We discuss forecasting in detail in Chapter 5.

The only inputs required for fitting the Beta CDF are the observed bid amounts and their associated time stamps. The resulting price path representation is characterized by only two parameters. The simplicity and parsimony of the Beta model distinguish it from alternative approaches. In the following, we describe an algorithm for fitting the Beta CDF in a way that again minimizes residuals in both bid amount and bid time dimensions simultaneously.

Beta-Fitting Algorithm: For a given auction, we estimate α and β from the observed bids as follows:

Step 1: Standardize Bid Amounts and Bid Times. Since the range (y) as well as the domain (x) of the Beta CDF is [0, 1], we first standardize the bid amounts and bid times by the following two transformations:

$$y \leftarrow \frac{\text{bid} - \min(\text{bid})}{\max(\text{bid}) - \min(\text{bid})}$$

and

$$x \leftarrow \frac{\text{time} - \min(\text{time})}{\max(\text{time}) - \min(\text{time})}.$$

x and y are now bid times and bid amounts standardized within [0, 1].

Step 2: Compute α_0 and β_0, the Initial Values of $\hat{\alpha}$ and $\hat{\beta}$. Since we treat x as a Beta-distributed random variable, it is reasonable to assume that the empirical average and the variance of x are close to their theoretical mean and variance. That is, $\text{mean}(x) \simeq \frac{\alpha}{\alpha+\beta}$ and $\text{var}(x) \simeq \frac{\alpha\beta}{(\alpha+\beta)^2(\alpha+\beta+1)}$. Therefore, the initial values of α and β are found by solving the minimization problem

$$(\alpha_0, \beta_0) = \left\{(\alpha^*, \beta^*)|\text{DIST}^A(\alpha^*, \beta^*) = \min(\text{DIST}^A(\alpha, \beta))\right\},$$

where

$$\text{DIST}^A(\alpha, \beta) = \left(\text{mean}(x) - \frac{\alpha}{\alpha+\beta}\right)^2 + \left(\text{var}(x) - \frac{\alpha\beta}{(\alpha+\beta)^2(\alpha+\beta+1)}\right)^2.$$

Step 3: Compute $\hat{\alpha}$ and $\hat{\beta}$. To capture both the bid levels and the bid times, the model minimizes error both in y and x directions simultaneously. Specifically, we choose to minimize the sum of the squared residuals in y and x directions. With the initial values α_0 and β_0 from Step 2, we solve for $\hat{\alpha}$ and $\hat{\beta}$ through the following minimization problem:

$$(\hat{\alpha}, \hat{\beta}) = \left\{(\alpha^*, \beta^*)|\text{DIST}^B(\alpha^*, \beta^*) = \min(\text{DIST}^B(\alpha, \beta))\right\},$$

where $\text{DIST}^B(\alpha, \beta) = \sum(y - \text{pbeta}(x, \alpha, \beta))^2 + \sum(x - \text{qbeta}(y, \alpha, \beta))^2$; pbeta and qbeta represent the cumulative distribution function and the inverse of the cumulative distribution function of the beta distribution, respectively.

The above algorithm is computationally very efficient. It takes, on average, 0.0329 s to fit the Beta model to one auction, which compares favorably to the four-member growth family (0.0305 s). Unsurprisingly, penalized splines fare better (0.0184 s) since they do not encounter any iterations. Conversely, fitting monotone splines, which do require iterative passes through the data, result in 50 times larger computing times (on average of 1.7049 s per auction). All computations are based on an IBM Lenovo T400 laptop computer with 2 GB memory and a dual core 2.26 GHz processor.

Properties of the Beta Model: The Beta model shares the main properties of competing methods (p-splines, monotone splines, and the four-family growth model), but it also has several additional properties that set it apart. Like all competing methods, the derivatives of the continuous Beta curves can be used to capture price dynamics. The Beta model produces monotonically nondecreasing curves, yet it is computationally fast. Unlike nonparametric approaches, fitting the Beta model does not involve any nuisance parameters.

The Beta model has two additional unique properties that make it especially advantageous in the online auction context: First, because both of its dimensions (bid time and bid amount) are derived from a probability function, the Beta summary statistics can be used to learn about the bid timing distribution. Second, there is an easy and straightforward way to measure pairwise distances between price paths. The latter is especially useful in the context of pairwise comparisons and dynamic forecasting. In particular, Zhang et al. (2010) use this property within a functional K-nearest neighbor (KNN) forecaster to produce more accurate price forecasts (see Chapter 5). We discuss each of the Beta model properties in detail next.

Representing Price Dynamics: The Beta CDF representation of the price paths means that price velocity, which is the first derivative of the price curve, is given by the Beta probability density function (PDF). In particular, at any given time T, the price velocity of an auction with shape parameters α and β can be computed as follows:

$$\text{Vel}(t, \alpha, \beta) = \frac{t^{\alpha-1}(1-t)^{\beta-1}}{B(\alpha, \beta)}, \tag{4.19}$$

TABLE 4.1 Relationship Between Beta Model and the Four Growth Models

Growth Models	Beta Model	
Exponential	$\alpha = 1$	$\beta < 1$
	$\alpha > 1$	$\beta \leq 1$
Logarithmic	$\alpha < 1$	$\beta \geq 1$
	$\alpha = 1$	$\beta > 1$
Logistic	$\alpha > 1$	$\beta > 1$
Reflected logistic	$\alpha < 1$	$\beta < 1$

where t is the normalized T on a scale of [0, 1] ($t = T/\text{duration}$) and $B(\alpha, \beta)$ is the beta function.

Higher order price dynamics can also be readily obtained by taking higher order derivatives. For example, price acceleration can be computed as

$$\text{Acc}(t, \alpha, \beta) = \frac{t^{\alpha-1}(1-t)^{\beta-1}}{B(\alpha, \beta)} \left(\frac{\alpha - 1}{t} - \frac{\beta - 1}{1 - t} \right).$$

As pointed out earlier, price dynamics carry important information about the auction process. Therefore, accurate approximations of price dynamics are beneficial across multiple applications.

Characterizing Growth Patterns: Similar to the four-member growth family, the Beta model provides a tool for characterizing price process types. In fact, the four growth models are special cases of the Beta model. For example, if both α and β are less than 1, then the price curve is similar to the reflected-logistic model. Table 4.1 lists the relationship between the Beta parameters and each of the four growth models. The implication of this relationship is that it allows us to easily characterize auctions in terms of their type of price dynamics, without the need for more specialized techniques such as functional clustering (e.g., Jank and Shmueli, 2008b, and Section 3.7) or via laborious visual examination (e.g., Hyde et al., 2006).

Characterizing Bid Timing: The estimated Beta parameters α and β can be used to compute summary statistics that capture bid timing information. Table 4.2 gives the

TABLE 4.2 Beta Distribution Summary Statistics and Their Auction Interpretation

	Formula	Explanation
Variance	$\frac{\alpha\beta}{(\alpha+\beta)^2(\alpha+\beta+1)}$	Dispersion of the bid arrivals
Mode	$\frac{\alpha-1}{\alpha+\beta-2}$	The peak of the velocity curve; price increases fastest at this point
Skewness	$\frac{2(\beta-\alpha)\sqrt{\alpha+\beta+1}}{(\alpha+\beta+2)\sqrt{\alpha\beta}}$	Asymmetry of the bid arrivals

formulas for the variance, mode, and skewness. The variance gives information about the dispersion of the bid arrivals; the mode, which is the peak of the price velocity curve, tells us about the time during the auction when the price moved fastest. Finally, skewness measures the level of asymmetry in the bid timings. Online auctions tend to see high bidding activity at the start and/or at the end.

Measuring Distances Between Price Paths: Unlike other models for the price path, the Beta model allows us to easily measure the *distance* between two paths. If both price paths are Beta curves, then the distance between them can be measured via the Kullback–Leibler (KL) distance (Kullback and Leibler, 1951). The KL distance is a noncommutative measure of the difference between two probability distributions. For two distributions X and Y, it measures how different Y is from X. The KL distance is widely used in the field of pattern recognition for feature selection (Basseville, 1989) and in physics for determining the states of atoms or other particles (Nalewajski and Parr, 2000).

In the case of the Beta model, the KL distance is especially simple. Consider X and Y Beta distributed with parameters (α, β) and (α', β'), respectively. The KL distance between X and Y is then given by a simple function of the Beta parameters alone (Raubera et al., 2008):

$$D_{\mathrm{KL}}(X, Y) = \ln \frac{B(\alpha', \beta')}{B(\alpha, \beta)} - (\alpha' - \alpha)\psi(\alpha) - (\beta' - \beta)\psi(\beta) + (\alpha' - \alpha + \beta' - \beta)\psi(\alpha + \beta), \tag{4.20}$$

where B and ψ are beta and digamma functions, respectively.

The importance of this property is that it allows us to incorporate dynamics in distance-based forecasting model, such as KNN methods. In fact, Zhang et al. (2010) use this property to develop a functional KNN algorithm and show that it can increase the forecasting accuracy especially when auctions are very heterogeneous (see Chapter 5).

4.1.6 Empirical Comparison of Price Path Models

In this section, we compare the Beta model with three alternatives: penalized splines, monotone splines, and the four-member growth family. Comparisons are made on two different dimensions: In terms of fit, we compare the ability of different models to generate accurate price representations of observed auction data, and in terms of prediction, we compare the forecast accuracy of the four methods in predicting the final price of a set of ongoing auctions. We will see that in the forecasting context, it is especially important to have an adequate approximation of not only the auction's price path, but also its price dynamics.

Model Fit Comparison: To evaluate goodness of fit of a model, we examine the residual error in terms of both bid amount (y) and bid time (x). For this purpose, we define the residual error for the ith auction with n bids as

$$\text{resid}_i = \frac{1}{n}\sum_{k=1}^{n}[0.5(y_k - \hat{y}_k)^2 + 0.5(x_k - \hat{x}_k)^2],$$

where (x_k, y_k) and $(\hat{x}_k, \hat{y}_k)$ are the observed and fitted values, respectively. Note that since both penalized splines and monotone splines minimize only errors in terms of bid amount (y), we set $x_k = \hat{x}_k$, which may result in an overly optimistic view of these two methods.

We apply all four methods to a set of auction data. For each auction, we estimate a p-spline, a monotone spline, the best fit of the four-member growth family, and the Beta model. For each auction, we first normalize the observed bids p and associated times t into a [0, 1] scale. We then fit each of the models[2] to the normalized data. Normalization results in an equal weighing of the residuals in both bid time and bid amount dimensions since they are measured on equal scales.

The distributions of the absolute residuals are shown in Figure 4.8. We can see that the Beta model (top left panel) results in the second-best fit (average error = 0.0125), surpassed only slightly by the fit of monotone splines (bottom right panel; average error = 0.0112). Both p-splines and the four-family growth models result in a much worse representation of the data (average error of 0.0326 and 0.0434, respectively). But also recall the much longer estimation time for monotone splines: it took a total of 2650 s (or 44 min) to fit all the 380 auctions in our data; this compares to only 19 s for the Beta model!

Forecasting Accuracy Comparison: We now compare the four methods in terms of their capability of producing accurate forecasts of an auction's final price. Information of the final price ahead of time has advantages for all auction participants. Bidders can use this information to make more informed bidding decisions (Jank and Zhang, 2009a). Sellers can use predictions to identify times when the market is more favorable to sell their products (e.g., higher demand, lower supply). We pay particular attention to the role of price dynamics: While recent research has shown that the incorporation of dynamics into real-time forecasting models for the price of an online auction can improve the forecasting accuracy tremendously, what has not been investigated to date is whether the method used to *estimate* these dynamics has an impact on the forecast accuracy. It is quite plausible that different estimates for the price curve will yield different estimates for the dynamics and, as a consequence, will result in varying forecasting performance. We investigate this next.

We have already shown that the Beta model can be computed very quickly, much faster than monotone splines and almost as fast as p-splines. Hence, the Beta model lends itself for real-time forecasting scenarios where updates need to be computed in split seconds for many hundreds or thousands of auctions at a time. What still remains to be shown is if these fast computing times also lead to accurate forecasts. We have argued earlier that p-splines, despite being very fast to compute, do not capture the

[2] For p-splines and monotone splines, the smoothing parameters are determined using leave-one-out cross-validation.

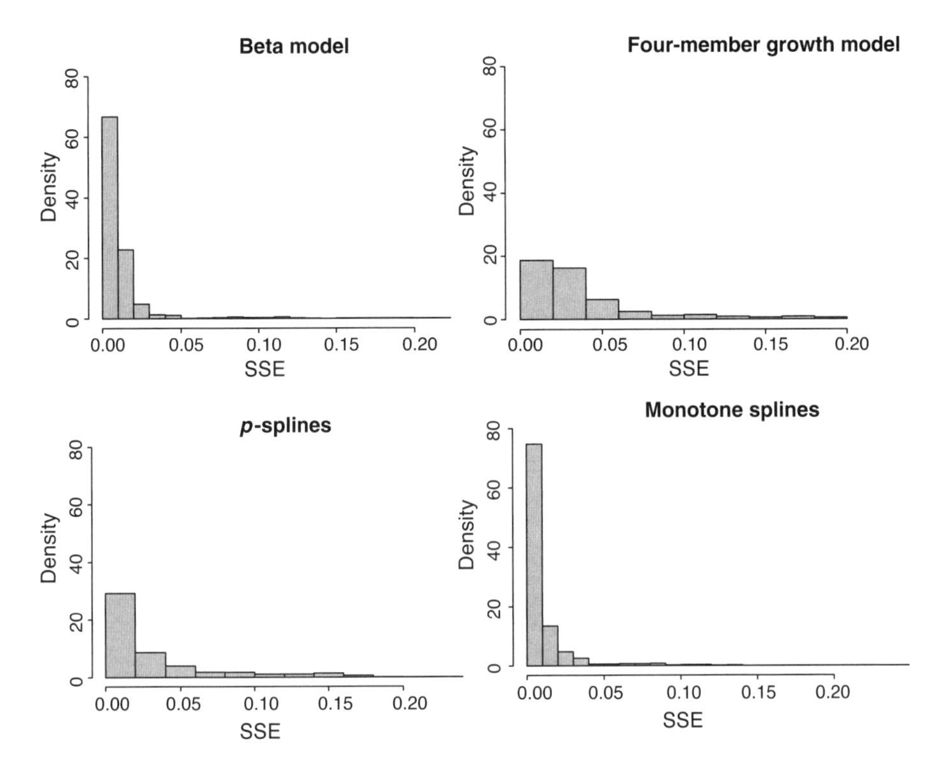

FIGURE 4.8 Model residuals: Beta model (top left), four-member growth models (top right), p-splines (bottom left), and monotone splines (bottom right).

monotone auction process and thus may not be as useful for forecasting purposes. In contrast, while monotone splines are, at least in theory, a natural choice for predicting the continuation of an auction process, they are slow to compute. We will thus now compare the Beta model against its competitors in terms of predictive accuracy.

We conduct our investigation in the following way. For a given forecasting model (e.g., a linear model or a nonlinear model), we assemble four sets of predictors, one for each of our four price path models. All four sets contain the same input information (e.g., they all contain the opening price, the auction duration, or the price velocity); their only difference is *how* we compute estimates for the price velocity. Then, by applying this set of four predictors to the same forecasting model, we observe differences in forecasting accuracy (measured on a holdout set). We can hence conclude that the observed differences must be due to the differences in how we computed the price dynamics. More details are provided below.

Forecasting Models: We compare the effect of price dynamics within four different types of forecasting models—a linear model, a generalized additive model, a neural network, and a regression tree. We start by describing a linear forecasting model. Consider an ongoing auction at time point T. For instance, for a 7-day auction (the most common auction length on eBay), $T = 5$ would imply that the auction has been ongoing for 5 days. Our goal is to forecast, at time T, the final price of an auction; in the above example, our goal would be to forecast at day 5 the final auction price. The linear forecaster is given by

$$\text{FinalPrice}_i = \boldsymbol{\beta}_1'\mathbf{X} + \beta_2'\text{Price}_{Ti} + \beta_3'\text{Velocity}_{Ti}, \tag{4.21}$$

where the design matrix $\mathbf{X}$ includes control variables that characterize the seller, the product, and the auction features. Such variables include the opening price, the auction duration, shipping fees, the seller's feedback score, whether or not the auction features a picture, whether the seller has an eBay store, whether the seller is a power seller, the number of bids placed on the auction, and the average bidder rating (both measured at time T). Note that the information in $\mathbf{X}$ is identical across all forecasting models.

While $\mathbf{X}$ is identical and held constant, we vary the information in Price_{Ti} and Velocity_{Ti} that denote estimates for the price and its velocity at time T, respectively, using price path model i, $i \in$ {Beta model, four-member growth family, penalized spline, monotone spline}. We thus obtain four different forecasts for the final price, each arising from a different price path model.

A comment on the estimation of the price dynamics is in order. Note that Price_{Ti} and Velocity_{Ti} denote the current price and its velocity *at time* T, that is, during the ongoing auction. To obtain these estimates, we estimate the price path model using the bids placed only *up to time* T.

In addition to the linear model in (4.21), we also investigate three nonlinear forecasting model formulations. These include a generalized additive model (GAM), a neural network, and a regression tree. The rationale behind these additional model formulations is that differences in observed forecasting accuracies may be due to nonlinearities or higher order interaction terms between the price dynamics and the response, and investigating more flexible modeling alternatives allows us to quantify this difference. For details on forecasting method, see Chapter 5.

Results: Forecasting performance of all forecasters is evaluated on a holdout set. In particular, we randomly split the set of auctions into training and holdout sets, each consisting of 50% of the auctions. Model parameters are estimated using the training set, and then the predictive accuracy is measured by computing the mean absolute percentage error (MAPE) in the holdout set:

$$\text{MAPE} = \frac{1}{N}\sum_{i=1}^{N}\left|\frac{y_i - \widehat{y}_i}{y_i}\right|, \tag{4.22}$$

where y_i and $\widehat{y}_i$ denote the true and the estimated final price in auction i (in the holdout set), respectively.

Figure 4.9 shows the results. The top panel corresponds to the linear model formulation (left) and generalized additive model (right); the bottom panel corresponds to the neural network (left) and the regression tree (right). The y-axis corresponds to the forecasting error; the x-axis corresponds to the forecasting window. By "forecasting

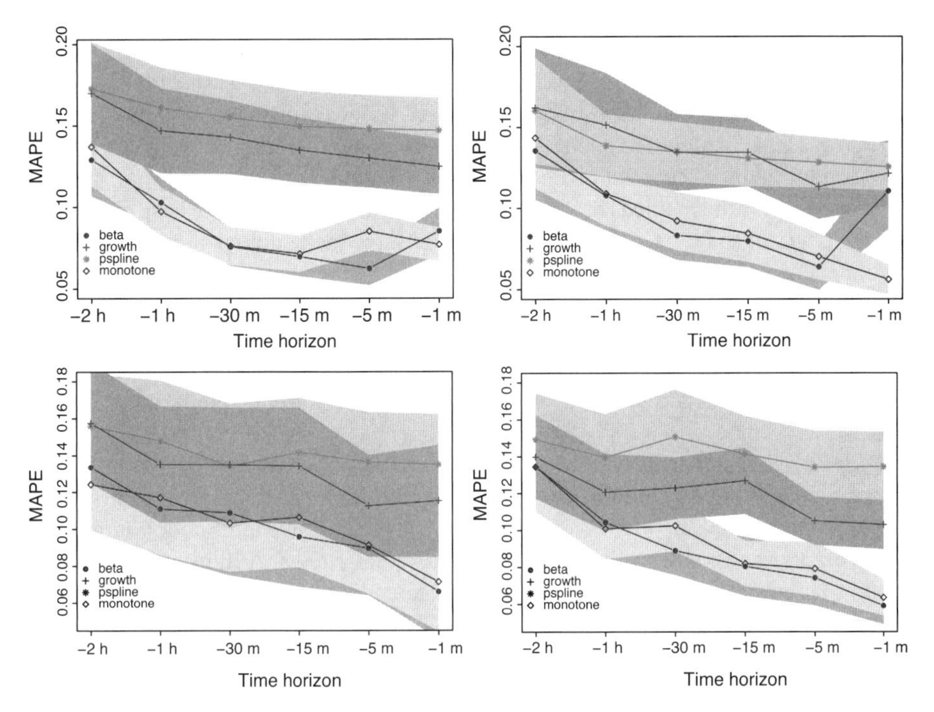

FIGURE 4.9 Comparison of the forecasting accuracy of different price path models by forecaster. The top panel corresponds to linear forecasters (left) and GAM (right); the bottom panel corresponds to neural networks (left) and regression trees (right). The x-axis corresponds to the forecasting window; the gray bands correspond to 95% error confidence bounds.

window" we mean the amount of time *before the end of the auction* at which we compute the forecast. For instance, $T = -2\,\text{h}$ means that we are forecasting the price 2 h before the auction closes. As expected, we can see that the longer we wait (i.e., as T approaches zero), the better will be our forecast. We can also see that, regardless of the forecasting model, dynamics based on the Beta model and the monotone spline always outperform those based on the penalized spline or the four-member growth model. While this difference is most pronounced in the linear forecasting model (top left panel), it is statistically insignificant for the neural network (bottom left panel), suggesting that the relationship between dynamics and the response is highly nonlinear. Also, remember that while the performance of the Beta model and the monotone splines appears very similar, it takes over 50 times more computing time to estimate the monotone spline. For instance, in our data that consist of 698 auctions, it takes only 23 s to fit all Beta models, but over 1190 s (~20 min) to fit all monotone splines. Given that the goal is to perform real-time forecasting and given that forecasts need to be updated 5 min or sometimes 1 min before the close of the auction, 20 min computing time is not very practical. We can thus conclude that the Beta model is a very attractive alternative to existing methods when the goal is to obtain fast and accurate estimates of the price path.

4.1.7 Case Study: Studying the Winner's Curse

In this section, we illustrate the usefulness of having available measures for velocity and acceleration to take a fresh look at core economic principles. In particular, we propose a new method for quantifying the degree and direction of value affiliation in auctions taking functional approach, that is, via modeling the *bid velocity*. To this end, we employ the methods for representing the price process that we have discussed in this section. In fact, while we do employ the same methodological principles, we will not capture the price process over time, but rather the price process as a function of the number of bidders. The difference is of substantive nature: while we will again use smoothing methods to arrive at an estimate for dynamics, the interpretation will be different compared to the interpretation of, for example, the price velocity, as we will now smooth as a function of the number of bidders. More details are provided below.

Introduction and Motivation: Auctions are characterized by asymmetric information, and therefore determining the appropriate information structure is the first step in beginning to develop models of auctions. Typically, the two information structures used are the well-known private and common value settings. A third, more general, setting allows for the existence of both private and common value components in bidder's information about the object being sold (see the general setup described in Klemperer (1999, p. 58) and Krishna (2002, Section 6.1)). It is well established that these information structures lead to very different bidding patterns and correspondingly different auction outcomes (Milgrom and Weber, 1982). However, establishing whether a given auction setting falls under the pure private value setting or the common value setting is nontrivial and has been treated as a matter of the researcher's judgment. Current research (Borle et al., 2003) that polls a variety of auction experts on whether a given object falls under a private or a common value setting suggests significant polarity in definitions for given items, of whether they are common value or private value. Given the importance of establishing this basic building block for doing any theoretical or empirical work with auction data, we suggest an empirical procedure—based only on the availability of estimates for the velocity and acceleration—that can be applied as a pre-theory test to detect a private or affiliated-value setting.

The current state of the art on detecting a private- versus common-value setting has been the use of the winner's curse test in the context of real online auction data by Bajari and Hortacsu (2003). Their approach, implicitly conditioned on the fact that bidders in common value setting are engaging in theoretically predicted optimal bidding behavior by shading their bids to avoid a winner's curse, looks for the presence of bid shading as an indication of a common value auction. To test whether there is bid shading, they rely on Milgrom and Weber's formulation:

> ... the "winner's curse" test suggested by the model of Milgrom and Weber (1982)... The possibility of a winner's curse is greater in an N-person auction as opposed to an (N-1)-person auction; therefore, the empirical prediction is that the average bid in an N-person second-price or ascending auction is going to be lower than the average bid in an $(N-1)$-person auction.

Here, we propose an alternate approach that isolates the effect of the number of bidders on the rate of bid increase. Our interest is to determine whether there in any affiliation in the observed bid values as the auction progresses. In particular, we propose a method that estimates this relationship while capturing the dynamic nature of online auctions, the unknown and changing number of bidders in an auction, and the bid timing. The lack of affiliation would strongly imply a private value setting, where bidders are not influenced by what other bidders are doing. In contrast, a presence of affiliation can be interpreted in different ways depending on the direction of the relationship. A positive affiliation suggests that bidders are revising their values upward in response to other bidders' bids, perhaps due to nonoptimal bidding (from a rational utility maximizing Bayesian Nash equilibrium strategy prediction perspective). This could potentially lead to winner's curse. A negative relationship would suggest bid shading in the presence of an anticipated winner's curse in a common value auction.

Bid Shading and the Winner's Curse: Bid shading means bidding lower than your actual valuation of the auctioned item. According to auction theory, a major reason for strategically bid shading is to avoid the "winner's curse." Winner's curse is a phenomenon that occurs in auctions for common value items. These are auctions where the item being auctioned is valued the same by each participant, although the participants may not know the precise value. In such a setting, each bid is assumed to be a realization from the value distribution, and the average bid is assumed to be the estimate of the common value. If bidders bid their estimate, then the highest bid is an overestimate of the common value, leading to the "winner's curse" phenomenon. To avoid this curse, bidders in such auctions must shade their bid, that is, shrink their estimate. According to auction theory, the amount of shrinkage depends on the number of bidders in the auction: the more bidders, the more shading is required. This can be shown as follows: consider a random sample of n bids, $X_1, X_2, \ldots, X_n$, from a distribution F. The distribution of the maximum bid $\text{Max}_n = \text{Max}(X_1, X_2, \ldots, X_n)$ is then F^n, with expected value

$$E\{\text{Max}(X_1, X_2, \ldots, X_n)\} = \int_x xnF^{n-1}(x)f(x)dx, \tag{4.23}$$

where $f(x) = F'(x)$. The deviation of $E(X)$ from $E(\text{Max}_n)$ is therefore proportional to n. To illustrate this, consider two distributions that are popular choices in auction theory for bid distributions: the uniform and normal distributions. We chose two distributions that have $E(X) = 50$ and similar spread: $X \sim U(0, 100)$ and $Y \sim N(50, 17^2)$. In the uniform case, there is a closed form for (4.23), which is $E(\text{Max}_n) = 100n/(n+1)$. For the normal case, we use Harter's tables (Harter, 1961) that are based on numerical integration. The deviation $E(\text{Max}_n) - E(X)$ as a function of n is shown in Figure 4.10. In both cases, the relationship is monotone and nonlinear.

The Effect of Number of Bidders on Bids: According to auction theory, in a private value auction, where each bidder has his/her own valuation of the item and does not care about other bidders' valuation, we expect to see no relationship between bids and the number of bidders. In contrast, if the number of bidders appears to affect bids,

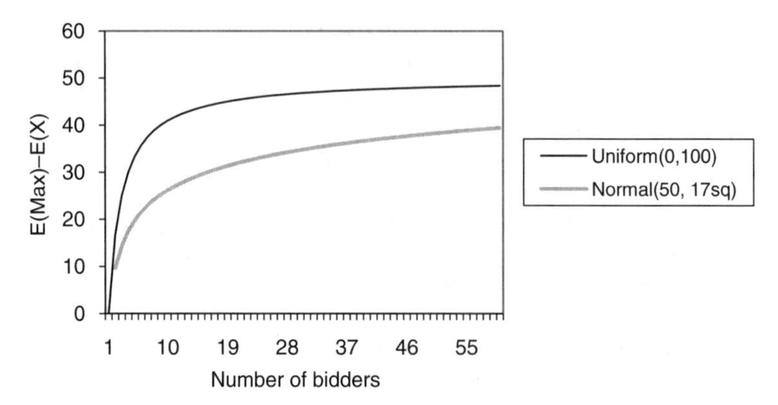

FIGURE 4.10 Deviation of $E(\text{Max}_n)$ from $E(X)$ as a function of the number of bidders (n) for uniform and normal distributions.

it means that bidders do care about how many competitors they have. If we consider once again the winner's curse, then we expect rational bidders to deflate their bids proportionally to the number of bidders (bid shading). Thus, bid shading would be indicative of affiliation between bidders' valuations. Of course, if bid shading does not take place in such a setting, then we would observe the winner's curse! However, another possibility that relates bids to the number of bidders, which is not prescribed by auction theory but is popularly known to occur in online auctions (Salls, 2005), is an inflation of bids in response to the number of bidders. In contrast to bid shading, bids would get inflated as more bidders are present. As bid shading, it indicates that bidders' valuations are affiliated.

The method that we propose is aimed at evaluating whether bids are affected by the number of bidders in the auction or not. The absence of such a relationship would be indicative of unaffiliated "private value"-type valuations. The presence of a relationship can be indicative of either bid shading (if bids are negatively impacted by the number of bidders) or the opposite effect of "auction heat," where bids actually are inflated as the number of bidders increases.

The Winner's Curse Empirical Test and Its Limitations: The implementation by Bajari and Hortacsu (2003) is based on fitting a linear regression to a sample of bids, where the response variable is the bid amount (normalized by the book value) and the explanatory variable is the number of bidders.[3] They find a negative slope and take this as suggestive evidence for the presence of bid shading, and thus infer a common value auction.

There are several challenges that arise when implementing the Milgrom and Weber model empirically in online auctions for the purpose of testing for bid shading. First, the number of bidders in online auctions (such as eBay) is typically unknown until the close of the auction, and it changes during the auction. This means that we cannot use

[3] In practice, they use a proxy for the number of bidders, which is the opening bid normalized by the book value.

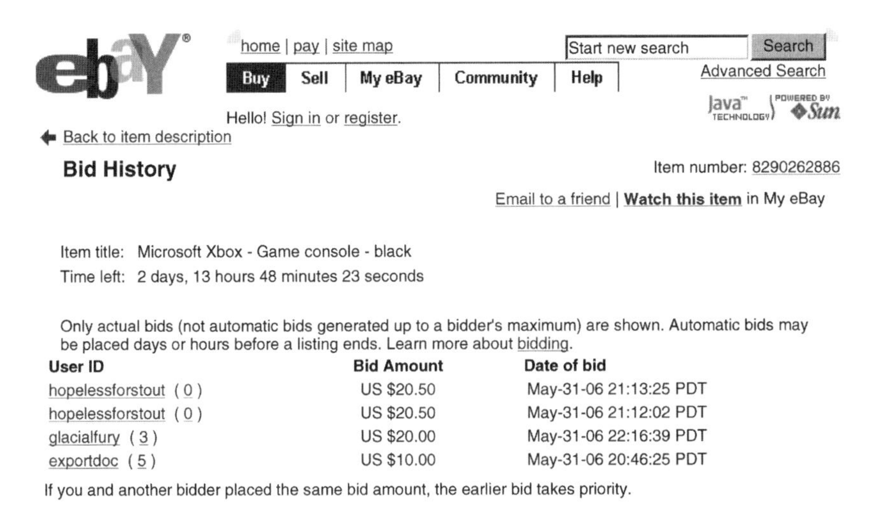

FIGURE 4.11 Bid history of an open auction for an XBox console.

the total number of bidders in an auction in our model because at the time of bidding this information was unavailable to the bidders. However, if we assume that each bidder bids as though she/he is the last bidder in the auction (and shades or inflates the bid accordingly), then we *can* compare bids of same order bidders (e.g., bids by all fifth bidders). In other words, we look at the *current number of bidders* rather than the total number of bidders. This is illustrated in Figure 4.11, which displays a bid history for an ongoing auction. At this point in the auction, there are three bidders who placed bids, the latest bid by *glacialfury*.[4] This bidder would then consider himself the third bidder in an auction with three bidders. This assumption is reasonable except for one factor that is not taken into account, which is the time that the bid was placed. In online auctions, which typically take place over a few days, the bid timing is an important factor. We assumed above that the nth bidder perceives himself as the last bidder in an n-bidder auction. However, bidders who arrive earlier in the auction are more likely to be followed by additional bidders than those who arrive toward the auction end. For example, a fifth bidder who places an early bid is more likely to be followed by additional bidders than a fifth bidder who bids at the last minute of the auction. We must therefore account for the bid timing. Bajari and Hortacsu (2003) addressed this by using data only from the last few minutes of the auction. Our approach allows one to use bids from the entire auction, while factoring out the bid timing element.

Another challenge is in treating multiple bids from a set of auctions as a random sample, in order to see whether bids get shaded as the number of bidders increases. Clearly, bids within the same auction are temporally dependent. Fitting a regression

[4] Note that the latest bid of $20 is lower than the previous bidder's $20.50 bid. This is possible because on eBay only the current second highest bid is shown.

model that assumes independence is therefore likely to yield biased results. Our proposed solution treats each auction as an observation rather than each bid.

Finally, we prefer to avoid the assumption of a linear relationship between bids and the number of bidders because there is no theory suggesting such a linear form. In fact, Figure 4.10 shows that the deviation between the expected maximum value and the expected value tends to be nonlinear in the number of bidders, and therefore the amount of bid shading should be increasing at a decreasing rate. To avoid the need for any parametric assumptions on the valuation distributions (which cannot be checked or validated with standard publicly available bid data), we propose a different formulation. We assert that bid shading *slows down the speed of the price increase*, while auction fever *speeds up the price increase rate*. In other words, we look at the effect of the number of bidders on the *rate of change* in bid magnitude. Our measure of interest is therefore the bid velocity.

Quantifying the Affiliation: A Functional Approach: We now introduce a novel method for directly measuring the degree of bid shading or auction fever by accounting for the dynamic nature of online auctions. The method is based on a functional data analytic approach, that is, based on the methods for representing the price process and its dynamics that we have discussed in this section. Its advantage is that it uses the entire set of bids, not just those in the last minutes of the auction; it maintains the relationship between the bid amount and the bid timing; and it captures the temporal dependencies between bids in the same auction.

We start by representing each auction as a functional observation, which relates its bids with the current number of bidders throughout the auction. In particular, we fit a smooth curve to the time series of bids, where the x-axis is the current number of bidders. An illustration is shown in Figure 4.12. Here, we perform curve fitting via penalized smoothing splines. Using penalized smoothing splines (rather than monotone splines or the Beta model) is actually an advantage in this application as we are modeling the proxy bids that, in contrast to the live bids, are not necessarily

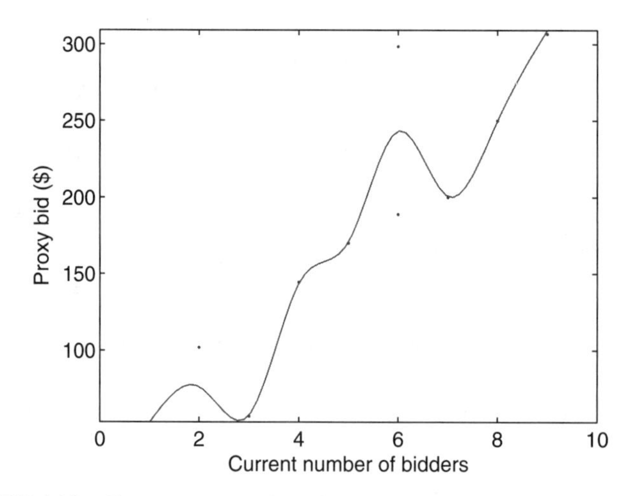

FIGURE 4.12 Curve representation of the sequence of bids in a single auction.

monotonic increasing. Moreover, as pointed out earlier, in contrast to the previous section we now smooth over the current number of bidders (rather than over time). The difference is of substantive nature: the resulting interpretation is different compared to the interpretation of, for example, the price velocity.

The middle panel in Figure 4.13 shows a sample of curves that were fitted to each of 92 seven-day auctions for a Microsoft X-Box game console. Rather than treating the 1852 bids from all 92 auctions as a single sample (as shown in the top panel of Figure 4.13), our sample consists of the 92 functional observations, where each curve is considered an observation. The thick curve is the point-wise average curve and represents the "typical" bid curve. We can see that the average curve is approximately linear in the current number of bidders, but it is also significantly affected by the declining number of auctions with many bidders (and especially those ending with very high bids). The object of our interest is, however, not the bid curves but rather the rate of change in the bid curves. This reflects the price increase per additional bidder and is captured by the first derivative of the bid curves, which we call the bid velocity curves. The bottom panel of Figure 4.13 shows the corresponding bid velocity curves for the 92 auctions. Although the single curves are volatile, this volatility is suppressed at average level, as can be seen in the average bid velocity curve (the thick line). In fact, the average curve is nearly constant in the number of bidders, and we might therefore conclude that the rate of bid increases is not affected by the number of bidders (i.e., bidders' valuations are unaffiliated). However, recall the effect of bid timing where earlier bids are more likely to be followed by additional bids. Our next step is therefore to factor out the bid timing before proceeding to fit functional auction curves and examining their velocities.

Accounting for Bid Timing: To account for the effect of bid timing on the bid amounts, we model the relationship between the two by regressing the bid amount on bid timing and then using the residuals as quantities that are free of bid timing. To do this, we fit a nonlinear regression to the bivariate sample of {bid amounts, bid times}. We use a kernel smoother to account for the very uneven distribution of points and, in particular, the high-density area toward the auction end. Figure 4.14 shows a scatterplot of bids versus bid times and the fitted kernel smoother. As expected, later bids tend to be higher than early bids. By subtracting the predicted bid amounts (based on their bid time) from the actual amounts, we obtain residuals that are "time free." This serves as a preliminary stage before applying the smoothing method described above. In other words, we fit curves to the residual bids of each auction and then examine the derivative of these curves. Figure 4.15 shows the residual bid curves (top) and residual bid velocity curves (bottom) for the 92 auctions. Compared to the flat average velocity curve before accounting for bid timing, the average residual bid velocity curve is slightly increasing in the number of bidders. This positive relationship can be indicative of auction fever, or bid inflation.

Results for Additional Products: For the purpose of comparison, we applied the method to three other types of products: 116 auctions for the book *Harry Potter and the Half-Blood Prince*, 194 auctions for Palm M515 PDAs, and 97 auctions for Cartier premium wristwatches. All auctions transacted on eBay and lasted 7 days. Figure 4.16 shows the residual bid velocity curves for each of the three items. In all

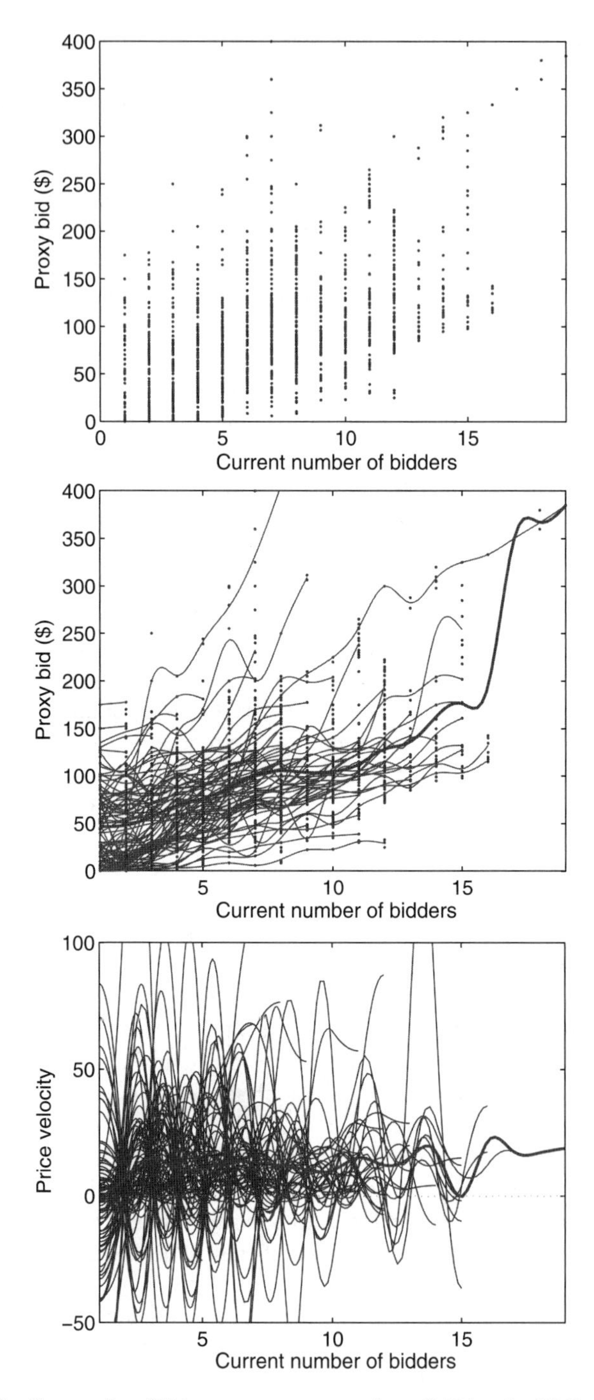

FIGURE 4.13 Scatterplot of bids versus current number of bidders for 92 Cartier wristwatch auctions (top), fitted auction curves (middle), and bid velocity curves (bottom).

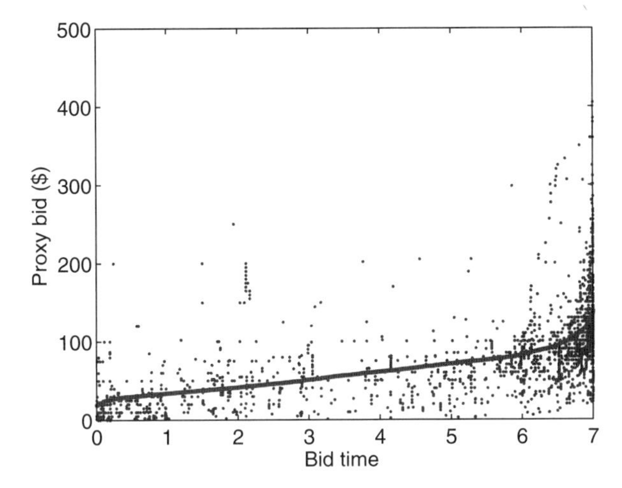

FIGURE 4.14 Nonlinear regression of bid amounts on bid times using a kernel smoother.

cases, the average velocity curve appears to be increasing in the number of bidders, indicating the presence of auction fever (bid inflation). However, there appear to be subtle differences in the nature of the increase. Of the three items, it appears that the average degree of affiliation is flattest for the Palm M515 auctions, suggesting bidding behavior closer to a private value setting. In contrast, both the Harry Potter book and the Cartier wristwatch auctions show significant deviation from zero affiliation and suggest that individual bidders are influenced by the bids placed by other bidders.

Discussion: Identifying which informational setting (private or common value) is a key building block of the vast auction theory literature has traditionally been left to the judgment of the researcher. Alarmingly, current research (Borle et al., 2003) that polls a variety of auction experts on whether a given object falls under a private or a common value setting suggests significant polarity. We propose a new method, based on functional data analysis, that detects and when present quantifies the degree and direction of affiliation in bid data. Our approach overcomes methodological limitations of using a linear regression approach (Bajari and Hortacsu, 2003) to implement the winner's curse test. These limitations include the lack of consideration of the endogenous nature of the number of bidders, particularly in online auctions, to the lack of independence of bids within the same auction. We recommend that researchers analyzing empirical data use our proposed FDA-based test prior to analyzing auction data to determine whether a private or a common value setting is appropriate.

4.1.8 Conclusion

The price path and its dynamics are useful not only in real-time forecasting applications, but also for studying bid profiles, as well as different factors that affect or interact with them.

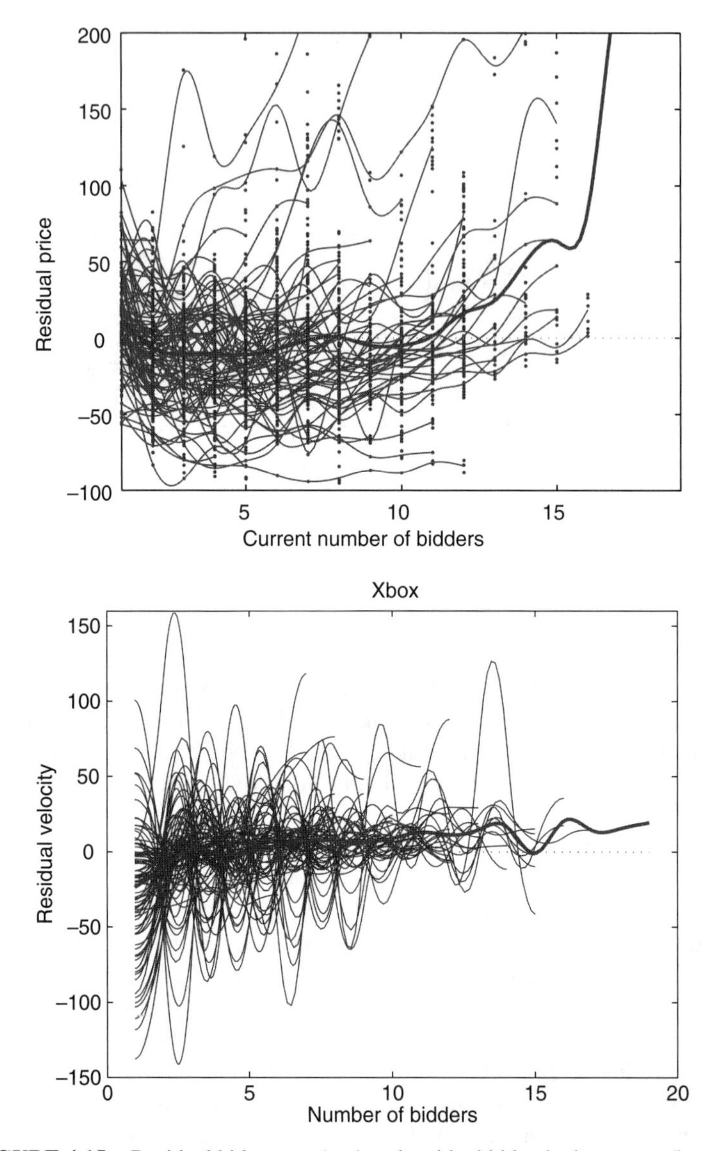

FIGURE 4.15 Residual bid curves (top) and residual bid velocity curves (bottom).

In terms of assessing the effect of various factors on price dynamics, there has been much interest in understanding the effects of product features (e.g., make or model), seller characteristics (e.g., a seller's reputation), and auction features (e.g. the length of the auction or its opening bid) on the outcome of an auction (i.e., its final price). However, while all these factors do affect an auction's outcome, they also affect its

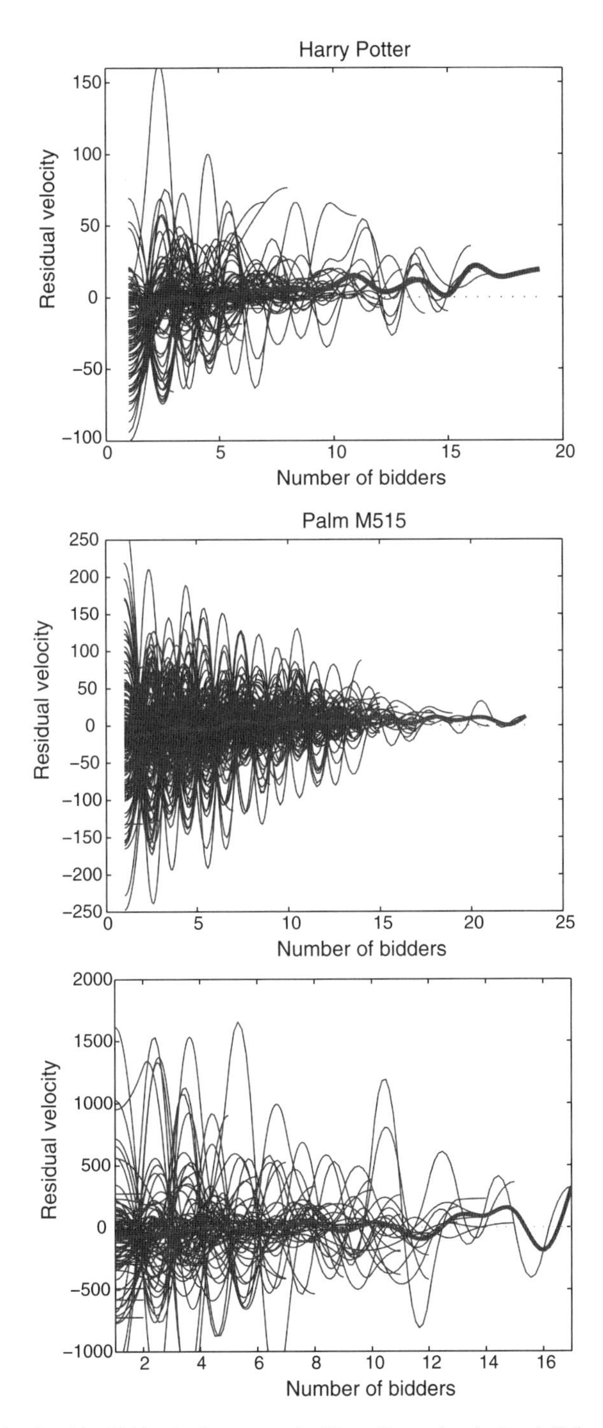

FIGURE 4.16 Residual bid velocity curves for Harry Potter books (top), Palm PDA (middle), and Cartier wristwatches (bottom).

dynamics, which in turn affects the final price (Bapna et al., 2008b). Shmueli and Jank (2008), for instance, illustrate the effect of auction features (such as the opening bid) on the auction dynamics and find that higher opening bids result in lower price dynamics. Jank et al. (2008b) expand upon this finding and develop model-based regression trees to relate differential equation models to auction features. Thus, a better understanding of price dynamics will also lead to a better understanding of the precise role of auction features or product descriptions and their effect on the outcome of an online auction.

Another component that is receiving increased interest is competition (Haruvy et al., 2008). This includes competition between bidders within the same online auction, across multiple auctions, or even beyond the online auction marketplace. The price path and its dynamics can reflect unobservable auction behavior such as the degree of competition between bidders, both within the same auction and across different auctions (Dass et al., 2009). Therefore, adequately capturing and modeling the price path can help understand the effects of competition.

There exist alternate models for capturing the price path of an auction. Among them is the semiparametric mixed model proposed by Reithinger et al. (2008). Assume that the price curve is modeled as

$$\text{Price}_i(\text{t}) = \alpha_0 + \alpha(\text{t}) + b_{i0} + \epsilon_i(t), \tag{4.24}$$

where b_{i0} is a random effect with $b_{i0} \sim N(0, \sigma_b^2)$ and α_0 is the (fixed) intercept for the model. Note that in (4.24) we assume a common slope function $\alpha(\text{t})$ for all auctions. We also assume that the intercepts of all auctions vary randomly with mean α_0 and variance σ_b^2. The residual error ϵ_i is assumed to be $N(\mathbf{0}, \sigma_\epsilon^2 R)$, where R is a known residual structure, for example, autoregressive first order. In this sense, all auctions are *conditionally* independent given the level (the random intercept). In this fashion, we can model all auctions within one parsimonious model and yet obtain a price curve estimate for each auction individually. Reithinger et al. (2008) estimate this model via a boosting approach that automatically estimates the smoothing parameters.

Figure 4.17 shows the estimated mean slope function (together with pointwise confidence bounds) of $\alpha(t)$ for a sample of 7-day auction. It shows two inflection points around days 1 and 5 and is monotonically increasing, as expected from an ascending auction. Moreover, the slope is steepest at the beginning and at the end of the auction, which is consistent with the phenomena of early bidding and bid sniping observed in the online auction literature (Bapna et al., 2003; Shmueli et al., 2007). It is also intriguing that, in contrast to most smoothers, the width of the confidence interval is decreasing toward the end. This is due to the fact that most bids arrive toward the end of the auction, thus providing greater accuracy in the estimation of the population trend.

Figure 4.18 shows the resulting individual curve estimates for 12 sample auctions. The solid lines correspond to the mixed model fit; for comparison, we also show the fit of ordinary penalized smoothing splines (dashed lines). We can see that the penalized

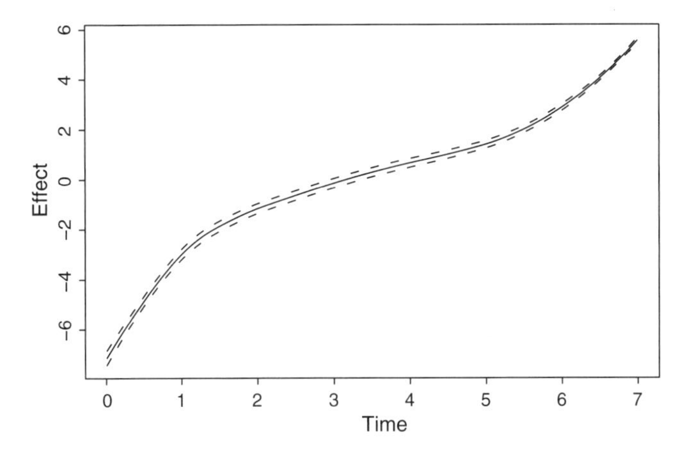

FIGURE 4.17 Mean spline function $\alpha(t)$ (solid line) and ± 1 std pointwise confidence band (dashed lines) estimated for a sample of auctions.

smoothing splines can result in poor curve representations: in some auctions, there is a lack of curvature (e.g., #27), while in others there is excess curvature (e.g., #23, 24, 34); yet in other auctions they do not produce any estimates at all due to data sparseness (e.g., #35), or data unevenness may result in very unrepresentative curves (e.g., #23, 24, 34). Moreover, many of the curves produced by penalized splines are unsatisfactory from a conceptual point of view: for instance, in auction #24 or 34, penalized splines result in an estimated price path that is not strictly monotonic increasing, which violates the assumption underlying ascending auction formats.

The estimates produced by the mixed model approach are much more appealing. Mixed model smoothing takes the mean slope as blueprint for all auctions and allows for variation from the mean through the random effect b. For instance, while the solid lines in Figure 4.18 all resemble a common mean slope, they differ in steepness and the timing of early and late bidding. These differences from the mean are driven by the amount (and distribution) of the observed data. For instance, auction #27 has a considerable number of bids that are distributed evenly across the entire auction length. As a result, the estimated curve is quite different from the mean slope function. On the other hand, auction #35 only has two observations. While penalized smoothers break down with too little information (and do not produce any curve estimates at all), the mixed model approach is still able to produce a reliable (and conceptually meaningful) result by borrowing information from the mean slope.

It is interesting to note that all of the price curves created by the mixed model approach are monotonically increasing. This is intriguing since the mixed model in equation (4.24) does not incorporate any monotonicity constraints. However, it appears to have "learned" this feature from the pooled data. This makes it a very flexible and powerful approach suitable for many different data scenarios.

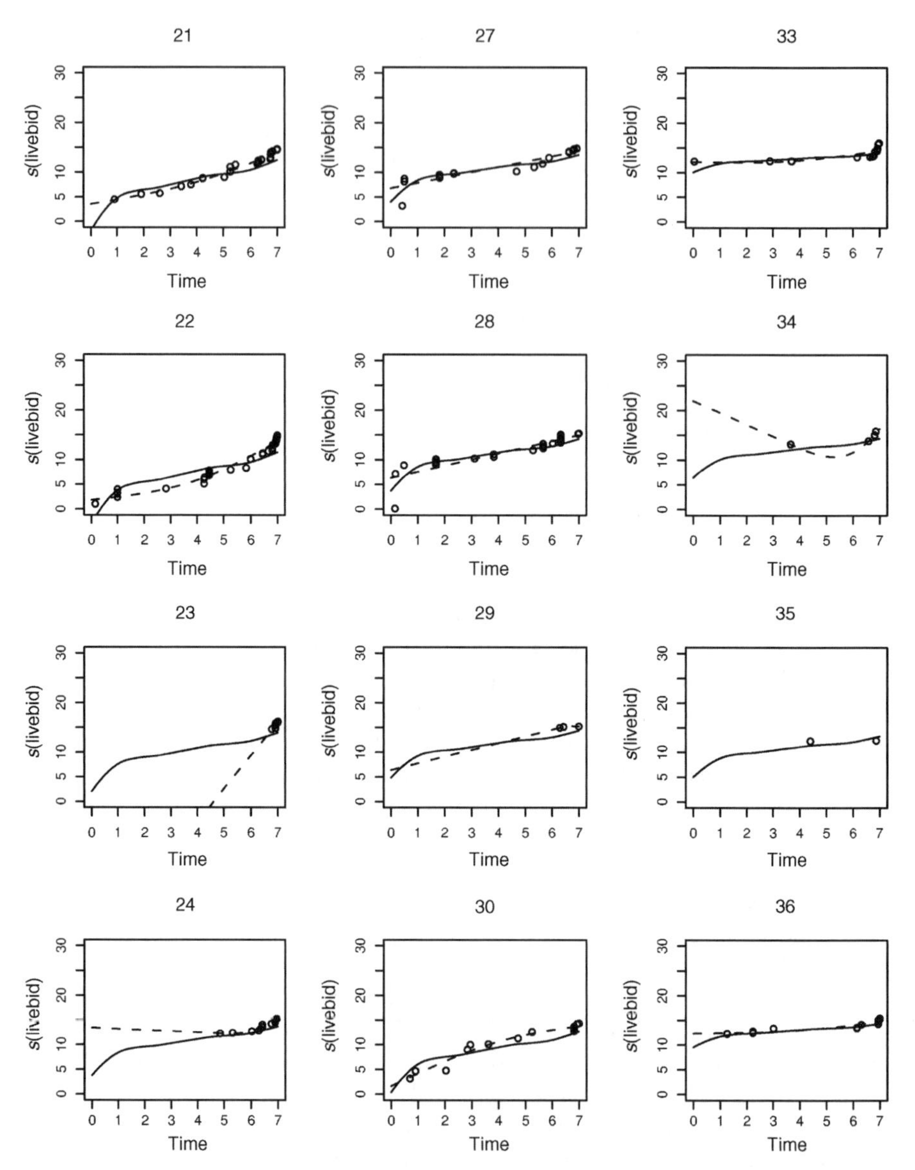

FIGURE 4.18 The mixed model fit for 12 sample auctions (solid lines); the separately fitted penalized splines are given by the dotted lines.

4.2 MODELING THE RELATION BETWEEN PRICE DYNAMICS AND AUCTION INFORMATION

In this section, we focus on the dynamics generated during an online auction and different ways for modeling them. Dynamics have only recently been found to affect

COLOR PLATES

home | pay | register | sign in | services | site map | help

Browse | Search | Sell | My eBay | Community

Powered By IBM

Search tips

Search titles and descriptions

eBay.com Bid History for

Palm Pilot m515 Color Handheld PDA 515 NEW NR (Item # 3074620884)

Currently	**US $200.50**	First Bid	**US $0.99**
Quantity	**1**	# of bids	**17**
Time left	**Auction has ended.**		
Started	Jan-28-04 20:30:00 PST		
Ends	Feb-02-04 20:30:00 PST		
Seller (Rating)	**uscagent (121 ☆)**		

View page with email addresses (Accessible by Seller only) Learn more.

Bidding History (Highest bids first)

User ID	Bid Amount	Date of Bid
golspice (26 ☆)	US $200.50	Feb-02-04 20:28:02 PST
audent (1)	US $198.00	Feb-02-04 17:18:14 PST
jay4blues (0)	US $196.50	Feb-02-04 14:57:54 PST
audent (1)	US $193.00	Feb-02-04 17:17:59 PST
audent (1)	US $188.00	Feb-02-04 17:17:33 PST
eagle2sc (11 ☆)	US $180.25	Feb-02-04 11:48:20 PST
audent (1)	US $179.00	Jan-31-04 09:18:51 PST
istariken (38 ☆)	US $175.25	Jan-30-04 20:45:20 PST
audent (1)	US $175.00	Jan-31-04 09:18:37 PST
audent (1)	US $169.55	Jan-31-04 09:18:16 PST
audent (1)	US $159.00	Jan-31-04 09:17:40 PST
amanda_s_brooks (14 ☆)	US $150.00	Jan-30-04 09:12:58 PST
audent (1)	US $80.05	Jan-28-04 21:54:09 PST
cscott24 (23 ☆)	US $80.00	Jan-28-04 20:35:05 PST
powergerbil (3)	US $45.01	Jan-28-04 21:05:14 PST
powergerbil (3)	US $42.99	Jan-28-04 21:04:44 PST
powergerbil (3)	US $40.00	Jan-28-04 21:04:08 PST

FIGURE 3.1 Bid history of a completed eBay auction for a Palm Pilot M515 PDA.

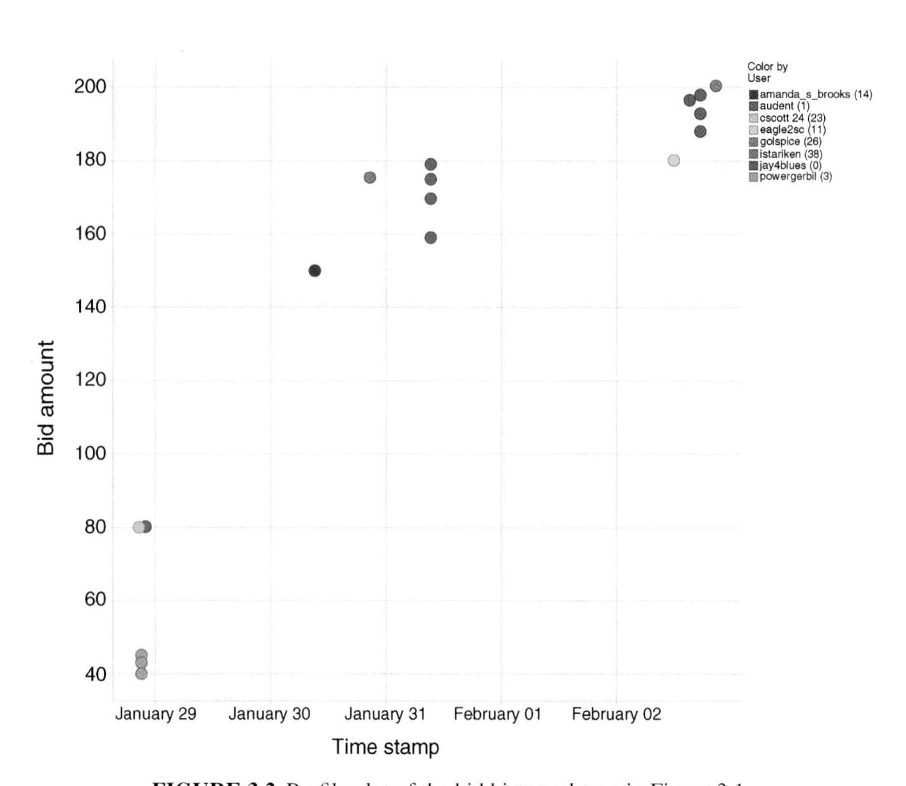

FIGURE 3.2 Profile plot of the bid history shown in Figure 3.1.

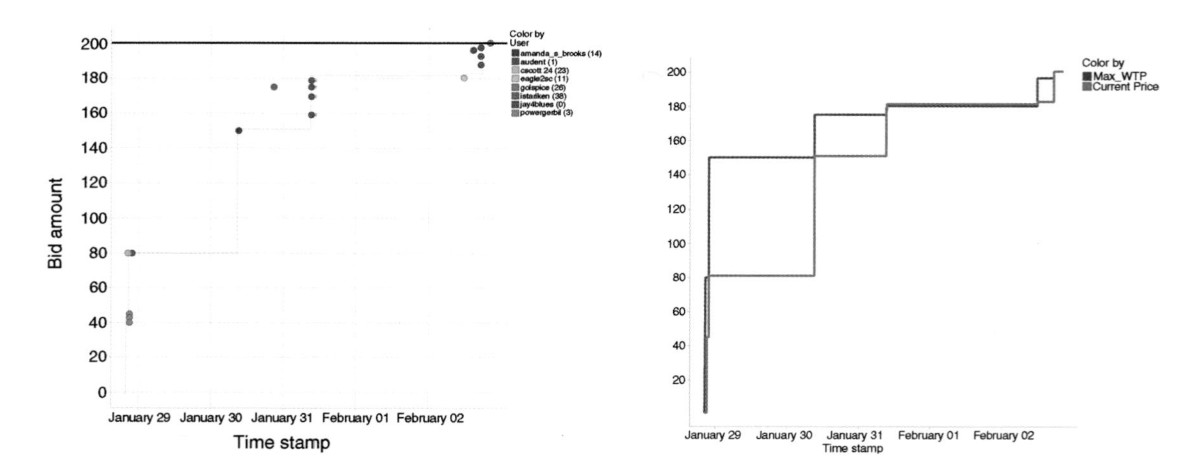

FIGURE 3.3 *Left panel*: profile plot and current price step function (as seen during the live auction). The final price is marked by a horizontal line. *Right panel*: comparing the current price function with the current highest bid.

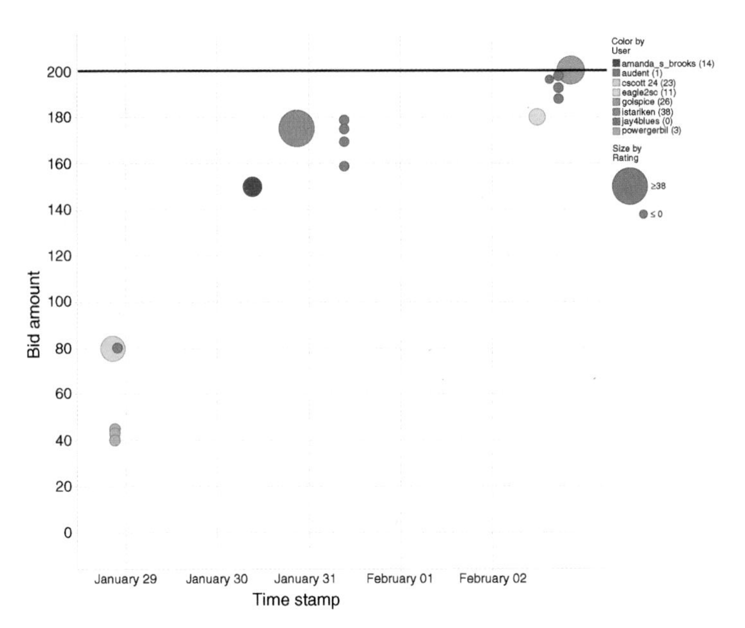

FIGURE 3.4 Profile plot with bidder rating information conveyed as circle size. The final price is marked by a horizontal line.

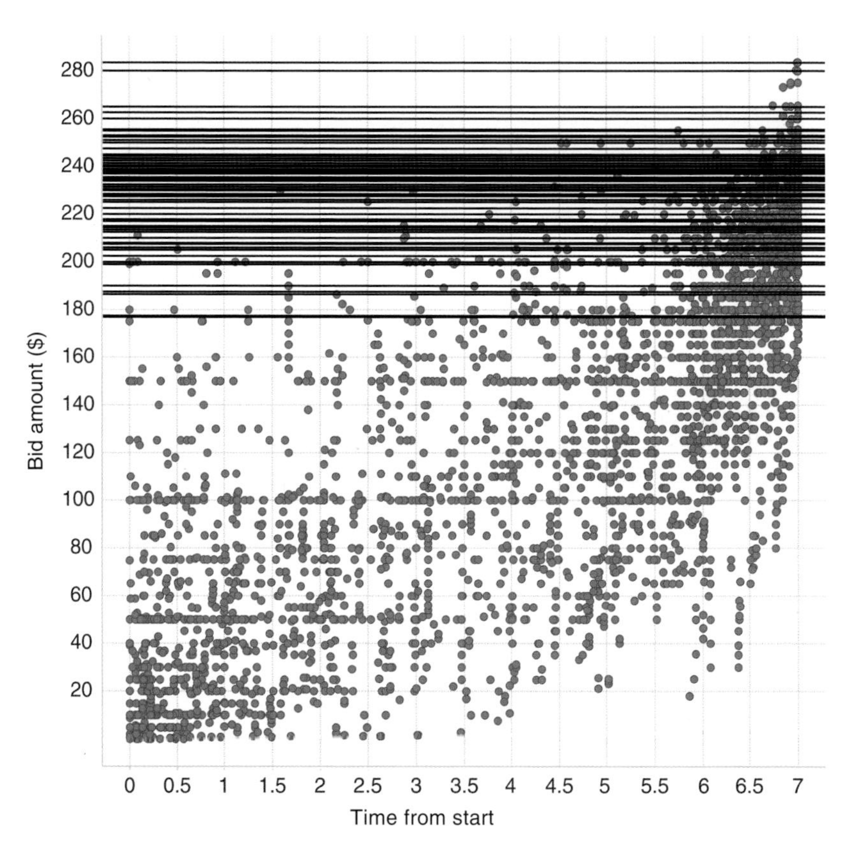

FIGURE 3.5 Profile plot for over 100 seven-day auctions. The final prices are marked by horizontal lines.

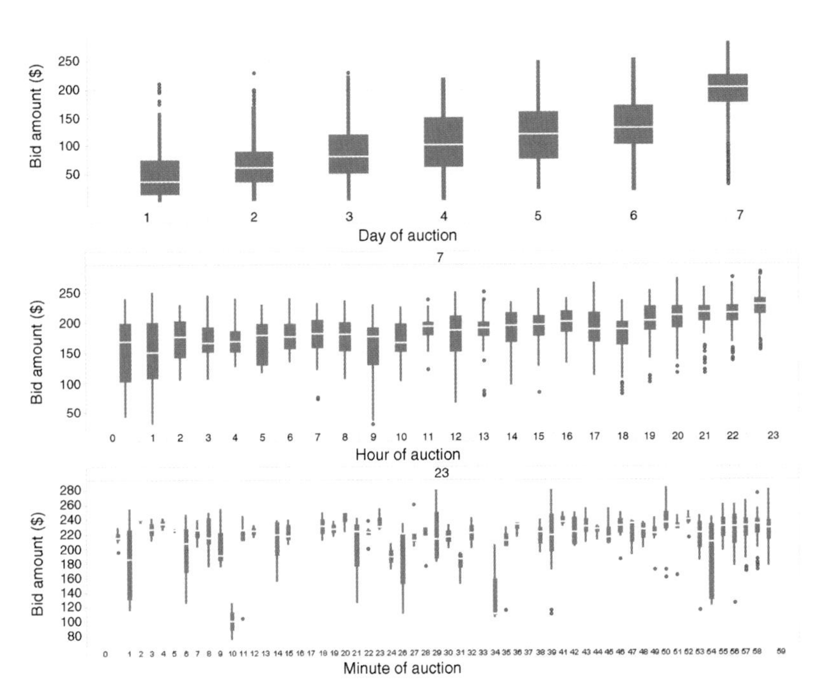

FIGURE 3.6 Bid distributions by varying the temporal hierarchy. Top panel displays daily distributions. Middle panel displays hourly distributions on the last day of the auction. Bottom panel displays minutely distributions for the last hour of the auction.

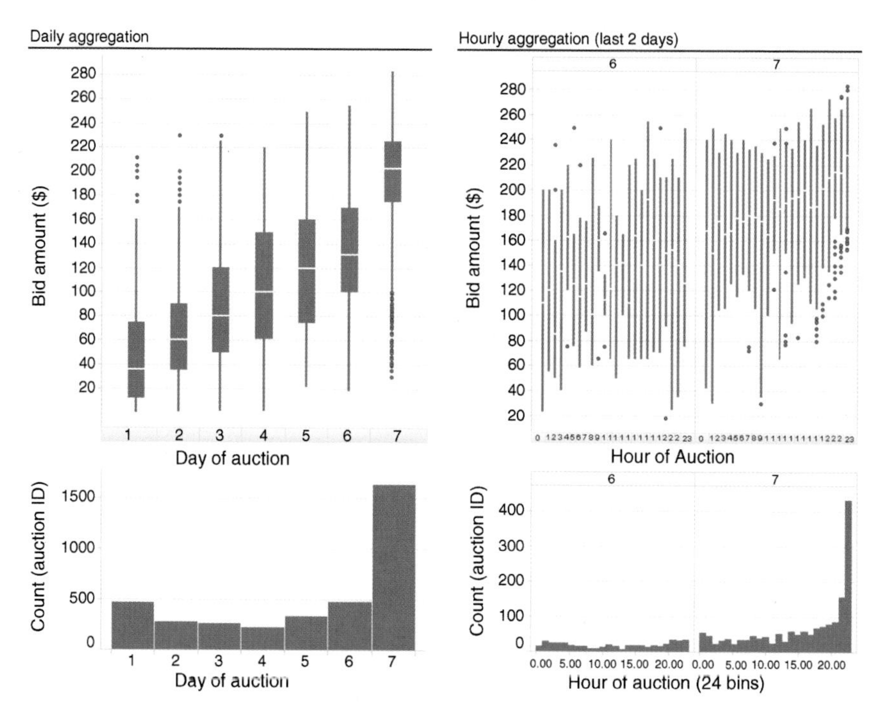

FIGURE 3.7 Bid distribution and intensity by varying the temporal hierarchy. *Left panel*: daily aggregation. *Right panel*: hourly aggregation for last 2 days of auction.

the outcome of an online auction (Bapna et al., 2008b). Jank and Shmueli (2008b) find that auctions selling identical products fall into one of the three segments of price dynamics, namely, "steady auctions" that experience a constant flow of dynamics, "low-energy auctions" with late dynamics, and "bazaar auctions" that see the largest push of dynamics. Shmueli and Jank (2008) illustrate the effect of auction parameters (such as the opening bid) on an auction's dynamics and find that higher opening bids result in lower price dynamics. Wang et al. (2008b) show that an auction's price dynamics can be characterized well using a single class of functional differential equation models and Jank et al. (2008b) extend upon this idea and develop model-based regression trees to relate differential equation models to auction covariates. Moreover, Wang et al. (2008a) show that the inclusion of price dynamics into forecasting models significantly improves the predictive capability of an auction (see also Bapna et al., 2002b). More recent studies focus on competing auctions (Dass et al., 2009; Jank and Zhang, 2009a) and find that dynamics not only can improve the predictability of an auction, but can also capture otherwise unobservable information such as bidding competition.

Recall that by "dynamics" we refer to the price velocity or the price acceleration. The price velocity measures how fast the price is changing and it can be estimated via the first derivative of the corresponding price curve (e.g., the first derivative of $f(t)$ in equation (4.1)). Similarly, the price acceleration measures changes in the price velocity and it can be computed via the second derivative of the price curve. Our starting point in this section is the prior observation that dynamics can be quite heterogeneous—even if we consider auctions for the same item, sold during the same period of time, using the same auction format as in Figure 4.5. In Chapter 3, we described curve clustering for the purpose of segmenting a data set of auctions by heterogeneity in price curves and dynamics. In this section, we present various models for capturing this heterogeneity and modeling its relationship to explanatory variables. We discuss *functional regression models* that allow us to link the dynamics generated during an auction with explanatory variables such as the auction format or the seller's reputation. We also discuss *differential equation models* that are a natural approach for modeling process dynamics. In that context, we first introduce a class of differential equation models that capture a wide range of auction dynamics. One methodological difficulty with differential equation models is that it is not quite obvious how to include predictor variables. To this end, we conclude with discussing a novel approach via *differential equation trees*.

4.2.1 Regression Models for Price Dynamics

We have shown in Chapter 3 that auction dynamics exist and that they can be quite different from one auction (cluster) to another. What we do not know yet is *why* some auctions experience fundamentally different price dynamics compared to other auctions. In other words, while we have found that differences exist, we cannot yet explain the reasons for their existence or relate them to other information about the auction. To this end, we now discuss functional regression models that allow us to investigate relationships between auction dynamics and associated explanatory variables. We start by discussing the main principles of functional regression analysis and then illustrate the method on a small case study. Note that recent work by Yang et al. (2010) on quantifying the dependence of pairs of functional observations could well be used as an input into our modeling efforts. Moreover, while we discuss functional regression analysis in the context of continuous response curves only, it can also be generalized to different types of responses (e.g., James and Hastie, 2001; James, 2002; Spitzner et al., 2003) and modified to help interpretations (James et al., 2009).

Functional Regression Analysis One of the goals of statistical modeling is to study the change of a response variable in reaction to changes in explanatory variables. In traditional statistical models, both the response variable and the explanatory variable have either scalar or vector values, representing univariate or multivariate data. In functional modeling, however, these variables can be more general (i.e., functional) objects. In the context of auction modeling, the response variable could be the price evolution $f(t)$ that describes the price progress over time. Explanatory variables could be auction characteristics such as the auction duration, the product category, a seller's rating, or a bidder's rating. The functional approach allows that, in addition to the price path, we can also model the price dynamics. Such a model enables us to study those factors that influence the price velocity $f'(t)$ and the price acceleration $f''(t)$ and subsequently lead to a better understanding of the price formation process.

We first describe the general functional regression model and its estimation process. We use vector notation similar to that of ordinary least squares. Let $\mathbf{Y}(t) = (y_1(t), y_2(t), \ldots, y_J(t))$ be a $J \times 1$ vector of functional objects where J denotes the total number of auctions. For instance, if we model the current price, then we set $y_j(t) = f_j(t)$. Alternatively, if want to find a model for the price velocity, we set $y_i(t) = f_i'(t)$, and so on. Let $\mathbf{Z}$ denote the $J \times (q+1)$ design matrix

$$\mathbf{Z} - \begin{pmatrix} 1, z_{11}, \ldots, z_{1q} \\ \vdots \\ 1, z_{J1}, \ldots, z_{Jq} \end{pmatrix}. \tag{4.25}$$

For instance, if the first covariate is the starting price of the auction, then we set z_{j1} = starting price for auction number j. If the second covariate is the seller's rating, then we set z_{j2} = seller rating for auction number j, and so forth. While the model formulation so far strongly resembles ordinary least squares, one of the main differences is that we use parameter *curves* rather than parameter *vectors*. Define a q vector of parameter curves $\boldsymbol{\beta}(t) = (\beta_0(t), \beta_1(t), \beta_2(t), \ldots, \beta_q(t))$. At every time point t, $\beta_1(t)$ measures the influence of the first covariate on the average response curve $y(t)$. For instance, if we set $y(t) = f''(t)$ to model the bid acceleration (where t is the day of the auction, and $\beta_1(t)$ denotes the parameter curve corresponding to the starting price), then $\beta_1(2)$ measures the average unit change in the price acceleration for a unit increase in the starting price *on the second day of the auction* (holding all other factors constant). Similarly, $\beta_1(6)$ measures this unit change *on the sixth day of the auction*. Thus, the flexibility of the functional approach stems from the fact that functional regression models capture the change of the covariates' influence on the response over time. This is in contrast to traditional models where the parameters remain constant. These varying parameter models are very useful in the online auction context since the relationship between, say, the starting price and the current price can be expected to change over the course of the auction.

While the functional regression model allows a better understanding of the change in the price formation process (and its dynamics), it also allows for new insight into the factors that lead to that change. Consider variables like the bidder rating or the number of bidders. Both of these variables are dynamic. That is, they differ from static variables like the starting price or the seller's rating in that the information changes with every new incoming bid. Consider Figure 4.19 that shows the *current average bidder rating* for a sample auction. Functional regression models can account for the changing nature of variables by introducing *dynamic covariates*. Dynamic covariates can reveal additional insight into the bid formation process that would otherwise be lost.

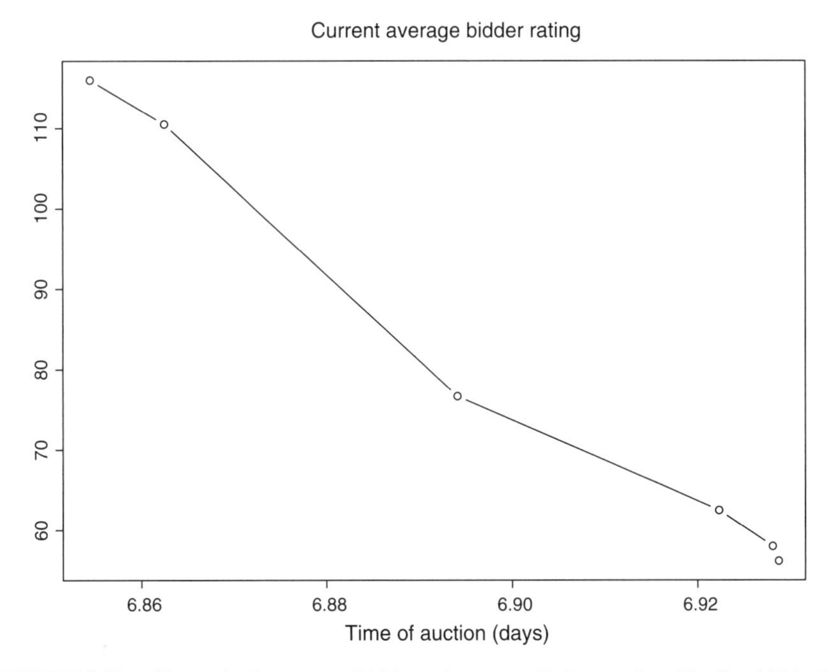

FIGURE 4.19 Change in the average bidder rating over a 7-day auction. The first bidder has a rating of 116 (leftmost circle). The second bidder rates at 105 that results in an average rating of $(116 + 105)/2 = 110.5$ (second leftmost circle). To arrive at a continuous representation, we linearly interpolate (solid line).

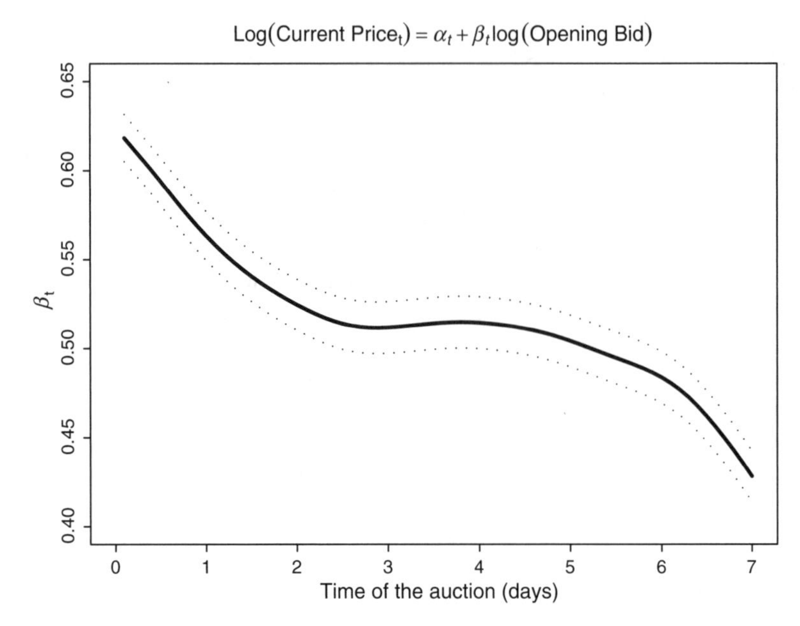

FIGURE 4.20 The estimated parameter curve (solid line) for a functional regression model of the opening bid on the price path. The dashed lines represent 95% confidence bounds.

Estimation of the parameter curves proceeds as follows. We attempt to find $\boldsymbol{\beta}(t)$ such that the expected value of $\mathbf{Y}(t)$ equals $\mathbf{Z}\boldsymbol{\beta}(t)$ for each value of t. This problem can be written similar to the least squares minimization objective of ordinary regression. The objective function

$$\text{ISSE}(\boldsymbol{\beta}) = \int \|\mathbf{Y}(t) - \mathbf{Z}\boldsymbol{\beta}(t)\|^2 dt \tag{4.26}$$

defines the *integrated error sum of squares* (ISSE), where $\|\cdot\|$ denotes the Euclidian norm. The goal is to find $\boldsymbol{\beta}(t)$ that minimizes ISSE. Ramsay and Silverman (2005) point out that since there is no particular restriction on the way in which $\boldsymbol{\beta}(t)$ varies as a function of t, one can minimize ISSE by minimizing $\|\mathbf{Y}(t) - \mathbf{Z}\boldsymbol{\beta}(t)\|^2$ on a suitable grid of values $t_1, t_2, \ldots, t_n$. This yields a sequence of parameter estimates $\hat{\boldsymbol{\beta}}(t_1), \ldots, \hat{\boldsymbol{\beta}}(t_n)$. One then reconstructs the continuous parameter vector $\hat{\boldsymbol{\beta}}(t)$ by simply interpolating between the values $\hat{\boldsymbol{\beta}}(t_1), \ldots, \hat{\boldsymbol{\beta}}(t_n)$.[5]

Interpretation One of the challenges of functional regression modeling is the careful interpretation of the results. Consider Figure 4.20 for illustration. We see that the *parameter curve* for the starting price follows a decreasing, S-shaped path. This has several implications. Overall, the parameter curve is positive during the entire

[5] An alternative—and equivalent—way is to operate on the spline coefficients rather than on a grid (Ramsay and Silverman, 2005).

auction duration, indicating a positive relationship between the starting price and the current price. In other words, higher starting prices are associated with higher prices at any time during the auction. Note, however, that the parameter curve is at its peak at the auction start ($\hat{\beta}(0) \approx 0.62$) and then decreases toward day 7, implying that the *strength* of the relationship between the starting price and the current price is continuously weakening. At the auction end, the parameter estimate has decreased to a value of only $\hat{\beta}(7) \approx 0.43$. The steep decline in the coefficient toward the auction end implies that the information contained in the starting price loses its usefulness for explaining the auction price as the auction progresses.

Case Study We illustrate functional regression analysis on a small case study using a random sample of 1009 auctions that took place on one single day on eBay's auction page (Bapna et al., 2008b). We first define the following covariates for our functional regression model:

z_{j1} = (log) starting price for auction j
z_{j2} = (log) item's final price (or selling price) for auction j
z_{j3} = (log) seller reputation (+5) for auction j
z_{j4} = (log) current average bidder experience (+5) for auction j
z_{j5} = (log) current number of bidders for auction j
z_{j6} = a dummy variable indicating U.S. currency in auction j
z_{j7} = a dummy variable indicating usage of secret reserve price in auction j
z_{j8} = duration of auction j (in days)
z_{j9} = a dummy variable for the product category type in auction j

The covariate z_{j1} is simply the natural log of the starting price for auction j. Similarly, z_{j2} denotes the item's final price[6] on a log scale. The covariate z_{j3} denotes the log of the seller rating.[7] Since some sellers have negative ratings in the range $(-4, \ldots, -1)$, we add 5 to each seller's rating before taking logs, thus assuring that the log transformation is well defined.

We also include information on the bidder experience. As with the seller ratings, we compute the log of the average bidder rating after adding 5. However, note that in contrast to the seller rating, the average bidder rating does not remain constant throughout the auction. In fact, the average rating of currently participating bidders most likely changes with every new incoming bid. We therefore use a *dynamic* covariate that takes this change into account. Using an evenly spaced grid of points across the auction duration, say $t_1, t_2, \ldots, t_n$, the *current* mean bidder rating at t_i is calculated as the average rating of all bidders that participate at or before time t_i. Taking logs, we denote this covariate by z_{j4}. Thus, z_{j4} measures the average experience level of currently participating bidders.

As with the bidder ratings, the number of bidders also changes with every new incoming bid. To measure the effect of the *current* number of bidders on the price

[6] By item final price, we mean the selling price; that is, the price that the highest bidder pays.

[7] We measure seller reputation as the total number of positive feedback minus the total number of negative feedback. Alternative ways exist, for example, the ratio of positive to total feedback.

formation, we create another dynamic covariate z_{j5}. Using the same grid as above, z_{j5} denotes the (log of) the total number of bidders that participate at or before t_i. In this sense, z_{j5} measures the effect of the current competition level.

To capture the effect of currency on the auction outcome, we include a dummy variable z_{j6}, which assumes the value 1 for auctions in U.S. currency and the value zero for auctions in non-U.S. currency (GBP or the Euro). Thus, z_{j6} measures the geographical market differences in the bidding dynamics between the United States and Europe. A similar dummy variable (z_{j7}) is created for the secret reserve price that is set equal to 1 if the seller uses this option.

Another important factor of price formation is auction duration, measured by the covariate z_{j8}. Most auctions on eBay range from 1 to 10 days. To measure the effect of duration, auctions of different length have to be incorporated into the same functional regression model. Varying-length auctions result in varying-length smoothing splines $f(t)$. To align splines of different length, we standardize auction duration into unit time intervals. After this standardization, every auction has starting time 0 and ending time 1.

We use curve clustering (as described in Chapter 3) to group eBay categories into two segments (denoted as A and B) where each segment consists of more homogeneous price dynamics. The allocation of categories into segments is shown in Table 4.3. The dummy variable z_{j9} assumes the value 1 for products in category B.

Finally, we also investigate interactions between several variables. While the usage of interaction terms is well understood in traditional regression models, there have not been many applications of the interaction concept in the context of functional data analysis. Generalizing the interaction concept to functional modeling is not straightforward, in part because its interpretation can be prohibitively complicated. The following results show that the functional interaction term proves very useful, especially in the auction setting.

Results: After fitting the above model to our data, we obtain the parameter estimates depicted in Figures 4.21–4.23. We discuss these estimates by the type of information they convey.

TABLE 4.3 Two Distinct Clusters of eBay Categories

Category A	Category B
Collectibles	Computers
Toys & Hobbies	Jewelry
Coins & Stamps	Consumer Electronics
Books	Antiques
Sports	Clothing & Accessories
Pottery & Glas	Automotive
Music & Movies & Games	
Travel & Tickets	
Home & Garden	
Health & Beauty	
Business & Industrial	

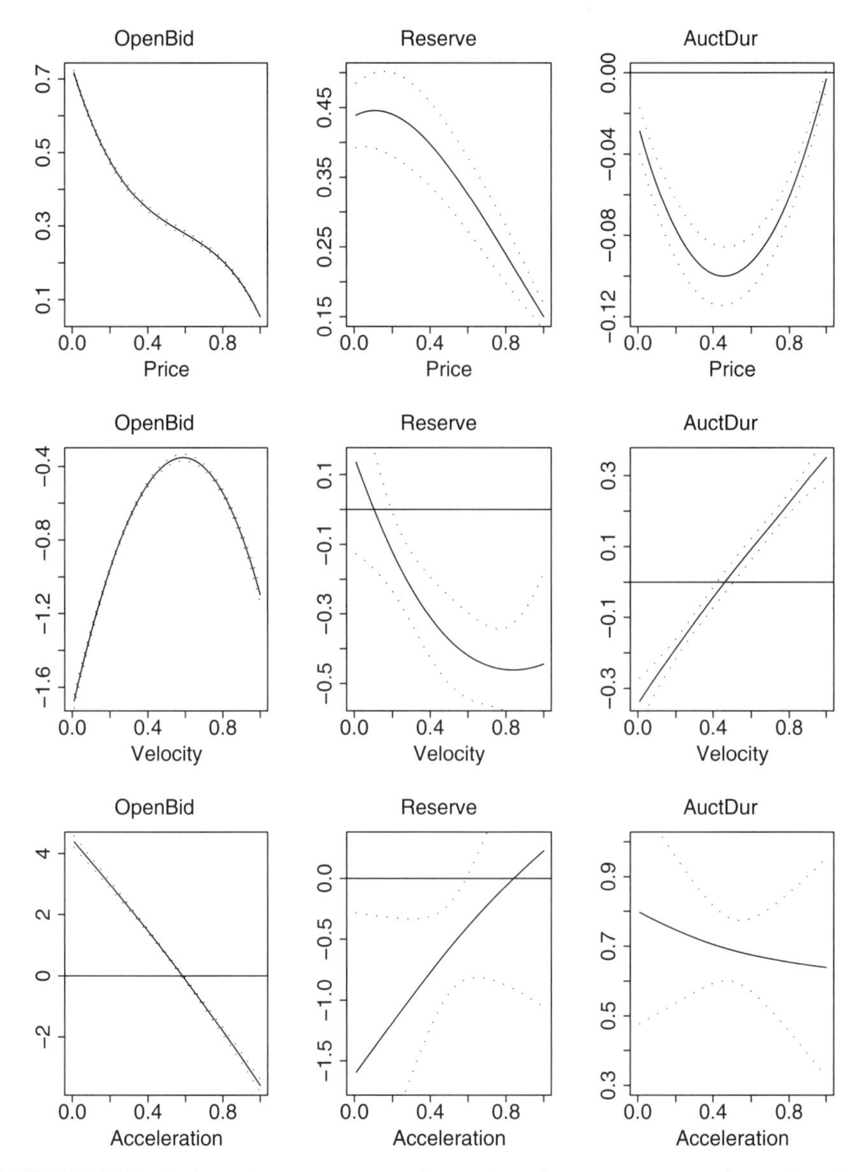

FIGURE 4.21 Estimated parameter curves for starting price, secret reserve price, and auction duration.

Seller's Mechanism Design Choices—Starting price, Reserve, and Duration: Consider the leftmost panel in Figure 4.21 that shows the estimated parameter curves associated with the (log) starting price. The top panel shows the parameter curve for a regression on the current price, $f(t)$, while the middle and the bottom panels correspond to regressions on the price velocity $(f'(t))$ and price acceleration $(f''(t))$.

The parameter curve in the top plot is positive throughout the entire auction, starting at about 0.7 and ending just below 0.1. A positive parameter implies a positive relationship between the starting price and the current price. Note that this relationship is statistically significant as indicated by the very tight confidence bounds (dashed lines) that remain above zero throughout.[8] The implication is that the higher the starting price, the higher the current price. However, although the parameter remains positive, it gradually declines in magnitude toward the auction end. The decreasing magnitude implies that the impact of the size of the starting price on the price formation process reduces throughout the auction. In other words, the information contained in the starting price loses its usefulness for explaining the current price. Consistent with auction theory, we observe that the starting price influences entry into the auction and has higher signaling value in the early stages of the price formation process. We know, for instance, that more experienced sellers make better mechanism design choices that attract a higher number of bidders (Bapna et al., 2008a). This effect tends to lessen as the auction rounds progress and the competitive elements of the auction on hand take over. This also motivates us to consider the interaction between the number of bidders and the starting price, as discussed further below.

The impact of the starting price on the price dynamics can be seen in the middle and bottom panels. The middle panel shows that the parameter curve associated with the price velocity is negative throughout the auction. This negative relationship means that higher starting prices result in *slower price increases*. The higher the starting price, the smaller the difference to an item's valuation. Bidders, unclear about the exact valuation, can be expected to place smaller increments above the starting price, resulting in lower *auction dynamics* and thus a slower price formation.

Yet even more information about the price formation process can be extracted from the relationship between the starting price and the price acceleration in the bottom panel. It is most noteworthy that the parameter curve changes its sign from positive to negative about midway through the auction and decreases further to the end. The negative parameter estimate at the auction end implies that high starting prices are associated with a high *negative* price acceleration, or *price deceleration*. Put differently, auctions with high starting prices experience a slowdown in the price increase at the end of the auction and vice versa. This finding is interesting in light of the commonly encountered phenomenon of *bid sniping* (Roth and Ockenfels, 2002).

It is observed that the usage of a hidden reserve price has a positive impact on the price level, but during the second half of the auction such a tool used by the seller results in slower price increases than in non-reserve auctions.

The relationship between auction duration and price formation is strongest during the middle of the auction and has a negative sign. While eBay's auctions experience heavy participation at the beginning and at the end, the middle of the auction is typically marked by a "draught" with only few incoming bids. Our analysis shows that this draught is stronger in longer lasting auctions. On the other hand, participation

[8] We used 95% confidence levels for all confidence bounds.

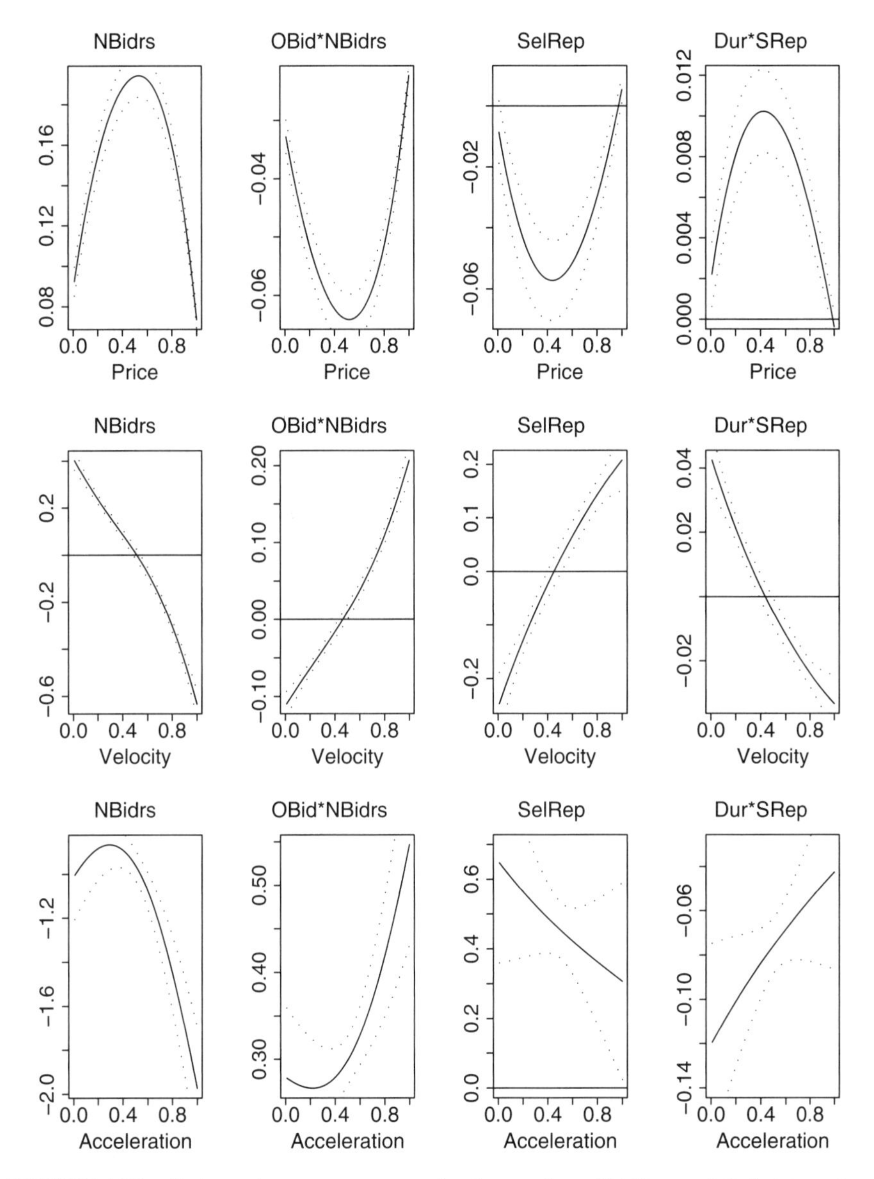

FIGURE 4.22 Estimated parameter curves for the number of bidders and their interaction with the starting price, seller reputation, and its interaction with auction duration.

is typically heavy at the auction end and our analysis of the price velocity shows that at the end the auction dynamics are larger for longer lasting auctions.

Competition—Number of Bidders and Interaction with Opening Bid: The first panel in Figure 4.22 shows the effect of the *current* number of bidders. The top plot shows that, not surprisingly, the number of bidders is positively associated with the current price. A larger number of bidders result in increased competition for the item and thus in a higher (final) price. Interestingly, though, the strength of this relationship is strongest in the middle of the auction (note the ∩-shape of the graph). The middle of the auction typically experiences the smallest amount of bidder participation, marked by "evaluators" who place only a single bid and do not return until the auction is over (Bapna et al., 2003). These evaluators typically place higher bids than the average bidder, resulting in a stronger price increase per bidder.

The middle and bottom plots reveal the impact of competition on price dynamics. Note the decreasing parameter curves for both the price velocity and the acceleration. Auctions on eBay typically experience most bidding activity at the end. This is commonly described as bid sniping where, during the last moments, bidders compete heavily with each other since the winner takes it all. A high competition level is therefore not uncommon at the auction end, in contrast to earlier stages of the auction. The coefficients of the price dynamics thus suggest that the same competition level early in the auction results in a much stronger price velocity and acceleration than at the auction end.

It is rewarding to carefully examine the interaction between the number of bidders and the starting price. The shape of the parameter curve is almost exactly the opposite of the main effect of the number of bidders. In particular, the interaction for the price evolution is negative! The implication is that although the current number of bidders has a *positive* effect on the price formation process, the magnitude of this effect is *reduced* for a large starting price. A similar moderating effect can be seen for the price dynamics.

Seller Reputation and Interaction with Auction Duration: We observe that seller ratings, which also proxy for experience, negatively correlate with price levels and are moderated by the auction duration. Since we observe a negative main effect and a positive interaction effect, we conclude that in longer auctions higher seller ratings result in higher price levels. Looking at the dynamics, we also observe that the effect weakens as duration increases.

Market Characteristics—Item Value, Currency, Category: The impact of the item's value is shown in the leftmost panel of Figure 4.23. As in the case of the starting price, the relationship between the value and the current price (top graph) is positive throughout the auction, implying that auctions with higher valued items attract higher bids. However, in contrast to the starting price, the strength of this relationship increases steadily. This is not surprising: At the beginning of the auction, it appears that bidders either have not yet formed a clear opinion about the item's valuation or are not yet willing to fully reveal their valuation. Either way, the price formation is not strongly associated with the item's value. This changes at the end of the auction. At the end, bidders receive an ever increasing amount of information from other participants and also possibly from outside sources. As the auction closes, they are also more willing to reveal their true valuation in order to win the item. It is reasonable to assume toward the end that eBay's second-price mechanism induces truth telling. Using the lens of William Vickrey's (1961) stylized model, eBay's mechanism is a hybrid between an open ascending English auction and a sealed-bid second-price auction. For such hybrid mechanisms, multiple equilibria are likely to exist and are

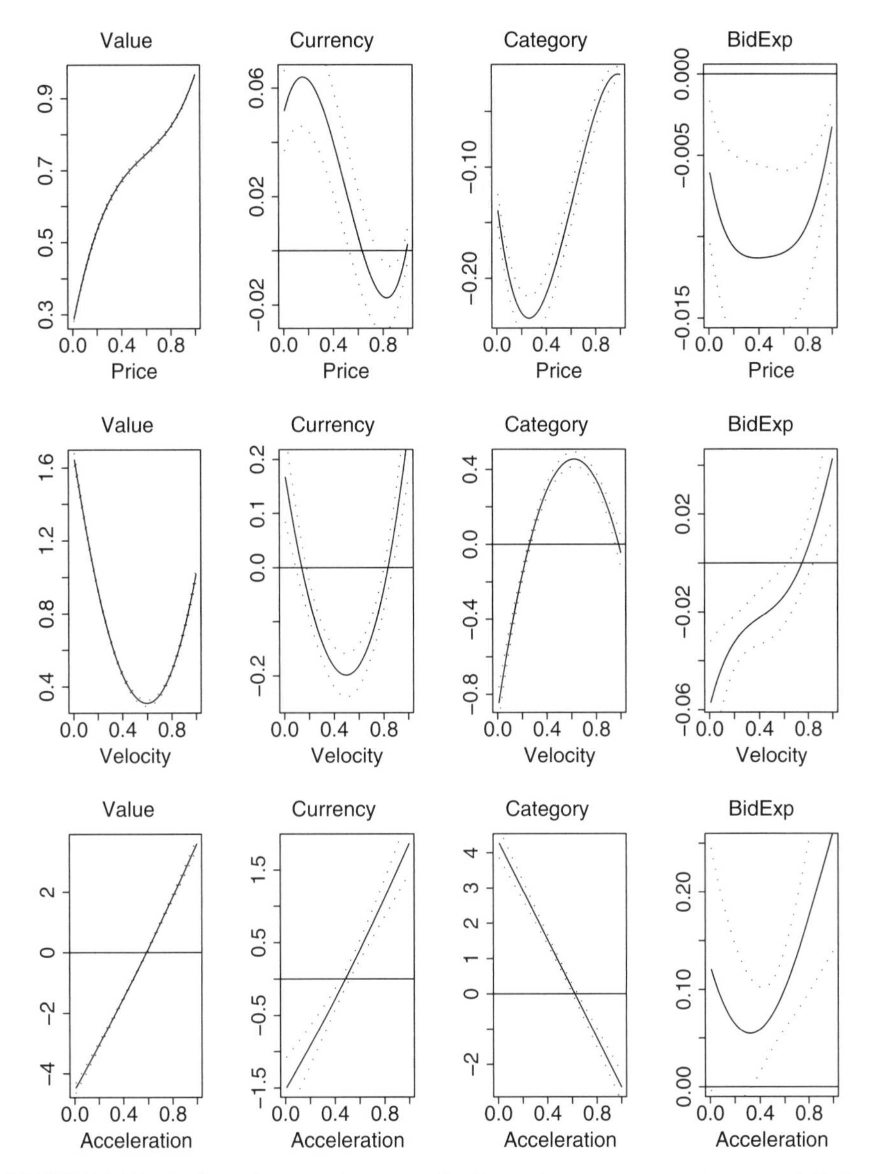

FIGURE 4.23 Estimated parameter curves for item value, currency, category, and bidder experience.

being currently explored (Bajari and Hortacsu, 2003). Toward the end of the auction, it is straightforward to prove that the absence of any response time to other bidders makes truth-telling sniping a weakly dominant strategy, and the eBay auction resembles a second-price sealed-bid auction. Thus the strength of the relationship increases.

As for the dynamics, we can see that toward the auction end, higher prices are associated with an increase in bidding dynamics. Indeed, while an item with a higher price strongly correlates with a faster price increase (middle graph), it is also associated with a faster *r*ate of increase (bottom graph). These two graphs suggest that for auctions with highly valuable items, fast price accelerations at closing can be expected. One explanation is that bidders for high-value items are more price sensitive and act more strategically in placing their bids compared to bidders for inexpensive goods. It also suggests that bid sniping is more prevalent in high-value auctions.

With respect to the currency effect, since we control for an item's final price, it is not surprising that the coefficient of the price path is near zero at the auction end. However, the dynamics are more interesting. Overall, U. S. auctions experience faster price movement at the auction end than their European counterparts. We also find that compared to European auctions, the price levels of U. S. auctions are on average 4% higher during the first half of the auction. These results indicate that the U. S. market, in contrast to Europe, is characterized by both early bidding and late bidding. While late bidding appears to be the dominant strategy in the European market, it is stronger in the United States. This may indicate a fundamental cross-cultural difference in bidding habits.

We find that prices of category B items are somewhat lower than category A items throughout the auction. The rate of price increase for category B items (jewelry, clothing, antiques, etc.) starts out slower than category A, but from mid-auction it catches up and subsequently increases at a faster rate.

Current Average Bidder Experience: Recall that in contrast to the seller rating, the average bidder rating does not remain constant throughout the auction. In fact, the average rating of currently participating bidders changes with every new incoming bid. We therefore use a *dynamic* covariate (average experience level) that takes this change into account. We observe a negative relationship with the current price, suggesting that experience pays off in terms of a lower price. However, once again the dynamics shed more light: It is evident that a higher experience level is suggestive of enhanced strategizing on the part of the bidders. We observe that early in the auction, high experience leads to slow price formation. This indicates that experienced bidders appear to hold back their valuations and do not bid early, a rationale consistent with the literature on sniping (Roth and Ockenfels, 2002).

Conclusion The most significant contribution of functional regression analysis is that it gives us a new *vocabulary* to describe and explain the dynamics of online auctions. This vocabulary was not available previously and it allows us to characterize auctions as dynamic systems. For instance, our case study shows that higher starting prices are associated with *slower* price increases, especially during the *early* stages of the auction. Not surprisingly, this effect is moderated by the number of bidders. A positive interaction suggests that the same starting price results in *faster* price increase during the *later* stages of high-competition auctions. This finding suggests that there is significant scope for enhancing the price formation contribution of the early stages of eBay's auction mechanism. We find that the incremental impact of an additional bidder's arrival on the rate of price increase is smaller at the end of the auction. This suggests that towards the closing, sniping stages of the auction, eBay's progressive mechanism verifiably turns into a sealed-bid second-price mechanism. We find that "stakes" do matter and that the rate of price increase is higher for more expensive items, especially at the start and the end of an auction. We also observed that higher seller ratings (which correlate with experience) positively influence the price dynamics, but the effect is weaker in auctions with longer durations. In an interesting effect across markets, we find that compared to European auctions, U. S. auctions get 4% higher price levels during the first half of the auction. After that the price levels are comparable. The second-order dynamics of this effect suggests that the main difference is in the middle of the auction. While U. S. auctions have faster price increases during the beginning and the end of the auction, they are slower than European auctions during the auction middle. Another interesting finding is that price level is negatively related to auction duration when the seller has low rating! However, in auctions with high-rated sellers, longer auctions achieve higher price levels throughout the auction, and especially at the start and the end. This suggests that new and inexperienced sellers are better off using shorter auction durations initially, and then switch to longer durations as their reputation increases. Taken together, functional regression analysis provides a behind-the-scenes look at the process that leads to end results that are visible to all.

Next, we describe yet another way of modeling price dynamics via differential equation models. The main difference is that while functional regression models can only account for external factors, differential equation models also capture effects inherent in the system.

4.2.2 Differential Equation Models for Price Dynamics

As seen in the previous section, functional regression analysis gives us a tool to link price dynamics with potential explanatory variables such as the auction format, seller characteristics, or bidding behavior. However, regression is not the most natural approach for modeling the dynamics of an evolving system. Moreover, it also does not capture the dynamics generated from "within the system." In the context of auctions, one could argue that some of the changes in price dynamics are caused from within the auction process. For instance, an auction that experiences fast price increases for some period of time will likely see a slowdown of price in later stages. Reasons for this slowdown could include bidders abandoning the auction in favor of other auctions with a lower (and slower moving) price. While the root cause is often unobservable, it is captured in the interplay (or dependence structure) of dynamics. Classic regression models cannot capture how dynamics depend on one another. To that end, we discuss a more natural approach for modeling dynamics via *functional differential equation models.* We again start by discussing the main principles of functional differential equations and then illustrate their use within the online auction domain on a small case study.

Model Formulation and Estimation Differential equations are a common tool in physics and engineering for modeling the dynamics of a closed system. They are commonly used for describing processes such as population growth, mixing problems, mechanics, and electrical circuits (Coddington and Levinson, 1955). Applications of differential equations are also found in economics and finance. The Solow model (Solow, 1956), for example, utilizes differential equations to model the long-run evolution of the economy see also Mankiw et al. (1992). In finance, partial differential equations, are used for pricing financial derivatives (Karatzas and Shreve, 1998; Fouque et al., 2000; Almgren, 2002). Here, we focus on the functional version of differential equations, also referred to as *principal differential analysis* (Ramsay, 2000; Ramsay and Silverman, 2005; Müller and Yao, 2010; Ramsay et al., 2007).

Let y_i be the price function for auction i ($i = 1, \ldots, N$), and let $D^m y_i$ be the mth derivative of y_i. Our goal is to identify a linear differential operator (LDO) of the form

$$L = \omega_0 I + \omega_1 D + \cdots + \omega_{m-1} D^{m-1} + D^m \tag{4.27}$$

that satisfies the homogeneous linear differential equation $L y_i = 0$ for each observation y_i. In other words, we seek a linear differential equation model that satisfies

$$D^m y_i = -\omega_0 - \omega_1 D y_i - \cdots - \omega_{m-1} D^{m-1} y_i. \tag{4.28}$$

An important motivation for finding the operator L is substantive: applications in the physical sciences, engineering, biology, and elsewhere often make extensive use of differential equation models of the form $L y_i = f_i$. The function f_i is often called a *forcing* or *impulse* function, and it indicates the influence of exogenous agents on the system defined by $L y = 0$. Returning to the online auction setting, we reason that variation in price is due to variation in the forces resulting from bid placement and bid timing, and that these forces have a direct or proportional impact on the acceleration of the price process.

In practice, due to the prevalence of noise, it is virtually impossible to find a model that satisfies (4.28) *exactly*. Hence, principal differential analysis adopts a least squares approach to the fitting of the differential equation model. The fitting criterion is to minimize the sum of squared norms

$$\mathrm{SSE}_{\mathrm{PDA}}(L) = \sum_{i=1}^{N} \int [L y_i(t)]^2 dt \tag{4.29}$$

over all possible models L. Note that identifying L is equivalent to identifying the m weight functions ω_j that define the linear differential equation in (4.27).

There are generally two approaches for estimating the weight functions ω_j. The first, pointwise minimization, yields pointwise estimates of the weight functions ω_j

by minimizing the (pointwise) fitting criterion

$$\text{PSSE}_L(t) = \sum_i (Ly_i)^2(t) = \sum_i \left[\sum_{j=0}^{m-1} \omega_j(t)(D^j y_i)(t) + (D^m y_i)(t) \right]^2. \tag{4.30}$$

The pointwise approach can pose problems, especially if the ω_j's are estimated at a fine level of detail. An alternative approach, which is computationally more efficient, is to use basis expansions. Taking this approach, we approximate each weight function ω_j by a linear combination of basis functions ϕ_k, say, $\omega_j \approx \sum_k c_{jk}\phi_k$, where c_{jk} denote basis function coefficients. Using this expansion, one can approximate $\text{SSE}_{\text{PDA}}(L)$ in (4.29) as a quadratic form that can be minimized using standard numerical techniques. For more details, see Ramsay and Silverman (2005) or Ramsay (1996).

Model Fit An initial impression of the model fit can be obtained via visualization. If the model represents the data well, then the identified differential operator L should be effective at annihilating variation in the y_i, and this can be visualized by plotting the *empirical forcing functions* Ly_i. If the plotted Ly_i's are small and mainly noise-like, then the model provides good data fit. As a point of reference for the magnitude of the Ly_i's, we use the size of the $D^m y_i$'s, since these are the empirical forcing functions corresponding to $\omega_0 = \cdots = \omega_{m-1} = 0$.

To confirm visual impression, the quality of fit can also be gauged by more quantitative statistics. In the differential equation context, this can be done via the pointwise error sum of squares $\text{PSSE}_L(t)$ in (4.30). A logical baseline against which to compare PSSE_L is the error sum of squares in which we set $\omega_j = 0$ so that the comparison is simply with the sum of squares of the $D^m y_i$, which is analogous to the classical sum of squares in ANOVA:

$$\text{PSSE}_0(t) = \sum_i [(D^m y_i)(t)]^2. \tag{4.31}$$

Thus, we can assess the model fit of the differential equation by examining the pointwise squared multiple correlation function

$$\text{RSQ}(t) = \frac{\text{PSSE}_0(t) - \text{PSSE}_L(t)}{\text{PSSE}_0(t)} \tag{4.32}$$

and the pointwise F-ratio

$$F\text{-ratio}(t) = \frac{(\text{PSSE}_0(t) - \text{PSSE}_L(t))/m}{\text{PSSE}_0(t)/(N-m)}. \tag{4.33}$$

A Special Case: The Second-Order Linear Differential Equation We now discuss in further detail a special case of the above general differential equation model:

the second-order linear differential equation. This model is particularly useful in the context of auctions, where the focus is on the first two derivatives of the price curve. Consider the following differential equation:

$$Ly_i = \omega_0 y_i + \omega_1 Dy_i + D^2 y_i = 0. \tag{4.34}$$

Setting $\omega_0 = 0$, we get

$$Ly_i = \omega_1 Dy_i + D^2 y_i = 0, \tag{4.35}$$

which describes a strictly monotone, twice-differentiable function (Ramsay, 1998). Given that the auction price is monotonically increasing, equation (4.35) appears to be a reasonable candidate for online auctions. From here on, for ease of notation, we write $\omega = \omega_1$ and $\omega^* = -\omega$.

The coefficient function $\omega^* = -\omega = D^2 y / Dy$ measures the relative curvature of the monotone function in the sense that it assesses the size of the curvature of $D^2 y$ relative to the slope Dy. The special case of $\omega^* = -\alpha$ leads to $Y(t) = C_0 + C_1 \exp(\alpha t)$, whose exponent has constant curvature relative to α, while $\omega^* = 0$ defines a linear function. Thus, small or zero values of $\omega^*(t)$ correspond to locally linear functions, whereas very large values correspond to regions of sharp curvature. In mechanical systems, the latter type is generally caused by internal or external frictional forces or viscosity. In the context of online auctions, sharp curvature in the price process can be related to jump bids caused by bidders attempting to apply an external force ("deterring other bidders") to the bidding process. In contrast, $\omega^* = 0$ indicates a very slowly moving price process that is often observed during the middle of the auction ("bidding draught").

Case Study In this case study, we illustrate the above second-order linear differential equation model. We consider two different data sets: one for a high-value product (Microsoft Xbox game consoles, valued at approximately \$179.98 at the time of data collection) and the other for a low-value product (Harry Potter books, valued at approximately \$27.99 at the time of data collection); see also Wang et al. (2008a,b). The reason that we consider both high- and low-value products is to investigate the effect of an item's value on price dynamics. To that end, we show the results of three different analyses: one in which we analyze the combined data of both Harry Potter and Xbox auctions (we refer to this as the "mixed" data since it mixes high- and low-value items), and two additional analyses, one for each item separately. Graphically, we distinguish between these three analyses in the following way: solid lines mark the combined (mixed) data, dashed lines mark the Harry Potter data, and dotted lines mark the Xbox data.

Figure 4.24 shows a phase-plane plot (see also Figure 3.25) for the average second derivative of price (or price acceleration) versus the average first derivative (price velocity). The numbers along the curve indicate the day of the auction (for 7-day auctions). We can see that the price velocity is high at the beginning of the auction (days 0 and 1): At the auction start, it takes an instantaneous "burst of energy" to

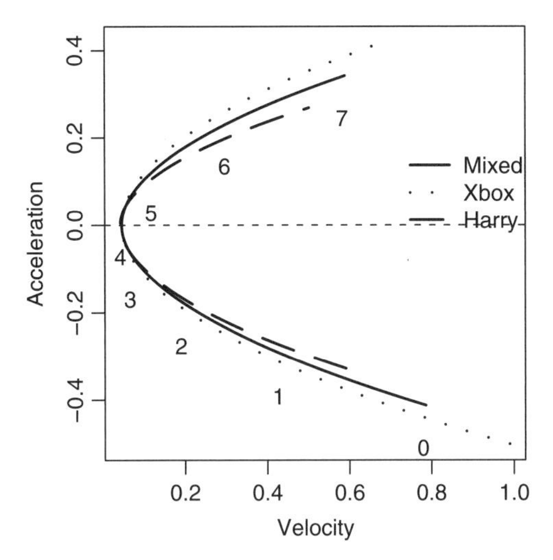

FIGURE 4.24 Phase-plane plot of the second derivative (acceleration) versus the first derivative (velocity). The solid line corresponds to the combined data for both Harry Potter and Xbox. The dashed and the dotted lines correspond to the individual Harry Potter and Xbox data, respectively.

overcome the opening bid (which can be considerably high). After this initial burst, the dynamics slow down: the price acceleration is negative and since changes in acceleration precede changes in velocity, we observe a consequent slowdown in the price speed. This slowdown continues until about day 4, after which the dynamics reverse: acceleration turns positive and leads to an increase in velocity. In fact, price increases quite rapidly until the auction end.

Several aspects are noteworthy in Figure 4.24. First, the phase-plane plot has a"C" shape, which is typical of an online auction: a phase of decrease in dynamics is followed by a transitional phase of change that is followed by a phase of increase in dynamics. Many auctions are characterized by a "C"-shaped phase-plane plot. However, while the shape is similar across many auctions, its magnitude can vary quite significantly, depending on different auction characteristics. Figure 4.24, for instance, shows that the "C" has a wider range for the (higher valued) Xbox auctions, indicating more price activity between the start and the end of the auction compared to (lower valued) Harry Potter auctions.

We now fit the data to the second-order linear differential equation in (4.35). As previously, we show the results for three different analyses: one for the combined Harry Potter and Xbox data and two analyses for each of the two items separately. The estimated coefficient function $\omega^* = -\omega$ is displayed in Figure 4.25. We observe that the results for the combined data are very similar to those for the two individual data sets. This is in line with the observed similarity of the phase-plane plots in Figure 4.24 and it implies that, overall, auction dynamics are similar, regardless of the item value.

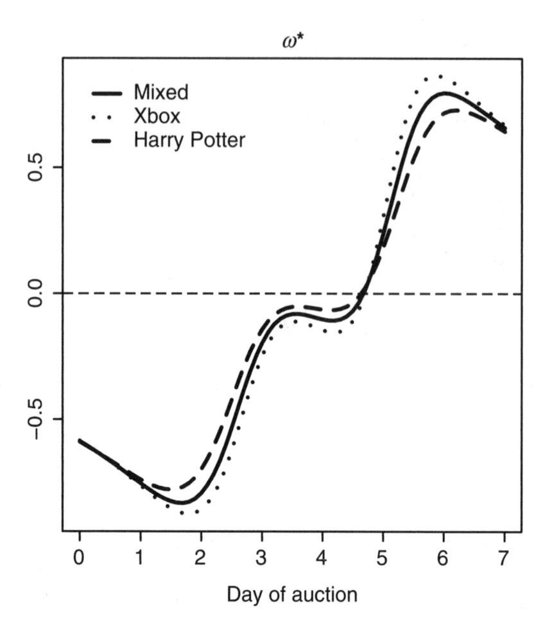

FIGURE 4.25 Estimated coefficient function of a monotone second-order linear differential equation model. The solid line corresponds to the combined Harry Potter and Xbox data; the dashed and the dotted lines correspond to the individual data sets.

A more careful investigation of Figure 4.25 reveals that ω^* has three levels of values: negative, zero, and finally positive. These capture and correspond to the three bidding phases during an auction: early activity, little mid-auction activity, and high late activity. The typical bidding behavior during an eBay auction consists of some early bidding, where bidders establish their time priority[9]. Then comes a period of "bidding draught," where there are hardly any bids placed,[10] and finally, during the last hours of the auction, bidding picks up again and dramatically peaks during the last auction minutes. This last-moment bidding is called "sniping," and there are various explanations as to why bidders engage in it (e.g., to avoid bidding wars, because last-moment bidding does not allow other bidders to respond).

The values of ω^* reflect the phases as follows: Recall that a value of zero indicates linear motion of the price process (i.e., no dynamics), whereas large positive or negative values are indicative of changes in the dynamics (depressing them or increasing them, respectively). The first phase (up to day 3) is characterized by a negative ω^*, with a dip on day 2. This dip marks the change from early bidding to "bidding draught" when velocity decreases and we see $\omega^* = 0$. From day 5 on $\omega^* = 0$ begins to increase, turning positive and peaking at day 6.

[9] When the two highest bids are tied, the earliest bidder is the winner.

[10] One possible reason is that bidders avoid revealing their willingness to pay too early to avoid early price increases.

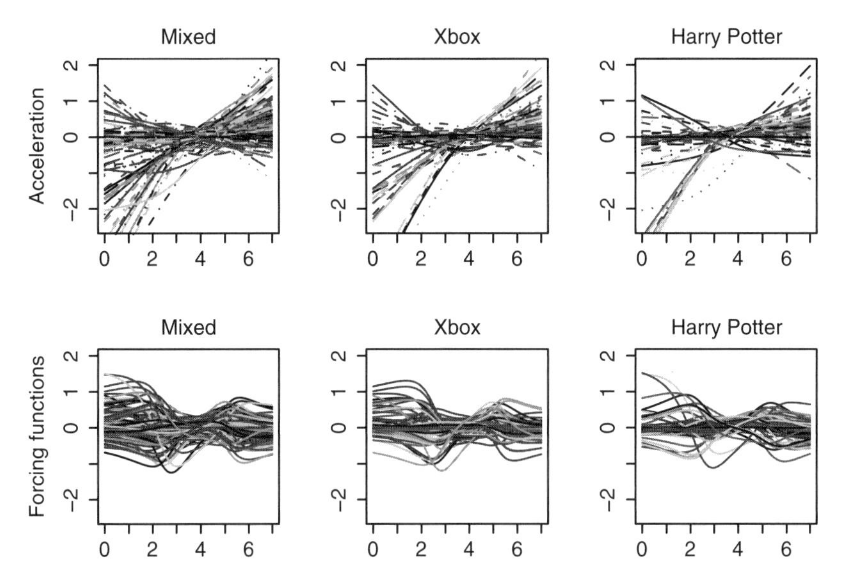

FIGURE 4.26 Comparison of the accelerations (second derivatives) and the observed forcing functions based on the monotone second order linear differential equation model.

The "double S" shape of the ω^* curve in Figure 4.25 is typical for an online auction (modeled by a linear differential equation), similar to the typical "C" shape observed in the phase-plane plots, and does not change much from one auction to another. What does change, however, is the *magnitude* of this double S, indicating larger or smaller price dynamics; its *steepness*, with steeper shapes resulting in a more abrupt change in dynamics; and the *changepoints*, that is, the locations where one bidding phase changes to another (see also Wang et al. (2008b)).

The fit of our differential equation model can be gauged from Figure 4.26 that shows the equivalence of a residual analysis. The top panel shows the observed price accelerations and the bottom panel shows the corresponding estimated forcing functions of the differential equation in (4.35). These two should be identical under perfect model fit (similar to the observed and fitted observations in regression). We can see that although the fit is not perfect, the range of the forcing functions is identical to that of the observed accelerations during most of the auction. The fit is especially close in the middle and the end of the auction; only the auction start does not seem to be captured as well by the differential equation model. We also see that the fit for the Harry Potter auctions is similar to that of the Xbox auctions.

Another quantification of model fit is given by the pointwise R-squared (RSQ) and F-ratio from (4.32) and (4.33). We show RSQ in Figure 4.27; the insight from the F-ratio is equivalent and is therefore omitted. RSQ is larger than 0.99 throughout the auction, indicating a very good global fit of the monotone second-order linear differential equation model (4.35). We see that the fit also varies: it is best at the beginning and end of the auction, but weaker during mid-auction. The fit is somewhat

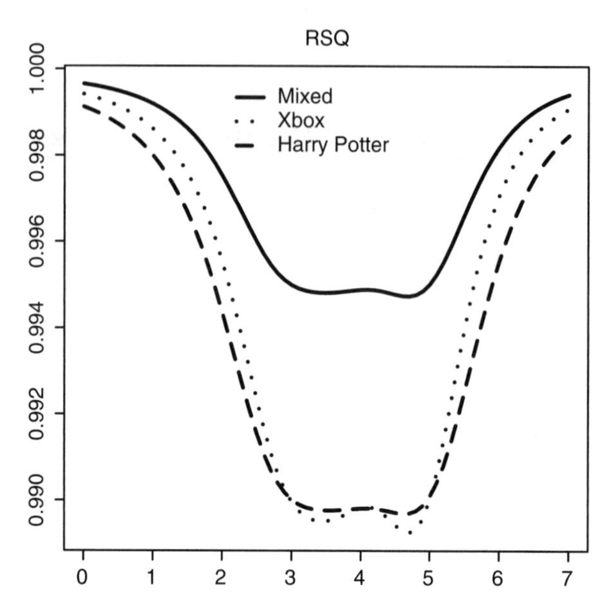

FIGURE 4.27 Model fit based on the residual sums of squares. The x-axis corresponds to the day of the auction.

weaker when considering Harry Potter and Xbox separately, which is likely due to the smaller sample size for the separate models.

In summary, we learn that a second-order linear differential equation model fits online auction data reasonably well. It captures the three phases of bidding and the interplay of dynamics that change over the course of the auction. We also see that the degree of model fit varies at different periods of the auction. This motivates our next step in which we want to investigate the reasons for this variation.

4.2.3 Differential Equation Trees

The previous section has shown that differential equation models can capture the changing dynamics in online auctions, but it has also shown that a significant amount of the variation in the dynamics is left unexplained. In the following, we set out to explain some of the residual variations using covariate information. In the auction context, plenty of covariate information is available (e.g., information about the auction format, the seller's characteristics, and the product or the bidding behavior). However, incorporating covariate information into differential equation models is not straightforward. Ramsay and Silverman (2005) suggest a way of incorporating covariate information via the forcing function, thereby creating a nonhomogeneous differential equation. We discuss an alternative innovative approach, borrowing ideas from recursive partitioning. In particular, we propose a novel tree-based approach for differential equation models. We discuss how to estimate differential equation trees and then illustrate their performance in a small case study.

Classification Trees Versus Differential Equation Trees Tree models give simple descriptions of often complex, nonlinear relationships between several predictors and a univariate or multivariate response. A classical reference is the monograph *Classification and Regression Trees* by Breiman et al. (1984). While tree models are generally very powerful, they encounter problems when the response is high dimensional. Yu and Lambert (1999) explore two ways of fitting trees to high-dimensional data. Both approaches proceeded by first reducing the dimensionality of the data and then fitting a standard multivariate tree to the reduced response. In the first approach, the dimensionality was reduced by representing the response as a linear combination of spline basis functions, while in the second one, the dimensionality was reduced using principal components analysis, retaining only the first few principal components.

Note that classical trees fit a scalar in each node of the tree. Since this tends to produce rather large and complex trees (e.g., Chan and Loh, 2004), research on incorporating (simple) parametric models into trees has recently received considerable attention. Such approaches are often referred to as *functional trees* (Gamma, 2004) with the most notable being *M5* (Quinlan, 1993). See also Loh (2002), Kim and Loh (2001), Chan and Loh (2004) for related approaches. One particularly noteworthy approach is that of Zeileis et al. (2005), who took the integration of parametric models into trees one step further by embedding recursive partitioning into the model estimation and variable selection framework. Within that framework, every leaf is associated with a conventionally fitted model such as a maximum likelihood model or a linear regression. The model's objective function is used for estimating both the parameters and the split points. The appeal of this approach is that the same objective function is used for partitioning and for parameter estimation. Building upon these ideas, Jank et al. (2008b) proposed a novel model-based functional differential equation tree that allows the incorporation of covariate information into dynamic models. We first briefly review the main ideas of model-based recursive partitioning and then use these ideas to derive our functional differential equation tree methodology.

Model-Based Recursive Partitioning Let $M(y, \theta)$ be a parametric model where $y = (y_1, \ldots, y_N)$ are (possibly vector-valued) observations and $\theta \in \Theta$ is a k-dimensional vector of parameters. We assume that $M(\cdot)$ can be estimated by minimizing some objective function, say, $\Psi(y, \theta)$, yielding the parameter estimate $\hat{\theta}$ where

$$\hat{\theta} = \underset{\theta \in \Theta}{\operatorname{argmin}} \Psi(y, \theta). \tag{4.36}$$

Estimators of this type are based on many well-known estimation techniques, the most popular ones being ordinary least squares (OLS) and maximum likelihood (ML). In the case of OLS, Ψ is given by the error sum of squares, and in the case of ML, it is the negative log-likelihood.

Let $(Z_1, \ldots, Z_L)$ be a set of partitioning variables (i.e., covariates). We assume that there exists a partition $\{\beta_b\}_{(b=1,\ldots,B)}$ of the space $Z = Z_1 \times \cdots \times Z_L$ into B cells (or segments) such that in each cell β_b, a model $M(y, \theta_b)$ with a cell-specific parameter θ_b holds.

The basic idea of model-based recursive partitioning is now that each node is associated with a single model. In the first step, the designated model $M(y, \theta)$ is fit to all observations by estimating $\hat{\theta}$ via minimization of the objective function Ψ; this yields a model in the top node. Then in the second step, a fluctuation test for parameter instability is performed to assess whether splitting of the node is necessary. Generally speaking, determining the necessity of a split based on a partitioning variable Z_l depends on comparing the estimated parameters of the post-split resulting daughters to determine whether they come from the same mean (and thus the split is unnecessary). If there is significant parameter instability with respect to any of the partitioning variables Z_l $(1 \leq l \leq L)$, then we select the variable Z_l that is associated with the highest parameter instability. In the third step, we compute the split point(s) of Z_l that locally optimize Ψ. Finally, we split the node into B locally optimal segments and repeat the procedure. If no more significant instabilities can be found, the recursion stops and returns a tree where each terminal node is associated with a model of type $M(y, \theta_b)$.

The steps of the algorithm are as follows:

1. Fit the model $M(y, \theta)$ to all observations in the current node by estimating $\hat{\theta}$ via minimization of the objective function Ψ.
2. Assess the stability of the parameters with respect to every ordering $Z_1, \ldots, Z_L$. If there is some overall instability, choose the variable Z_l associated with the highest parameter instability for partitioning; otherwise stop.
3. Search for the locally optimal split point(s) in Z_l by minimizing the objective function of the model ψ.
4. Split the node into daughter nodes and repeat the procedure.

The details for steps 1–3 can be found in Jank et al. (2008b).

Case Study We illustrate the method on the combined auction data for Harry Potter books and Microsoft Xbox game consoles discussed earlier (Wang et al., 2008a,b). For those data, we have different covariates such as the opening bid, the final price, the number of bids, and the seller and bidder ratings. We also have information on the item condition (used versus new), whether or not the seller set a secret reserve price, and whether or not the auction exhibited early bidding or jump bidding.

Figure 4.28 shows a series of *conditional* phase-plane plots, where the average derivatives are conditional on different levels of each auction characteristic. Here, pairs of plots can be compared to see the effect of the two different levels (e.g., new versus used items in the top left, or high versus low bidder rating in the bottom right). We can see that the general "C" shape identified in Figure 4.24 persists in all the conditional phase-plane plots; however, we also notice that the *magnitude* of the dynamics varies. Moreover, the different size of the "C" shapes also indicates that the magnitude of the *relationship* between velocity and acceleration differs.

In particular, we can learn from Figure 4.28 the following: For item value, higher valued items appear to have a larger range of dynamics compared to lower valued

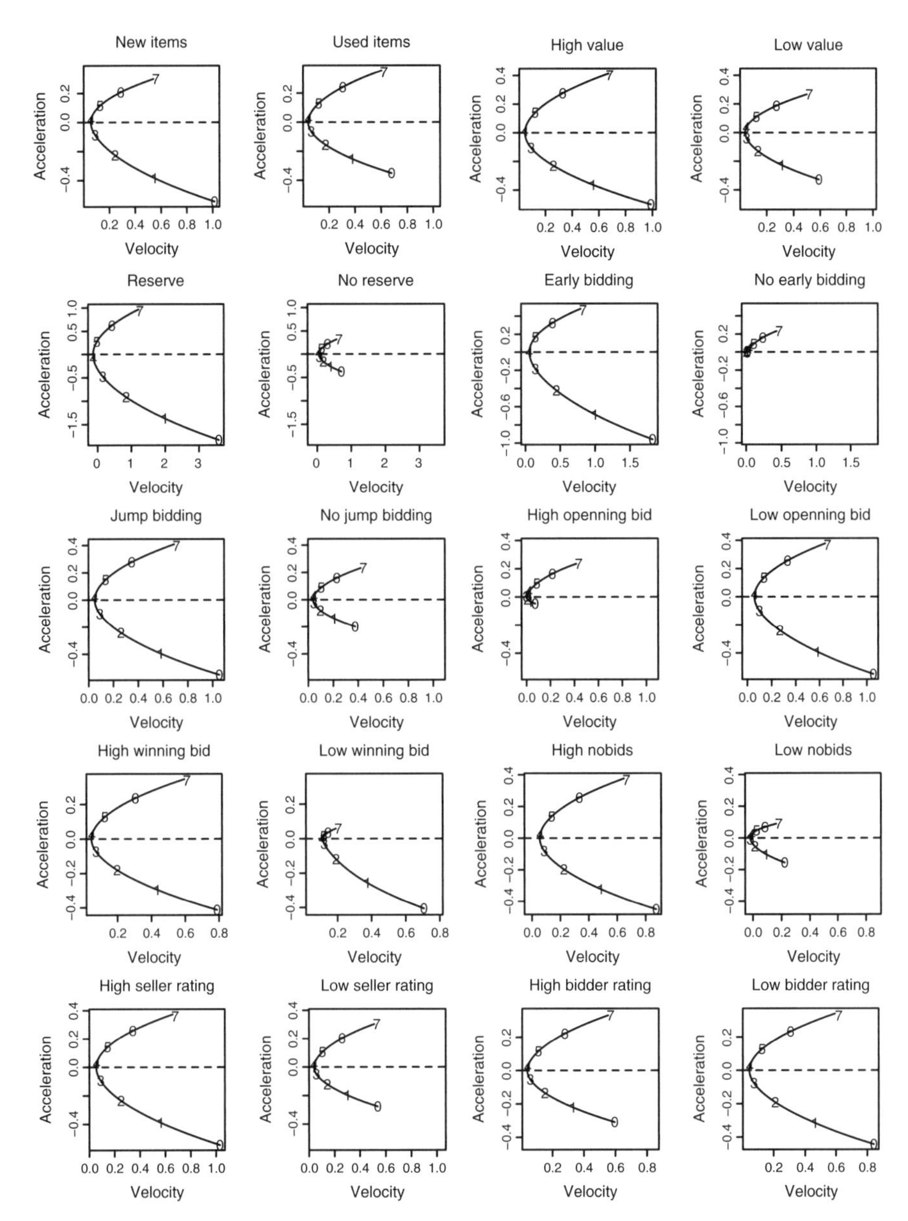

FIGURE 4.28 Conditional phase-plane plots, conditional on 10 different auction characteristics.

items. Early bidding seems to have a large effect on the dynamics not only at the auction start but also throughout the entire auction. For jump bidding, we see that auctions that experience jump bids have a different relationship between velocity and acceleration compared to auctions without jump bids. Additional observations are that the opening bid and the number of bidders both have an impact on the

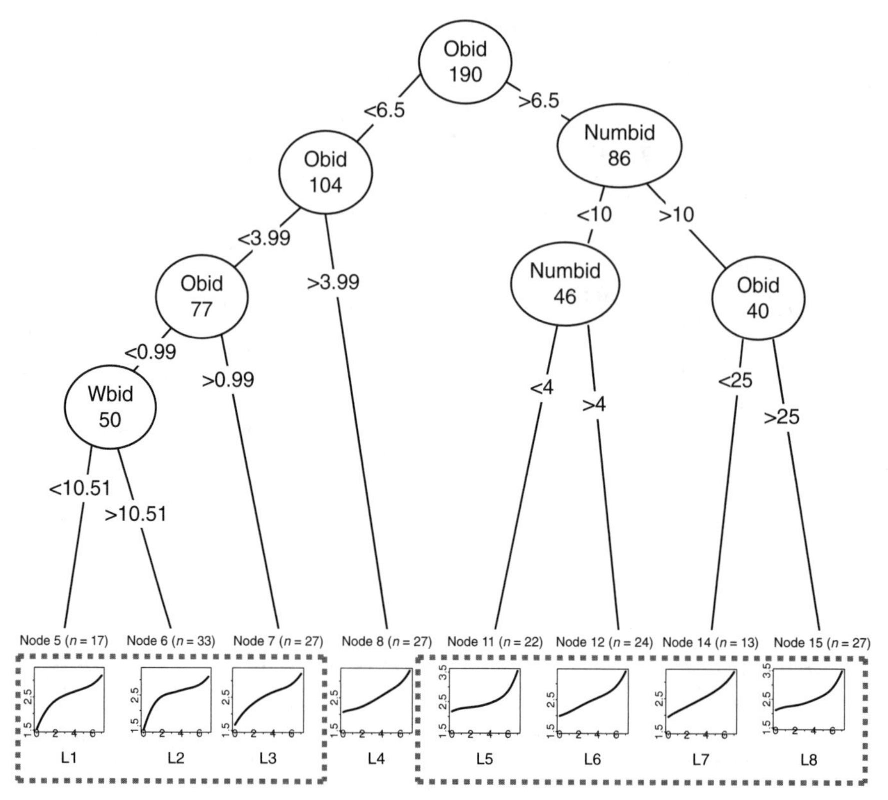

FIGURE 4.29 Estimated differential equation tree, using $\alpha = 0.01$ and min(number of observations in leaf node) > 10. "L1, ..., L8″ denote the eight terminal nodes.

dynamics. Interestingly, while different bidder ratings do not appear to make much of a difference, seller ratings do. Note however that several of these variables are correlated, for example, closing price and item value. Our goal is therefore to look at the effect of the combined information on price dynamics. For this purpose, we formulate a differential equation model that relates price dynamics to a set of predictor variables.

Figure 4.29 shows the fitted functional differential equation tree. This tree was obtained by employing a stopping criterion of a statistical significance level of $\alpha = 0.01$ and a minimum number of at least 10 observations within each final node. The resulting tree uses three different splitting variables to arrive at eight different price curves: the opening bid (Obid), the winning bid (Wbid), and the number of bids (Numbid). Returning to Figure 4.28, the tree confirms the differences between price dynamics observed for each of the variables. Moreover, as noted earlier, many of the variables are correlated, and therefore it is not surprising that some of them are absent from the tree. For instance, item value and item condition are associated with the winning bid; reserve price and early bidding are associated with the opening bid.

As in the phase-plane plots, the average bidder rating does not appear to lead to different price dynamics and is absent from the tree. In contrast, seller rating, which did appear to yield different phase-plane plots, is conspicuously missing from the tree. One possible reason for this is that seller rating (which proxies for experience) is indirectly associated with the opening bid, as the seller determines the opening bid.[11]

With respect to the resulting eight leaf nodes, the estimated price curves cover three main shapes: shape 1 is characterized by a fast initial price increase followed by an end-of-auction slowdown (L1–L3); shape 2 is characterized by almost linear increase (L4); and shape 3 shows little to moderate initial activity followed by late price spurts (L5–L8). We also note that these three main shapes differ by the opening bid only: For auctions with an opening bid lower than $3.99, the price curves follow the first shape; for those with opening bid higher than $6.5, the curves follow the third shape; and for auctions with opening bid between $3.99 and $6.5, the price curves follow the second shape, which is almost a linear increase. Further segmentation of these basic shapes is achieved via the winning bid and the number of bids.

4.2.4 Conclusion

In the previous section, we have discussed different methods for modeling the dynamics of online auctions. We described models for capturing the heterogeneity in the price process and relating it to explanatory variables via functional regression models. Functional regression models are similar in nature to classical regression models; however, their main difference (and advantage) is that they allow the modeling of auction dynamics. We then introduced differential equation models (and related to that, differential equation trees). Differential equation models differ from (functional) regression models in that they can capture the effects inherent in the system. In other words, dynamics are not only caused by external factors (such as the actions of individual bidders), but also change as a result of reaction (one bidder's action may cause other bidders' reactions), connectivity (sellers selling more than one auction, bidders bidding on more than one auction, bidders being loyal to certain sellers, etc.), and complexity of the online marketplace. Many of these effects are not directly measurable and hence hard to capture using traditional regression models. Differential equation models allow for a quantification of these effects and thus for a better understanding of the factors driving online auctions.

4.3 MODELING AUCTION COMPETITION

In this section, we discuss competition between online auctions. There are many different forms in which auctions can compete: The simplest form is competition

[11] Using a more lenient stopping criterion results in a larger tree. For example, using $\alpha = 0.1$ and a minimum number of at least five observations within a final node results in a tree that also incorporates the seller rating as splitting variable. Note that this tree is the same as earlier except for the addition of seller rating. We leave it up to future research to determine the optimal settings for the differential equation tree.

between similar items (Jank and Shmueli, 2007). Competition can then arise between simultaneous auctions for the same item, overlapping auctions, or sequential auctions. It can be between items by the same sellers or between competing sellers and it can also be between auctions with complementarities and substitutes. One can further expand the definition of competition by introducing different auction formats where format differences are defined as narrowly as adding a buy-it-now feature or as broadly as moving from an auction to a posted price or negotiation. Competition can also arise between different auction hosts vying for auction traffic. In such a platform, competition, network externalities, auction design, customization, and other considerations arise that have not been previously considered in the auction literature. For more discussion on auction competition, see Haruvy et al. (2008) and Haruvy and Popkowski Leszczyc (2009, 2010).

4.3.1 Motivation

In this section, we primarily focus on competition that arises between simultaneous, (partially) overlapping, or sequential auctions for the same (or similar) item, offered by the same (or different) seller. We also refer to these types of auctions as *concurrent* auctions since competition arises from the fact that similar items are being offered to the same set of bidders during the same period of time. More generally, competition arises from the interaction of concurrent agents, with one agent reacting to the action of others, resulting in an interplay of many events that take place simultaneously or within a short period.

Examples of concurrency are plentiful and they exist in both the offline and the online world. An example in the offline (or "brick-and-mortar") world is that of retail stores reacting to competitors' sales promotion by offering a promotion of their own. Action, in this case in the form of a sales promotion, can trigger a reaction and thus sales at different retail stores are correlated. A recent example is the price war over warranties between several large U. S. consumer electronic retail chains. Another example is that of two local grocery stores that use promotions to attract customers. Such events (i.e., promotions or price cuts) need not be identical: while one grocery store may offer a promotion on fresh fruit, the other might react with a promotion on fresh vegetables. The similarity of the two events can decrease even further if the second store decides to offer a promotion on bottled water.

The effect of competition in the previous example is limited since sales promotions for groceries affect only local markets. The number of actions and reactions is relatively small and limited to events from neighboring grocery stores only. But competition can also occur on a much larger scale, for instance, if it spans a wider geographic region. Examples are the promotional efforts of different car makers and their impact on nationwide sales, or price wars between large consumer electronic retail chains. The effect of competition increases even further in markets that are free of geographical boundaries, such as the Internet. Online markets are available day and night and accessible from anywhere in the world and thus attract a potentially unlimited number of competing sellers and buyers. This can make the effect of competition huge, with a large number of events influencing one another during any given time window.

Take the example of online auctions (such as on eBay). eBay offers almost inexhaustible resources for a wide variety of items. For example, an individual wishing to purchase a laptop through an eBay auction will find several hundred, probably thousand, relevant auctions taking place at any given time. While the individual may have a preference for an item with a certain set of features (e.g., a particular brand, a certain processor speed, and memory size), there will be additional auctions available for similar items, for example, those featuring comparable brands and similar product characteristics. Part of the reason for this vast amount of options is that markets such as those for electronics include an abundance of substitute products. Consequently, the decision-making process of bidders, and hence the final price of auctioned items, is affected not only by all auctions for items that have exactly the same specification, but also by auctions for *similar* items. This is particularly true for price-sensitive bidders who are willing to trade off feature preferences against a low price.

An additional factor that further complicates the study of concurrency is that bidders can use information on previous auctions to learn about ongoing and future auctions. eBay displays a huge archive of all recent completed auctions on its website, ready for users to inspect. Thus, bidders can inform themselves about similar auctions that took place in the past, their characteristics, and their outcome and use this knowledge to strategize about bidding in current or upcoming auctions. As a consequence, the price of an auction is also affected by auctions for the same or similar item *in the past.*

In short, there are two main components that drive competition in online auctions: item similarity and time. In particular, the price in one auction is driven by other simultaneous auctions for the identical item, by other simultaneous auctions for substitute items, and by auctions for identical or substitute items in the past. When we couple these two components with the daily abundance of items for sale on eBay (over one billion every day), the scale of competition becomes indeed huge.

The previous examples illustrate that competition, especially in the online world, is often characterized by two main components: a *temporal* component and a *spatial* component. By temporal component, we mean the effect of the time lag between a current event and the same (or similar) event in the past. The price of a car today will be affected by the price it sold for yesterday. The sales promotion offered by a grocery store today is influenced by the type, magnitude, and number of promotions that its competitors ran last week. The effect of time may be somewhat different in the online world where events often leave digital footprints. As pointed out earlier, eBay stores all completed auctions on its website and makes them publicly available. This makes it easy for users to inform themselves about key features such as quantity, quality, and price of items in previous auctions, and it can consequently influence their bidding strategies. The spatial component, on the other hand, addresses the effect of similar events. Returning to the laptop example, the price of a laptop with a certain set of features will be influenced by the price of other laptops with *similar* features. This is a plausible assumption since consumers are often willing to consider products with slightly different characteristics due to price sensitivity or insensitivity to particular features (e.g., whether or not the processor is manufactured by INTEL). This may be especially true for products such as cars, computers, or electronics that are comprised of a combination of many different features such as processor speed, memory, and screen size for laptops or horsepower, gas mileage, and color for cars. If we think of all possible combinations of features represented by the associated *feature space*, then two products are similar to one another if they lie in close proximity in that feature space. We can measure the proximity, or equivalently, the distance between two products, using an appropriate distance measure. The notion of a distance between two products in the feature space leads to the second component of concurrency that we call the *spatial* component.

We want to point out that our treatment of the similarity component of concurrency via spatial models is conceptually rare. Typically, the notion of "spatial models" is associated with geographically referenced data, for example, data with longitudinal and latitudinal information attached (Olea, 1999). Here, we choose to interpret the meaning of "space" in a slightly nontraditional way to conceptualize the more abstract notion of space among a set of features. This kind of interpretation is rare and only few studies point to the potential of this approach in an economic setting (Slade, 2005; Beck et al., 2006).

Finding statistical models for the temporal and spatial concurrency components is challenging for a variety of reasons. First, classical statistical time-series models tend to assume a constant time lag between measurements. For instance, financial performance data typically arrive every quarter, sometimes every month. In such a case, a "lag" is unambiguously defined as the previous time period. Now compare this to the online auction scenario. If we consider the series of closing times for a set of auctions, it is very unlikely to find an evenly spaced series. The reason is that on eBay the seller is the one who determines the ending time of his/her auction. While it might be more common to have an auction end at, for example, the afternoon or evening, the worldwide reach of the Internet makes an auction that ends in the afternoon on one part of the globe and end in the morning on another part. The implication for statistical analysis is that there exist no fixed time points at which data arrive. The result is a time series of auction endings (and associated closing prices) that is unequally spaced. This time series may at times be very dense (i.e., many auctions closing around the same time) and at other times very sparse (see, for example, the auction calendar and rug plot visualizations in Chapter 3). This poses new challenges to time series analysis and, in particular, to the definition of a "time lag." Simply applying the standard definition of a time lag to an unequally spaced time series can distort the lag-effect interpretation since the difference between two consecutive observations may refer to a long period for one pair of observations, but only to a short period for another pair. Statistical solutions for unequally spaced time series are rather thin (Engle and Russel, 1998; Pelt, 2005) and there exists no common recipe. In this section, we describe a new solution proposed by Jank and Shmueli (2007) that is particularly suited for the analysis of concurrency marked by unevenly spaced events. Specifically, we interpret a time lag as the information distributed over a certain time window in the past and we incorporate this information by including appropriate summary measures of this distribution.

Another challenge in modeling concurrency is the incorporation of the spatial component. While one can *model* the similarity between features using approaches similar to those used in the literature on geostatistics (Cressie, 1993), it is much harder to *define* a suitable and unifying measure of similarity. Products are typically characterized using a combination of quantitative and qualitative variables. For instance, while processor speed and memory size are both quantitative, product brand and color are qualitative. There exist an array of different similarity and distance measures for quantitative (or interval-scaled) variables, such as the popular Euclidian distance, the city-block distance, or the Mahalanobis distance.

Similarity measures for qualitative variables are typically lesser known. An overview over different approaches can be found in Bock (1974). Similarity for qualitative variables is often based on the co-occurrence and/or co-absence of observations in different variable categories. For instance, if we consider the single binary variable "owns a Rolex" versus "does not own a Rolex," then clearly two people who own a Rolex (co-occurrence) are more similar than two people who do not own a Rolex.[12] In such cases, measures such as Jaquard's coefficient are suitable. Alternatively, in the case of a binary variable such as "gender," it is clear that both co-occurrences (e.g., "female–female") and co-absences (e.g., "male–male") reflect similarity equally well. In such cases, measures such as the matching coefficient are more adequate. Our situation features binary variables that are all conceptually closer to a "gender" variable example, where both co-occurrences and co-absences are indicators of proximity.

After defining suitable similarity measures for quantitative and qualitative variables individually, one has to combine the individual measures into an overall measure that represents the entire quantitative–qualitative feature space. The popular approach is to combine the standardized distances through a weighted average (Hastie et al., 2001). One such measure is Gower's general coefficient. The weights are then a function of the type of distances used. For instance, using a city-block distance metric for the numerical variables would give them a weight equal to the inverse of the range of values for that variable. In our application, we chose to standardize the numerical variables to a (0,1) scale and use a similar distance metric for both numeric and binary variables. This gives equal weights to all variables and consequently does not *a priori* penalize any of the feature variables as being less important.

We incorporate the temporal and spatial components of concurrency into one unifying model using the ideas of semiparametric modeling (Ruppert et al., 2003). Semiparametric models are popular for their flexibility in that they allow standard parametric relationships coexist with nonparametric ones under the same roof. Parametric approaches are useful if the functional relationship between predictor and response is roughly known, while nonparametric approaches are purely data driven and thus useful if the exact relationship is unclear. Another advantage of the semiparametric model is that it can be represented in the form of mixed models (Wand, 2003),

[12] While for two persons who do not own a Rolex, one may own a Cartier, and the other may in fact not own a watch at all.

for which parameter estimation and inference are well established (e.g., McCulloch and Searle, 2000). This approach provides a very familiar environment for the joint estimation of parametric and nonparametric relationships.

4.3.2 Semiparametric and Mixed Models

We now discuss the individual components of our competition model. The model is very flexible and this flexibility is accomplished because we use a semiparametric modeling framework; we thus first discuss details of semiparametric modeling and how it relates to mixed models.

We consider the very flexible class of semiparametric models that combine the advantages of parametric and nonparametric models. Parametric models postulate a strict functional relationship between the predictors and the response. The functional form often allows new insight into the relation between variables, such as in growth models where new insight can be derived from the observation that the response grows at an exponential rate, rather than at a polynomial one. The strict functional relationship imposed by parametric models can sometimes be overly restrictive. This restriction can be alleviated using nonparametric approaches. Nonparametric models are very flexible and posit only minimal functional assumptions. As a consequence, they often provide a much better fit to the data. On the downside of this flexibility, however, nonparametric models do not typically allow insight into the precise functional relationship between variables. In the above example, nonparametric models would not allow to distinguish between exponential and polynomial growth. Semiparametric models can be thought of as a wedding between parametric and nonparametric models, and they combine their advantages.

We describe the general principles of semiparametric models next. For illustrative purposes, we consider a simple scenario with one response variable y and only two predictor variables x_1 and x_2. More details, also for more complex models, can be found in, for example, Ruppert et al. (2003). As in classical regression, we model the response in an additive fashion, say as

$$y = \beta_0 + \beta_1 x_1 + f(x_2) + \epsilon. \tag{4.37}$$

The function f describes the marginal relationship between x_2 and y and it is completely unspecified. It corresponds to the nonparametric part of the model. On the other hand, the relationship between x_1 and y is parametric and in particular a linear relationship. The error assumptions are typically standard $\epsilon \sim N(0, \sigma_\epsilon^2)$.

To estimate the function f from the data, one assumes a suitable set of basis functions. While many different choices are possible, a popular approach is to use *truncated line bases* (Wand, 2003) that yield

$$f(x) = \tilde{\beta}_0 + \tilde{\beta}_1 x + \sum_{k=1}^{K} u_k (x - \tau_k)_+, \tag{4.38}$$

where $\tau_1, \ldots, \tau_K$ denotes a suitable set of knots and the function $(x - \tau_k)_+$ equals $x - \tau_k$ if and only if $x > \tau_k$, and it is zero otherwise. Instead of the truncated line basis, one could also use the truncated polynomial basis that leads to a smoother fit, or radial basis functions that are often useful for higher dimensional smoothing (Ruppert et al., 2003). Note that the set of all parameters that needs to be estimated from the data is given by $(\tilde{\beta}_0, \tilde{\beta}_1, u_1, u_2, \ldots, u_K)$.

Using (4.38), the model in (4.37) becomes

$$y = \beta_0^* + \beta_1 x_1 + \tilde{\beta}_1 x_2 + \sum_{k=1}^{K} u_k (x_2 - \tau_k)_+ + \epsilon, \tag{4.39}$$

where we set $\beta_0^* := (\beta_0 + \tilde{\beta}_0)$ to avoid two confounded intercepts in the model. One can estimate (4.39) within the very familiar mixed model setting. Mixed models (e.g., McCulloch and Searle, 2000) have become a very popular tool set in the statistics literature and methods to fit mixed models are readily available. We can embed (4.39) within the mixed model framework by rewriting it in a different form. We use the following notation:

$$\boldsymbol{\beta} = [\beta_0^*, \beta_1, \tilde{\beta}_1], \qquad \mathbf{u} = [u_1, \ldots, u_K]. \tag{4.40}$$

In the mixed model context, $\boldsymbol{\beta}$ and $\mathbf{u}$ would be referred to as fixed and random effects, respectively. We define the design matrices for the fixed and random effects as

$$\mathbf{X} = [\mathbf{1}, \mathbf{x}_1, \mathbf{x}_2]_{n \times 3}, \qquad \mathbf{Z} = [(\mathbf{x}_2 - \tau_1), \ldots, (\mathbf{x}_2 - \tau_K)]_{n \times K}, \tag{4.41}$$

where $\mathbf{x}_1 = (x_{11}, \ldots, x_{1n})'$ and $\mathbf{x}_2 = (x_{21}, \ldots, x_{2n})'$ are vectors of covariates. Then, setting $\mathbf{y} = (y_1, \ldots, y_n)'$, we can write the model in (4.39) in the mixed model matrix notation

$$\mathbf{y} = \mathbf{X}\boldsymbol{\beta} + \mathbf{Z}\mathbf{u} + \boldsymbol{\epsilon}, \tag{4.42}$$

with the standard mixed model assumption on the error structure

$$E\begin{bmatrix} \mathbf{u} \\ \boldsymbol{\epsilon} \end{bmatrix} = \mathbf{0}, \qquad \text{Cov}\begin{bmatrix} \mathbf{u} \\ \boldsymbol{\epsilon} \end{bmatrix} = \begin{bmatrix} \sigma_u^2 \mathbf{I} & \mathbf{0} \\ \mathbf{0} & \sigma_\epsilon^2 \mathbf{I} \end{bmatrix}. \tag{4.43}$$

One can estimate the parameters in (4.42) using maximum likelihood or restricted maximum likelihood. These are readily implemented in standard software such as R's package *nlme* (CRAN, 2003).

Using a semiparametric approach allows not only flexibility in modeling concurrency but also, as our results show, superior predictive performance. The concurrency model consists of several model components. For each component, we apply either a parametric or a nonparametric modeling strategy, depending on our *a priori* knowledge of the functional relationship among the variables and also on the evidence provided by the data. Specifically, our concurrency model consists of three major components. The first component captures all auction related information such as the auction length, the opening bid, or the level of competition. This is the auction component (AC). In that sense, this first component captures only factors related to the *transaction* of the item and not to the item itself. The second component captures item features, the relationships between the features, and their impact on price. We argue that for this component it is convenient to assume a spatial model and therefore call this the spatial component (SC). And finally, the third component captures the impact of time on the auction price. We will refer to this as the temporal component (TC). In the following sections, we consider these three model components separately, explain their structure and inherent challenges, and propose solutions via semiparametric modeling. Later, we combine all three components into one grand concurrency model. For this grand model, we write, rather generically,

$$y = g_{\mathrm{AC}}(\mathbf{x}) + g_{\mathrm{SC}}(\mathbf{x}) + g_{\mathrm{TC}}(\mathbf{x}) + \epsilon, \tag{4.44}$$

where g_{AC}, g_{SC}, and g_{TC} denote the auction, the spatial, and the temporal concurrency components, respectively.

We illustrate our model on a sample set of data. Our data consist of 8128 laptop auctions that transacted on eBay during May and June of 2004 see also Jank and Shmueli (2007). The data comprise of several different brands (Table 4.4). Note that the average price is very similar across all brands and is around \$500. The price similarity could suggest that many items are substitutable and that bidders, looking for the best value among two items with a similar price, choose the one with the better features. Prices are highly right-skewed with only few auctions selling at a very high price (see left panel of Figure 4.30). Transforming prices to the log scale results in a much more symmetric price distribution (see right panel of Figure 4.30). We use the (natural) log of price throughout the remaining part of the chapter.

In addition to price information, we also have information on the features of laptops (e.g., high processor speed versus low processor speed), which can be expected to have significant impact on price. We have information on seven laptop features: Besides the laptop's brand, we have processor speed, screen size, memory size, whether or

TABLE 4.4 Laptop Prices by Brand

Brand	Count	Avg	Median	StDev	Min	Max
DELL	2617	494.38	449.00	199.44	200.00	999.99
HP	1932	515.04	463.00	217.11	200.00	999.00
IBM	1466	421.46	390.86	167.01	200.00	999.99
SONY	522	537.02	500.00	221.06	200.00	999.95
TOSHIBA	874	534.13	524.50	227.75	200.00	999.99
OTHER	717	533.35	504.99	230.66	200.00	995.00

The table lists the number of auctions per brand together with summary statistics of price.

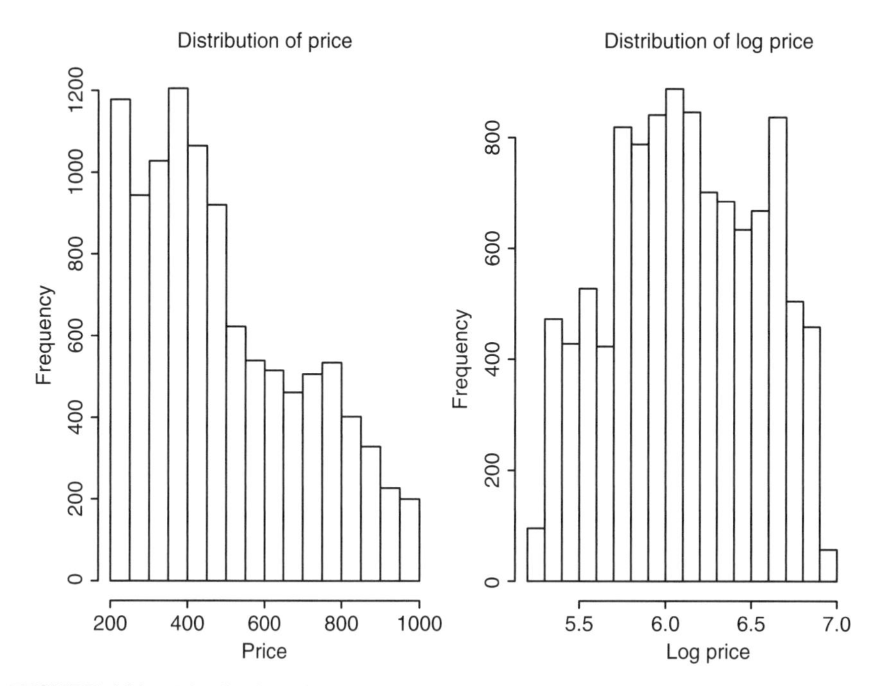

FIGURE 4.30 Distribution of price on the original and on the log scale.

not the chip is built by INTEL, whether or not a DVD drive is included, and whether or not the item is used. Table 4.5 summarizes this information.

There has been much work on understanding price in online auctions. One of the first studies by Lucking-Reiley et al. (2007) found that auction design, among other factors, can have a significant impact on the final auction price. In particular, they found that the magnitude of the opening bid, the use of a secret reserve price, and the length of the auction all have a positive effect on price. The number of bids, a measure for competition in the auction, can also affect price. In fact, Bapna et al. (2008a)

TABLE 4.5 Quantitative and Categorical Feature Related Predictors

Variable	Avg	Median	StDev	Min	Max
Processor speed	1070.54	800.00	734.27	100.00	3200.00
Screen size	13.88	14.00	1.77	7.00	20.00
Memory size	258.29	256.00	154.44	32.00	2000.00

Variable	No	Yes
INTEL chip?	675	7453
DVD drive?	3516	4612
Used item?	7027	1101

TABLE 4.6 Numerical and Categorical Auction Related Predictors

Variable	Avg	Median	StDev	Min	Max
Opening bid	193.51	80.00	244.37	0.01	999.99
Auction length	4.49	5.00	2.46	1.00	10.00
Number of bids	16.28	15.00	12.77	1.00	115.00

Variable	No	Yes
"Secret reserve price" used?	5791	2337
"Buy-it-now" used?	5451	2677

found a significant effect of the interaction between the number of bids and the opening bid (in addition to their main effects), suggesting that although generally higher competition leads to higher prices, a high opening bid will lead to lower competition and thus to a lower price.

Table 4.6 summarizes the distribution of the auction related variables (the opening bid, the auction length, the number of bids, the presence of a secret reserve price, and the use of a buy-it-now option). The buy-it-now option gives buyers the possibility to purchase the item immediately at a fixed (and posted) price. Since the buy-it-now price is visible to all bidders, it may influence their perceived valuation of the item and, therefore, the final auction price.

4.3.3 Auction Competition Component

We start by modeling the relationship between price and all auction related variables in the laptop data. These include the number of bids, the opening bid, the auction length, and the usage of a secret reserve price and a buy-it-now option. Figure 4.31 shows a scatterplot matrix of the pairwise relationships between these variables.

Secret reserve price and buy-it-now options are binary and thus enter our model in the form of dummy variables. Figure 4.31 also suggests a linear relationship between auction length ("days") and price, if there is any relationship at all. However, the relationship between price, opening bid, and number of bids is more intricate, and we discuss possible modeling alternatives next.

Figure 4.31 shows the relationship between price, opening bid, and number of bids on the log scale since logging reduces the effect of few extremes in all these three variables. Previous studies (e.g., Lucking-Reiley et al., 2007) have modeled the relationship between these three variables in a linear fashion, and Figure 4.31 does not show any evidence contradicting this functional relationship. However, it reveals another interesting feature: a very prominent curvature in the scatterplot between the opening bid and the number of bids! In fact, we pointed out earlier that the relationship between the price and the opening has been found to be moderated by the number of bids (Bapna et al., 2008a). This leads us to investigate a possible interaction term between the two variables (also displayed in Figure 4.31). In fact, the adjusted

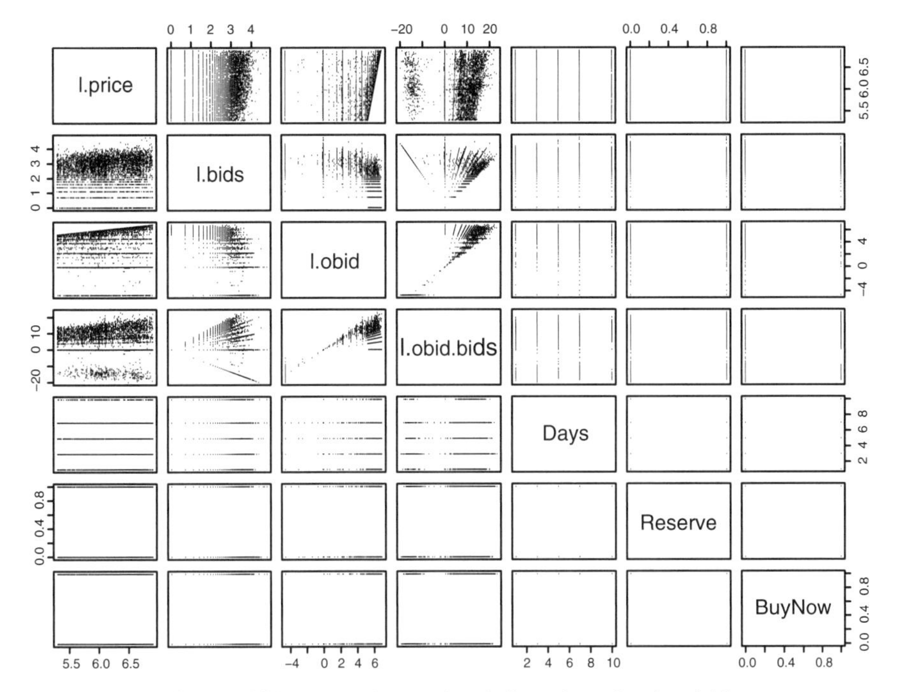

FIGURE 4.31 Scatterplot matrix of all auction related variables.

R-squared increases from 0.02 to 0.13 for the linear regression of price on the two variables opening bid and number of bids including the interaction between the two.

More evidence for the power of the interaction term can be seen in Figure 4.32 that shows "component-plus-residual" plots (e.g., Fox, 1997) for the model *with* the interaction term (left panel) versus a model *without* that term (right panel). Component-plus-residual plots graph $\hat{\epsilon} + \hat{\beta}_i x_i$, that is, the residual plus the contribution of x_i to the fitted regression equation, against the covariates x_i. Component-plus-residual plots are useful to detect nonlinearities and the absence of higher order terms in the regression equation. The dotted line corresponds to the least squares line fitted to the points on the graph, while the solid line corresponds to a lowess smoother. We can see that the inclusion of the interaction term renders the linearity assumption much more plausible. In fact, we take this as evidence that the relationship between price and all auction related variables can be modeled adequately in parametric, linear fashion. We thus define the auction related concurrency component g_{AC} in (4.44) as

$$g_{\text{AC}}(\mathbf{x}) = \beta_{\text{ob}} x_{\text{ob}} + \beta_{\text{nb}} x_{\text{nb}} + \beta_{\text{int}} x_{\text{ob}} x_{\text{nb}} + \beta_{\text{len}} x_{\text{len}} + \beta_{\text{rsv}} x_{\text{rsv}} + \beta_{\text{buy}} x_{\text{buy}}, \quad (4.45)$$

where $x_{\text{ob}}, x_{\text{nb}}, x_{\text{len}}$ denote observations on the opening bid, number of bids, and auction length, while x_{rsv}, and x_{buy} correspond to indicators for the secret reserve price and the buy-it-now option, respectively.

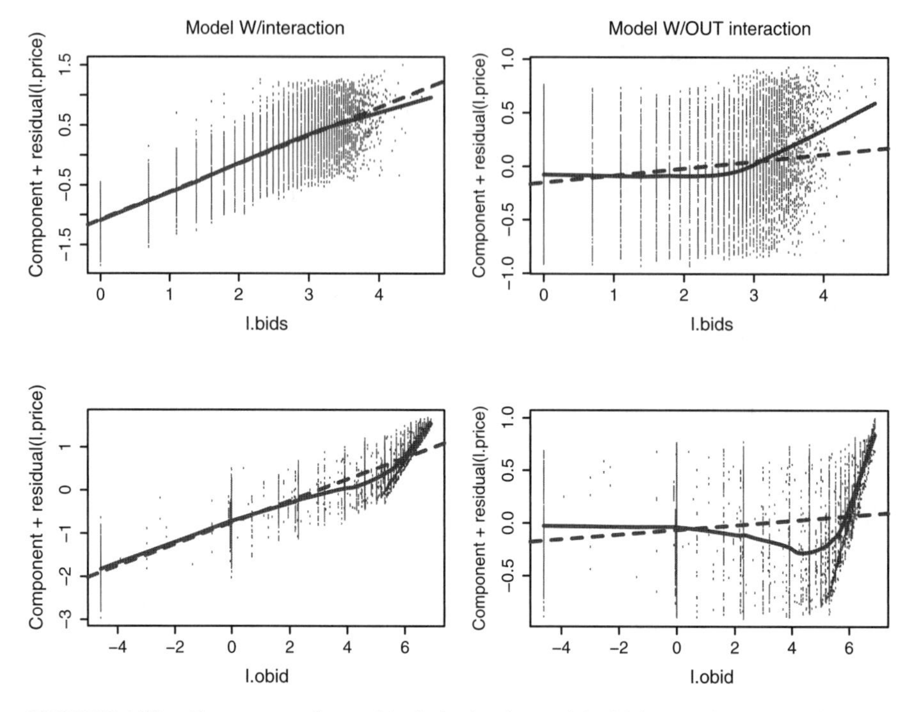

FIGURE 4.32 Component-plus-residual plot for the model with interaction term (left panel) and the model without the interaction term (right panel).

4.3.4 Spatial Concurrency Component

We use a spatial approach to model the interrelations between an item's features and their combined effect on price. Using a spatial approach has the advantage of making only minimal assumptions about the functional relationships across the features. In fact, the only assumption that we make is that for two sets of feature combinations, their effect on price is similar if they lie in close proximity to one another in the feature space. Conversely, a set of features may have a very different effect on price relative to another more distant set of features. Note that this approach is very flexible since no restrictive functional relationship is assumed. It therefore has the potential of detecting untypical but important relationships, some of which may exist only locally in the feature space, that are impossible to detect with a globally prespecified analytical function. For instance, it is quite possible that there exist "pockets" in the feature space with an extremely positive (or extremely negative) effect on price. Moreover, such pockets may not have a well-defined shape such as circular or rectangular. A spatial approach can detect such pockets and use them to its advantage.

Even if the relationship between the features and their effect on price were to follow *some* analytic function, a spatial approach can be advantageous since it can simplify the modeling task, given a sufficient amount of data. Recall that in our example we

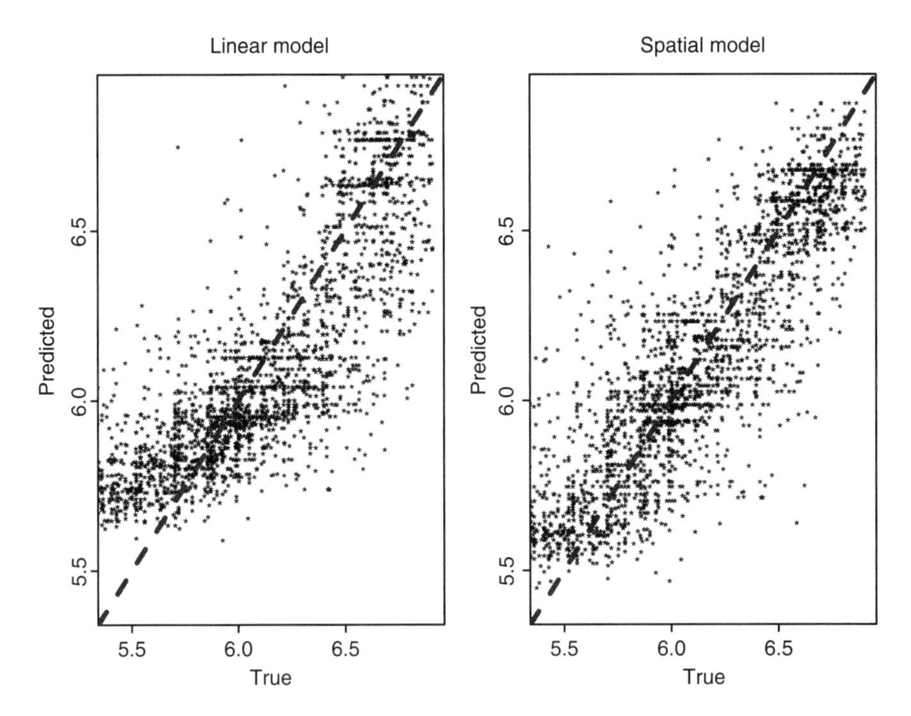

FIGURE 4.33 Predicted versus true values for a linear model (left panel) and the corresponding spatial model (right panel).

have a set of seven different features: processor speed, screen size, memory size, type of chip, inclusion of a DVD drive, item condition, and item brand. Using traditional linear or nonlinear models, it could certainly be very cumbersome to determine the exact functional relationship between the features and their effect on price. Traditional steps to determine that relationship include trying different transformations of the predictors and of the response, investigating second- or higher order effects via interaction terms among some (or all) of the predictors, creating new derived variables, and possibly many, many more steps. A spatial model can make many of these steps obsolete. Consider Figure 4.33 that shows a graph of true versus predicted values for our data set. In general, one would expect these values to scatter along the diagonal if the model provides an adequate fit to the data. The left panel in Figure 4.33 shows the corresponding graph of a linear regression model on log price for the seven feature variables described above. We can see that the scatter of predicted versus true displays curvature that may be due to nonlinearities in some of the variables, due to missing higher order terms, or due to other functional relationships that are not captured in the linear model. In contrast, the corresponding graph for a spatial model on the same variables (right panel) shows scatter right along the diagonal axis.

Feature-Based Distances Before *modeling* the spatial proximity of features, one has to first *define* exactly what is meant by similarity (or dissimilarity) between a set of features. Note that our seven features consist of variables that are measured on different scales. Some are measured on an interval scale (processor speed, memory size, and screen size), and others assume qualitative attributes (brand, type of chip, inclusion of a DVD drive, and item condition). To measure dissimilarity in the feature space, one first must define what is meant by the distance between two interval-scaled variables, between two categorical variables, and how these two distance measures can be combined into one overall measure of dissimilarity.

Note that one of the feature variables is "brand," which is nominal with six different levels ("Dell," "HP," "IBM," "Sony," "Toshiba," and "other"). We can represent "brand" in the form of five binary variables, similar to the use of dummy variables in linear regression. For instance, let x_1 denote the brand indicator variable for "Dell" and let x_2, x_3, x_4, x_5 denote indicators for the brands "HP", "IBM", "Sony", "Toshiba," respectively. In this form, we represent the feature "brand" in the space spanned by its first five levels x_1–x_5.

We proceed as follows. For interval-scaled variables, we use a version of the Minkowski metric. For binary variables, we define distance based on co-occurrences and co-absences. Although several of the binary variables indicate the *inclusion* of a certain feature, two items that *preclude* these features are in fact equally similar. For example, two laptops that exclude a DVD drive are equally similar as two laptops that include a DVD drive.

This results in the following overall distance measure. Let q denote the number of different product features and let $\mathbf{x} = (x_1, x_2, \ldots, x_q)$ denote the corresponding feature vector. Let $\mathbf{x}$ and $\mathbf{x}'$ denote two different feature vectors, corresponding to two different items. If the ith feature component $(1 \leq i \leq q)$ is binary, then we define the distance between $\mathbf{x}$ and $\mathbf{x}'$ on this feature as

$$d^i = \mathbf{1}(x_i \neq x_i'), \tag{4.46}$$

which equals zero if and only if the ith feature is identical on the two items and it equals one otherwise. On the other hand, if the ith feature is measured on the interval scale, then we define

$$d^i = \frac{|\tilde{x}_i - \tilde{x}_i'|}{\tilde{R}_i}, \tag{4.47}$$

where $\tilde{x}_i$ denotes the standardized value of x_i, standardized across all values of the ith feature, and $\tilde{R}_i$ denotes the range across all standardized values for the ith feature. This measure is similar to the scaled Minkowski metric (Jain et al., 1999). We then define the overall dissimilarity between $\mathbf{x}$ and $\mathbf{x}'$ on all q feature components as

$$d(\mathbf{x}, \mathbf{x}') = \frac{\sum_{i=1}^{q} d^i}{q}. \tag{4.48}$$

Note that by using $\tilde{R}_i$ in the denominator of (4.47), $d^i = 0$ if the ith feature has the same value on two different items, and it equals 1 if the feature values are at the opposite extremes. In that sense, the contribution of (4.46) and (4.47) to the overall dissimilarity measure $d(\mathbf{x}, \mathbf{x}')$ is comparable.

Semiparametric Spatial Model We now model the relationship between the features and their effect on price using spatial techniques. Having defined a meaningful measure for dissimilarity in the feature space, we do this using semiparametric models based on radial smoothing splines. Let $\mathbf{x}$ be a feature vector of dimension q. Let $\boldsymbol{\beta}_1$ denote a $(q \times 1)$ parameter vector and $\{\boldsymbol{\tau}_k\}_k$, $1 \leq k \leq K$, denote a set of knots, where each knot is also of dimension q. A radial smoothing spline (see, for example Ruppert et al., 2003) is defined as

$$f(\mathbf{x}) = \beta_0 + \boldsymbol{\beta}_1^T \mathbf{x} + \sum_{k=1}^{K} u_k C(r_k), \tag{4.49}$$

where, similar to equation (4.38), the set of all unknown parameters is $(\beta_0, \boldsymbol{\beta}_1^T, u_1, u_2, \ldots, u_K)$. We let $r_k = d(\mathbf{x}, \boldsymbol{\tau}_k)$, where $d(\cdot, \cdot)$ denotes the dissimilarity defined in (4.48). Note that $C(\cdot)$ denotes the covariance function. Many different choices exist for the covariance function such as the family of Matérn or power covariance functions (e.g., Cressie, 1993). We use $C(r) = r^2 \log |r|$, which corresponds to low-rank thin-plate splines with smoothness parameters set to 2 (French et al., 2001).

In a similar fashion to (4.42), we also write (4.49) in mixed model notation. This leads to the spatial concurrency component g_{SC} in (4.44) in the following way. Let $\mathbf{x}_{\text{Feat}}$ denote the features associated with response y. Write

$$\mathbf{x}_{\text{SC}} = [1, \mathbf{x}_{\text{Feat}}]_{(q+1)\times 1}, \qquad \mathbf{z}_{\text{SC}} = [C(r_k)]_{K\times 1}, \tag{4.50}$$

and then (4.49) can be written as

$$f(\mathbf{x}) = \mathbf{x}_{\text{SC}}^T \boldsymbol{\beta}_{\text{SC}} + \mathbf{z}_{\text{SC}}^T \mathbf{u}_{\text{SC}} \tag{4.51}$$

with $\boldsymbol{\beta}_{\text{SC}} = (\beta_0, \boldsymbol{\beta}_1^T)$ and $\mathbf{u}_{\text{SC}} = (u_1, \ldots, u_K)$. Note that $\boldsymbol{\beta}_1$ corresponds to the fixed effect parameter vector of $\mathbf{x}_{\text{Feat}}$. For an excellent introduction on how to fit model (4.51) using standard mixed model technology, and associated software code, see Ngo and Wand (2004). Note that while one could use the R package *SemiPar* by Wand et al. (2005) for similar spatial models, this package is currently limited to only two spatial dimensions and it also does not allow for data with qualitative features as is the case here.

Our spatial concurrency component thus becomes

$$g_{\text{SC}}(\mathbf{x}) = \mathbf{x}_{\text{SC}}^T \boldsymbol{\beta}_{\text{SC}} + \mathbf{z}_{\text{SC}}^T \mathbf{u}_{\text{SC}}, \tag{4.52}$$

with $\mathbf{x}_{\text{SC}}$, $\mathbf{z}_{\text{SC}}$, $\boldsymbol{\beta}_{\text{SC}}$, and $\mathbf{u}_{\text{SC}}$ defined as in (4.51).

TABLE 4.7 Estimated Coefficients for a Linear Regression Model on (Log) Price Using only the Seven Feature Variables as Predictors

Variable	Coeff	StErr	p-Value
Intercept	5.4619	0.0466	0.0000
Processor speed	0.0003	0.0000	0.0000
Screen size	−0.0005	0.0035	0.8809
Memory size	0.0007	0.0000	0.0000
DVD	0.0777	0.0078	0.0000
INTEL chip	0.0543	0.0125	0.0000
Item condition	0.0245	0.0117	0.0358
Dell	0.0685	0.0127	0.0000
HP	−0.0675	0.0133	0.0000
IBM	0.0394	0.0137	0.0041
Toshiba	−0.0711	0.0151	0.0000
Sony	0.0881	0.0172	0.0000

Spatial Concurrency Effects In the following, we illustrate the spatial concurrency model using the seven feature variables processor speed, screen size, memory size, type of chip, inclusion of a DVD drive, item condition, and item brand. Focusing the discussion here only on the spatial aspect of the overall model allows a more detailed inspection of its properties and its distinction from the traditional linear model. We discuss the overall concurrency model and its individual components later.

Table 4.7 shows the estimated coefficients and their significance levels for a linear regression model on (log) price, using only the seven feature variables as predictors. (Also, recall Figure 4.33 that shows the scatter of true versus predicted for this model.) We can see that, except for screen size, all features have a positive effect on price. The effect of screen size is negative, which may be a manifestation of the popularity of the so-called "tablet PCs" for which the viewpoint is often "the smaller the better." However, we also see that this effect is, quite surprisingly, insignificant. In fact, while all other variables are highly significant, only screen size has an insignificant effect on price.

Now consider the corresponding effects from the spatial model shown in Figure 4.34. We refer to these effects as "main effects" since they are computed by varying only one variable at a time and holding all other variables constant at their median values. In that sense, each graph in Figure 4.34 corresponds to the effect of only one variable, computed at the median value of all other variables. Recall that all quantitative variables are standardized, and thus the values on the x-axis are interpreted relative to the average of that variable. Moreover, the categorical variables DVD, INTEL chip, condition, and the five brand indicators assume only the values 0 or 1, and therefore only the endpoints of the corresponding graphs have meaningful interpretations.

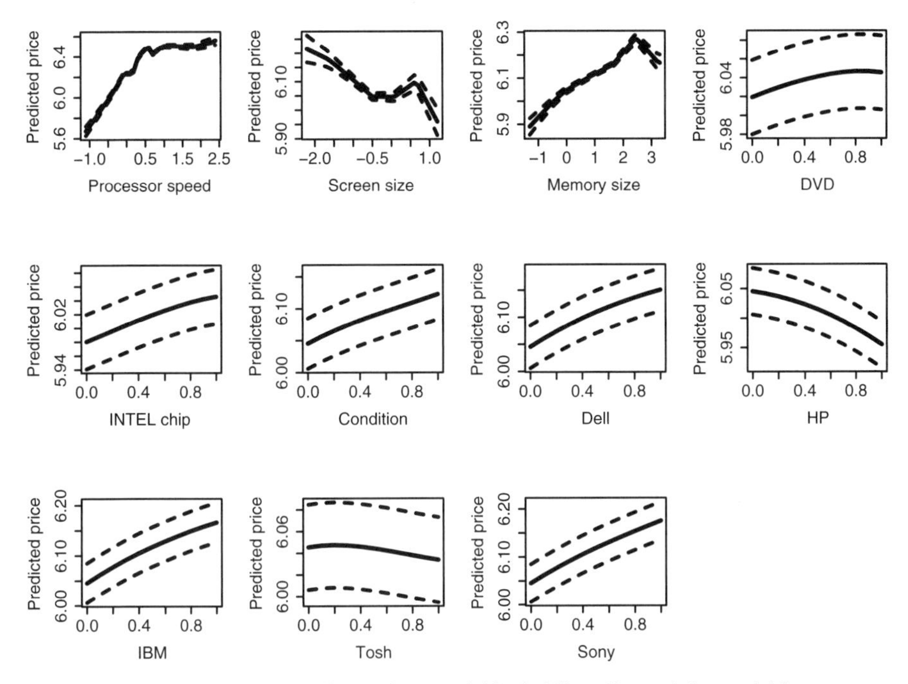

FIGURE 4.34 Spatial main effects of one variable, holding all remaining variables constant at their median values. The dotted lines correspond to 95% confidence intervals.

Consider the main effect of screen size in Figure 4.34. Note that the line is decreasing until it reaches zero, that is, the average screen size. That is, for all screen sizes smaller than the average, the effect on price is negative, which is in line with the result from the linear regression in Table 4.7. That is, tablet PCs and similar items achieve a higher price with a smaller screen. But note that the line *increases* between 0 and 0.5 and then again *decreases* for the largest screen sizes. It appears that for most screen sizes, the effect on price is negative. Only certain screen sizes, and in particular those that are slightly larger than the average screen size, result in higher prices. Note that this finding is very different from the linear regression model in Table 4.7. On the basis of the linear regression, one would have likely eliminated screen size as an important predictor.

Figure 4.34 reveals other insightful information. Note that the effect of processor speed on price is positive (which is similar to the finding from Table 4.7), but the magnitude of the effect decreases for processor speeds that are larger than 0.5 standard deviations above the average. In other words, for laptops with above average processor speed, adding additional processor capabilities does not increase the price significantly. Note though the spike right before 0.5: At certain configuration levels, adding additional processor speed adds tremendously to the item's price! A similar observation can be made for memory size, but its effect on price weakens only after memory size exceeds two standard deviations above the average. In fact, for laptop configurations with memory size above two standard deviations beyond the mean, the effect becomes negative and the price de-facto decreases. The effects of the features DVD, INTEL chip, condition, and the brand indicators are also similar to those in Table 4.7, at least in direction, but their effect magnitude and significance levels are different. For instance, it is quite interesting to note that in the linear regression model, Toshiba's prices are significantly different from the baseline, while this finding can no longer be sustained in the spatial model.

The main effects in Figure 4.34 paint only a very limited picture of the interrelations between all the different features in the spatial model. Since the main effects hold all other variables constant, they display only one effect at a time, and they capture the interrelation with other variables only at their median values. Ideally, we would want to display the interplay of *all* variables with one another, investigating all their possible values. Unfortunately, limitations in graphical display and human perception do not allow us to move beyond two, maybe three, variables at a time. Figures 4.35 and 4.36 show the interplay of *pairs* of variables. Specifically, Figure 4.35 shows the relationships between the three pairs of quantitative variables (processor speed, screen size, and memory size). Figure 4.36 displays the relationships between some of the remaining pairs. These two figures suggest that the pairwise relationships between all the variables and their combined effect on price are too complicated to be modeled by any known analytic function. The top panel of Figure 4.35, for instance, shows some interesting pockets of processor speed/screen size combinations for which price is higher than in other regions. Similar observations can be made in the other graphs of Figures 4.35 and 4.36.

In summary, Figures 4.34–4.36 suggest that the relationships among the features are too complex to be described satisfactorily by a (traditional) analytic function. The downside of the spatial approach is that it does not allow an easy interpretation of the variable effects. On the other hand, the spatial model turns out to have high predictive accuracy, we show later.

4.3.5 Temporal Concurrency Component

As pointed out earlier, modeling the temporal concurrency component involves creating a meaningful definition for a time lag. The "off-the-shelf" definition for time lags does not make much sense in online auctions, or, for that matter, in any situation where data arrive at irregular times. The reason for this is that the standard definition of a time lag assumes that the time between two consecutive events is always equal. This is the case in many situations such as sales data that become available every quarter, or measurements on experimental subjects that are performed every day. Note that in these two situations, the arrival of data is determined by one single source, a "referee" in some sense. In the first instance, the referee is the company that decides to make information available every quarter, and in the second instance it is the experimenter who decides to take measurements every day rather than every hour. Online auctions are different in the sense that there exists no such referee. In fact, the arrival of data in online auctions is determined by many different sources that act independently of each other, such as the seller who decides when to start and stop the

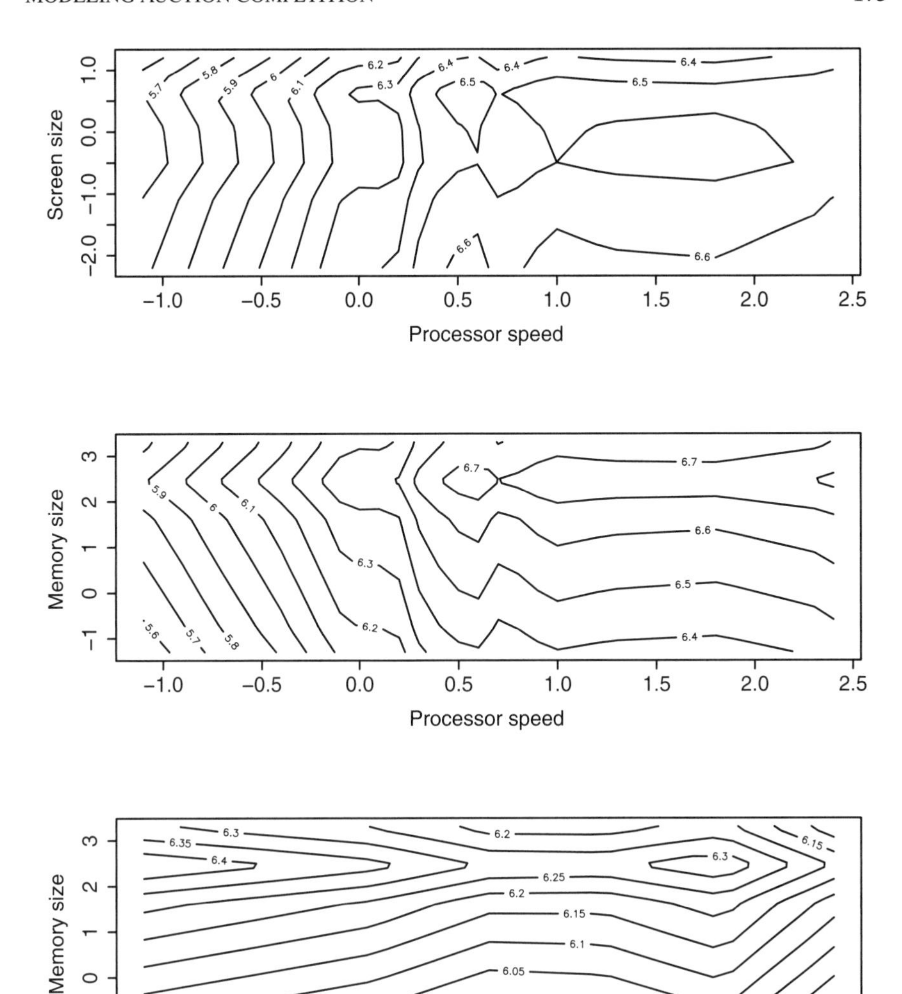

FIGURE 4.35 Pairwise spatial interaction effects, holding the remaining variables at their median values.

auction or the bidders who decide when and where to place their bids. The result is irregularly spaced data for which a traditional time lag is not very meaningful.

Consider Figure 4.37 that shows the distribution of a sample of auction endings over the course of 1 day. We can see that most auctions end later in the day, in the afternoon and evening. The arrival of ending times is very unevenly distributed though: while some auctions end at almost identical points in time (i.e., some points in Figure 4.37 are almost overlapping), there is a gap of several hours between others.

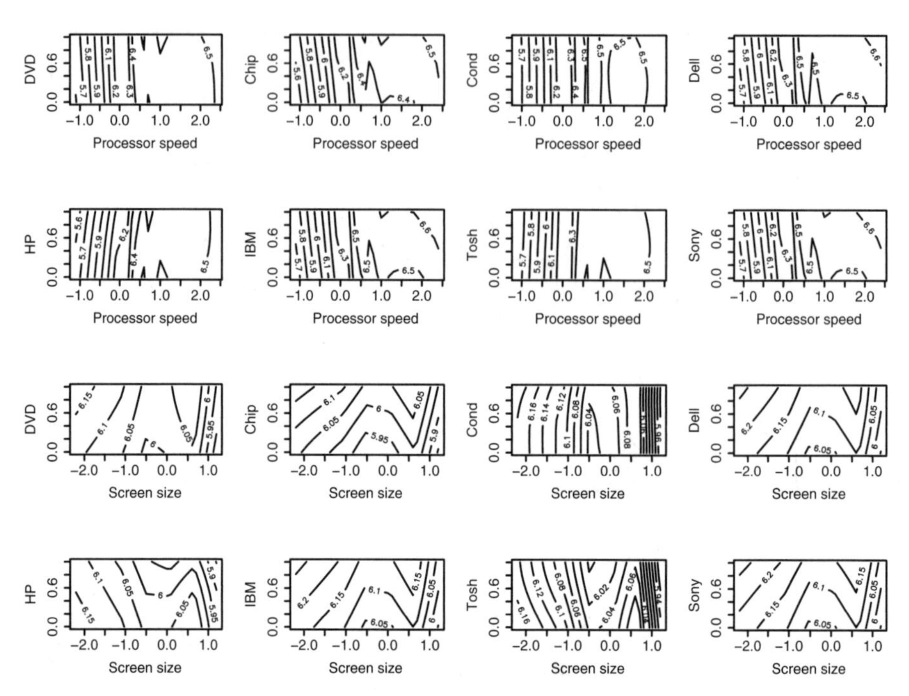

FIGURE 4.36 Pairwise spatial interaction effects, holding the remaining variables at their median values.

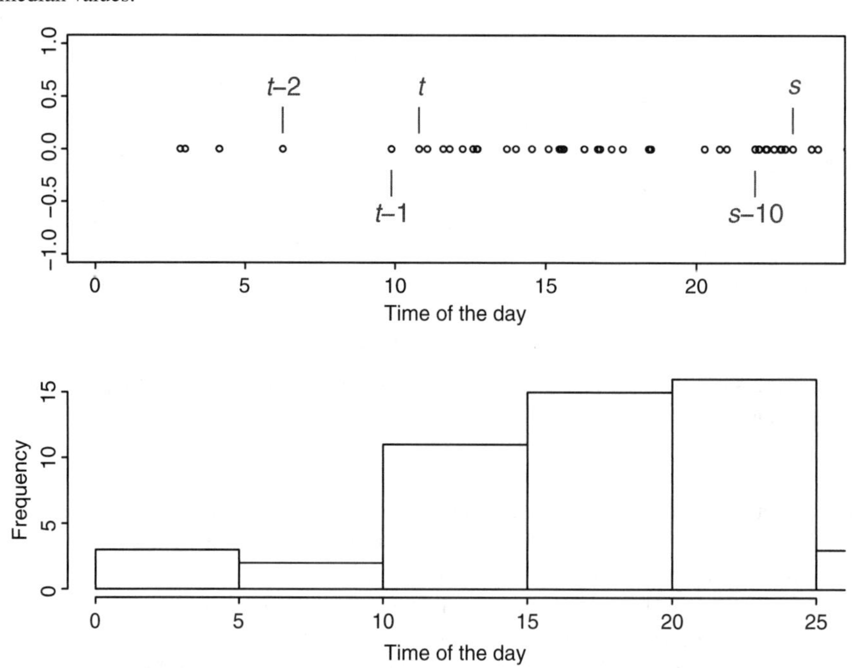

FIGURE 4.37 Distribution of ending times over the course of 1 day for a random sample of 50 auctions.

eBay users can monitor auctions and make their bidding decisions based on what they learn from them. It is likely that users will base their decision about bidding on a current or future auction not only on the price of the most recent auction, but also on prices from other auctions, further distant in the past. If, for instance, bidders wants to place a bid on the auction closing at time t in Figure 4.37, then their decision will be influenced not only by the most recent auction at $t-1$ but also by the auction that closed at $t-2$ and others further back. Using the traditional definition of a time lag, we would treat the time difference between the price at t and $t-1$ as equal to the time difference between $t-1$ and $t-2$; but this is clearly not appropriate here. Another problem with the traditional lag definition is that it will, more than likely, lead to information loss. Consider again Figure 4.37. If a bidder now wanted to place a bid on the auction closing at time s, then all the auctions closing within the last hour or so are likely to influence his decision process. Note though that there are 10 auctions that close within the last hour, the earliest of which is marked by $s-10$. Practical applications of traditional time series models, driven by model parsimony, would hardly ever include lags of order up to 10, or even higher! Thus, it is quite likely that a traditional time series model would ignore some of the earlier time lags and therefore would not capture all the important information in this situation.

In the following, we suggest a new approach for defining lag variables. This approach proves useful for investigating concurrency of price in online auctions, but its principles can find applications well beyond the online context, in other situations that are governed by irregularly spaced data. We first discuss the general method of defining lags and then illustrate its usefulness in the context of our example.

Time Lags for Irregularly Spaced Data Information on popular online auction sites is available in abundance, and it changes every hour, every minute, and with every new auction. Information is so plentiful that it is likely that users digest it in aggregate form rather than bit by bit. What we mean is that it is more likely that users conceptualize the information from auctions they observed over a certain period in a *summarized* form rather than each piece of auction information by itself. For instance, rather then remembering all individual auction prices over the last day, users might remember just the average price, its variability, its minimum and maximum, and even the price trend. This is also the type of information provided by websites such as Hammertap.com to assist sellers and bidders (see Chapter 2). Note that such summary statistics characterize components of the price *distribution* that may be easier for a user to remember and digest than all the individual prices. In that sense, we define lag variables that capture components of the price distribution for all auctions occurring during a certain period in the past.

Let y_t denote an observation at time t. In the following, we choose to interpret t literally as calendar time rather than as an event index. That is, y_t denotes an observation arriving on a certain day and time, such as 09/03/2005 at 5:24 p.m. In that sense, a lag of one looks backward one entire day. In this example, $t-1$= 09/02/2005 at 5:25 p.m. A lag of two would go back to 09/01/2005 at 5:25 p.m. One of the appeals of this approach is that it also allows lags of fractional order, for instance, a lag of 1.5 (which looks back to 09/02/2005 at 5:25 a.m). In the following, we focus our description only on lag-1 variables. The generalization to other lags, including fractional lags, is straightforward.

Let $Y_{t:t-1} := y_{t:t-1,1}, \ldots, y_{t:t-1,n}$ denote the set of all n observations between time t and time $t-1$; that is, all observations that occurred during the last 24 h. We want to capture important components of the distribution of $Y_{t:t-1}$ based on a few simple measures. To this end, define the vector-valued lag-1 variable $\mathbf{y}_{t-1}$ as

$$\mathbf{y}_{t-1} = \begin{pmatrix} g_1(Y_{t:t-1}) \\ g_2(Y_{t:t-1}) \\ \vdots \\ g_p(Y_{t:t-1}) \end{pmatrix}, \tag{4.53}$$

where the functions $g_i(Y_{t:t-1})$, $1 \leq i \leq p$, denote suitable summary statistics of $Y_{t:t-1}$ such as the mean, median, standard deviation, inner quartile range, minimum and maximum, and possibly other statistics.

Figure 4.38 revisits the sample of laptops from Figure 4.37 and shows their price distribution. Suppose we are interested in the effect of all prices from the previous day on the price at time t. There are $n = 50$ auctions between t and $t-1$. We may

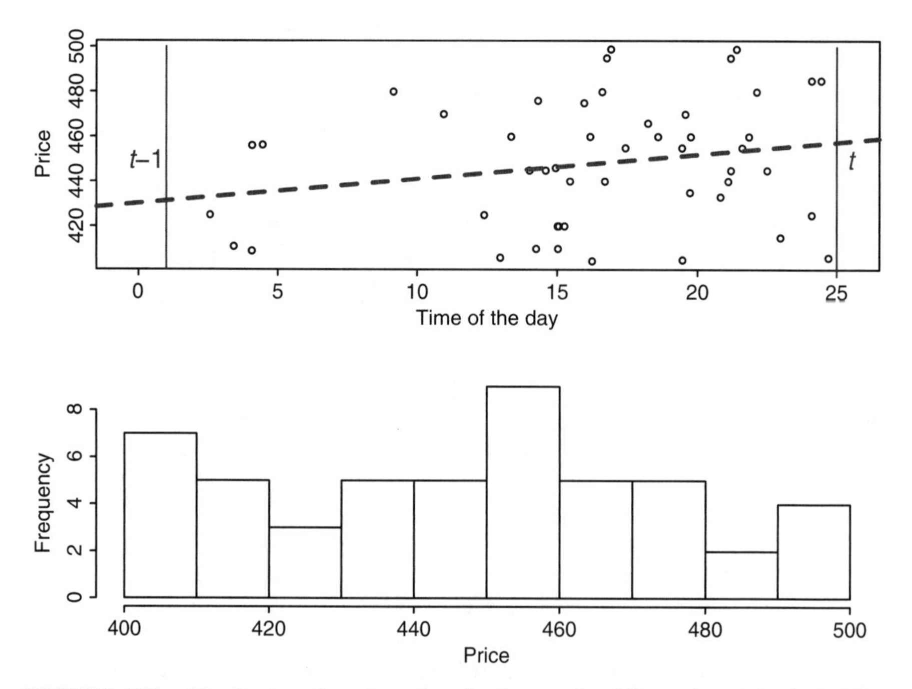

FIGURE 4.38 Distribution of auction prices for the sample of 50 auctions. The dashed line shows the slope of a least squares line fitted to the data.

choose to define the different components of $\mathbf{y}_{t-1}$ in (4.53) as

$$\mathbf{y}_{t-1} = \begin{pmatrix} \$455.0 \text{ (Median)} \\ \$450.6 \text{ (Mean)} \\ \$27.6 \text{ (StDev)} \\ \$404.5 \text{ (Min)} \\ \$499.0 \text{ (Max)} \end{pmatrix}, \tag{4.54}$$

where the first element of $\mathbf{y}_{t-1}$ (\$455.0) is the median of all 50 auction prices (and similarly for the other components of (4.54)). Summary statistics such as the mean, median, and standard deviation characterize a distribution, but they do not reveal a time trend in this period. The price at time t can be affected not only by the average price over the previous day, but also by a possible price trend. For instance, if prices have increased systematically over the last 24 h, this may result in a further increase of price at t. On the other hand, the effect may be different if prices have remained constant over the previous day. One simple way to measure a trend in $Y_{t:t-1}$ is via the regression slope. That is, consider the simple linear regression over time, $y_{t:t-1,i} = \alpha + \beta i + \epsilon_i$ $(1 \le i \le n)$, then $\beta = \beta(Y_{t:t-1})$ measures the price trend during the last 24 h. This adds an additional component, $g_i(Y_{t:t-1}) = \beta(Y_{t:t-1})$, to (4.53). The dashed line in Figure 4.38 shows the regression slope for the 50 auctions in this example. In a similar fashion, one could define components that capture nonlinear price trends and even more complicated functions of the data.

Similarity-Refined Time Lags Inclusion of a predictor variable into a regression model is meaningful only if that variable captures some of the variability in the response. This is no different for lag variables. One effective way to gauge the impact of the lag variables defined in (4.53) is via scatterplots.

Figure 4.39 shows scatterplots of y_t versus the components of $\mathbf{y}_{t-1}$ for all auctions in our data set. None of the six plots reveal a clear relationship between y_t and $\mathbf{y}_{t-1}$. This suggests that none of the components of $\mathbf{y}_{t-1}$ has a practical effect on y_t, and, therefore, $\mathbf{y}_{t-1}$ does not appear to be of much predictive value. While this is somewhat discouraging at first, one of the reasons for this phenomenon is that our data set is very diverse, with laptops differing vastly in their features and also in their prices. We therefore set out to obtain a more refined measure of temporal concurrency.

Specifically, we set out to obtain a refined lag variable $\tilde{\mathbf{y}}_{t-1}$ that, as previously, contains aggregate information from the prior day, but in contrast to our previous definition, it only contains information for those products that are *most similar* to one another. By similar we mean products with similar features such as similar processor speeds or similar memory sizes. The rationale for this approach is that bidders, making decisions about their bid, are more likely to look to prices from similar auctions for information gathering.

We calculate the refined lag variable $\tilde{\mathbf{y}}_{t-1}$ as follows. Let y_t denote the price of an item with feature vector $\mathbf{x}_t = (x_{t,1}, x_{t,2}, \ldots, x_{t,q})$. In our example, $\mathbf{x}_t$ contains features

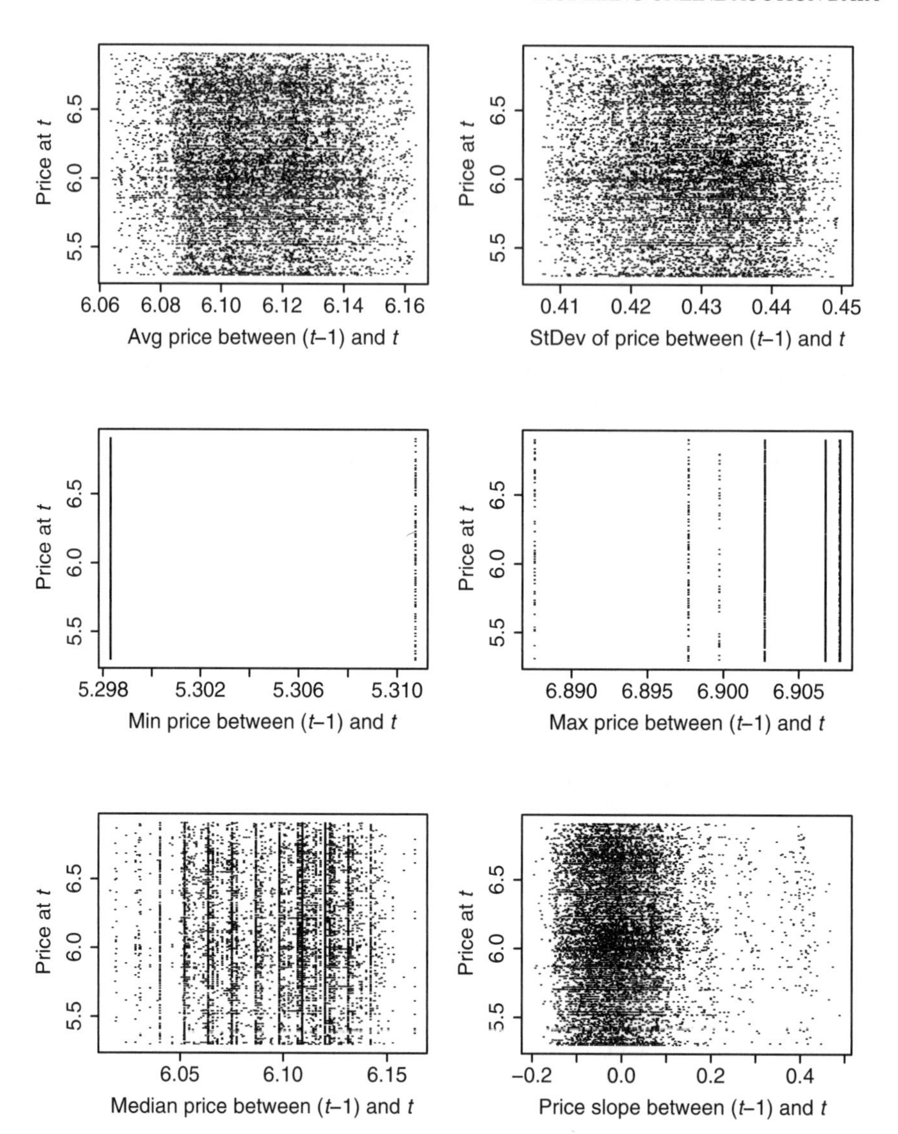

FIGURE 4.39 Scatterplot of price at time t, y_t, versus price lag summaries $\mathbf{y}_{t-1}$, corresponding to the mean, standard deviation, minimum, maximum, median, and slope of the price trend of all auctions from the previous day.

such as processor speed, memory size, or whether or not a DVD drive is included. As before, $Y_{t:t-1}$ denotes the set of all prices from the last day. Let

$$Y_{(t:t-1)} = \{y_{t:t-1,(1)}, \ldots, y_{t:t-1,(n)}\} \tag{4.55}$$

denote the *feature-ordered* set. That is, $y_{t:t-1,(1)}$ denotes the price for an item whose features, $\mathbf{x}_{t:t-1,(1)}$, are *most similar* to $\mathbf{x}_t$. Conversely, the item with price $y_{t:t-1,(n)}$

has features $\mathbf{x}_{t:t-1,(n)}$ that are least similar to $\mathbf{x}_t$. We compute the dissimilarity $d_{t,i} := d(\mathbf{x}_{t:t-1,i}, \mathbf{x}_t)$ in the same way as earlier using the definition in (4.48). We then obtain the feature-ordered set $Y_{(t:t-1)}$ by ordering $Y_{t:t-1}$ based on ranking the similarity scores $d_{t,i}$.

The refined lag variable $\tilde{\mathbf{y}}_{t-1}$ contains information only for those products that are most similar to one another. To that end, we select only those elements from $Y_{(t:t-1)}$ whose features are most similar to that of y_t; that is, we select its top $\alpha\%$. Let $Y^{\alpha}_{(t:t-1)}$ denote the set of top $\alpha\%$ elements from $Y_{(t:t-1)}$. We compute similar aggregate measure as previously; however, we restrict these measures to the set $Y^{\alpha}_{(t:t-1)}$. This results in the refined lag variable

$$\tilde{\mathbf{y}}^{\text{ms}}_{t-1} = \begin{pmatrix} g_1(Y^{\alpha}_{(t:t-1)}) \\ g_2(Y^{\alpha}_{(t:t-1)}) \\ \vdots \\ g_p(Y^{\alpha}_{(t:t-1)}) \end{pmatrix}, \tag{4.56}$$

where the superscript "ms" stands for "most similar." Applying the same principles, we can also calculate $\tilde{\mathbf{y}}^{\text{ls}}_{t-1}$, the lag variable corresponding to "least similar items," for example, by choosing the lowest $\alpha\%$ in $Y_{(t:t-1)}$. In our application, we use $\alpha = 33$, which means that $\tilde{\mathbf{y}}^{\text{ms}}_{t-1}$ contains information from the top third of the most similar items while $\tilde{\mathbf{y}}^{\text{ls}}_{t-1}$ contains the bottom third. This leaves the middle third that we denote by $\tilde{\mathbf{y}}^{\text{as}}_{t-1}$, or items with "average similarity."

Figures 4.40 and 4.41 show scatterplots of the refined lag variables. Note that these scatterplots are similar to the ones in Figure 4.39, but now they break up each lag variable into its three components: most similar, average similar, and least similar items! We can see that the refined lag variables carry a considerably larger amount of information about y_t. It is interesting, for example, that average prices for the most similar items are positively correlated with y_t, while the least similar items' average price appears to exhibit a negative correlation. We can also see new, previously unobserved patterns for the other lag variables such as the price variability and the minimum or the maximum price. Price trends, displayed in the bottom panel of Figure 4.41, appear to have only a weak correlation with y_t.

Selection of Refined Time Lags Figures 4.40 and 4.41 display a total of 18 refined lag variables. Inclusion of all these 18 variables would not be meaningful under common model parsimony considerations. Thus, we face the challenging task of selecting a suitable subset of lag variables, ideally a subset or some reduced dimension of the full set with high explanatory power.

We use principal components analysis to achieve this goal. Figure 4.42 shows a screeplot based on a principal component analysis of all 18 refined lag variables. We can see that the first principal component (PC) carries most of the information (78%). In fact, the first three PCs account for over 87% of the total variation across all 18 lag variables. The factor loadings for the first three PCs can be seen in Table 4.8.

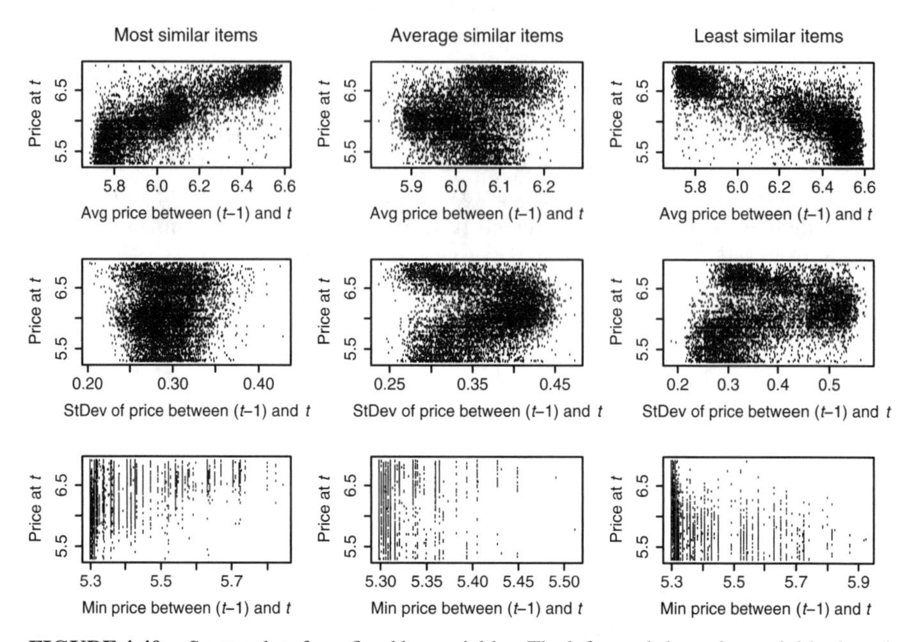

FIGURE 4.40 Scatterplots for refined lag variables. The left panel shows lag variables based on most similar items, and the right panel shows lags based on least similar items.

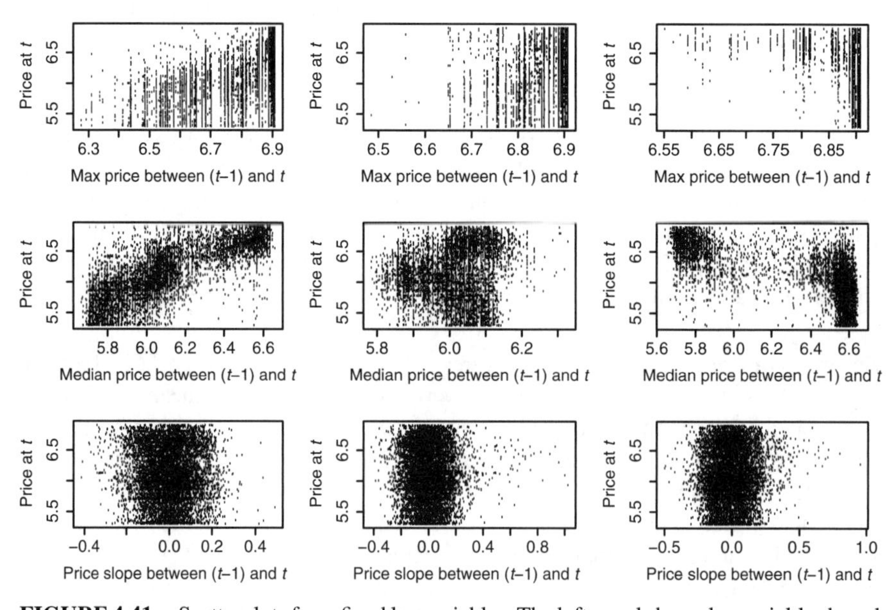

FIGURE 4.41 Scatterplots for refined lag variables. The left panel shows lag variables based on most similar items, and the right panel shows lags based on the least similar items.

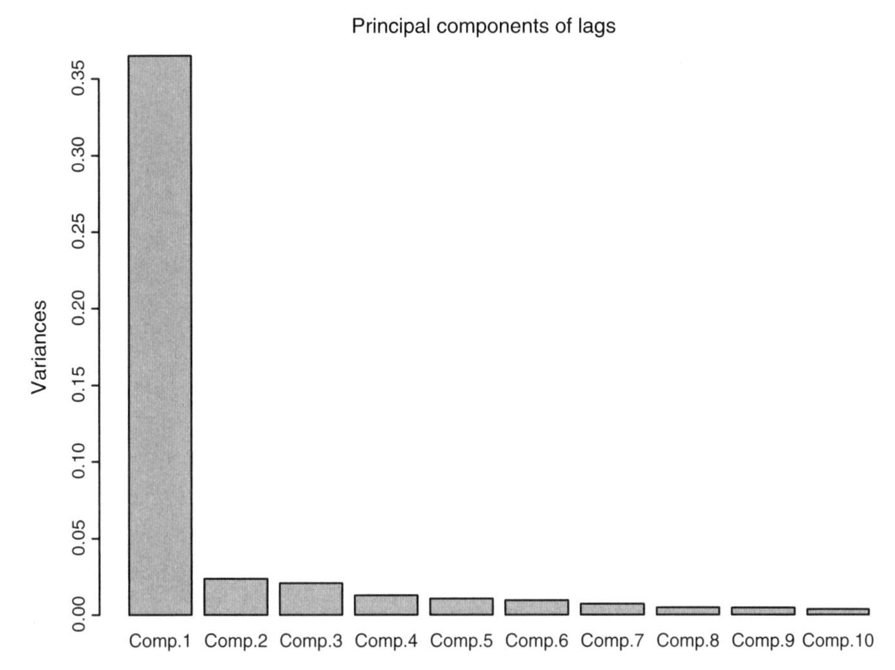

FIGURE 4.42 Screeplot of a principal component decomposition of all 18 lag variables.

TABLE 4.8 Factor Loadings of First Three Principal Components

Variable		PC1	PC2	PC3
	Most similar	−0.41	0.29	−0.10
Mean	Avg similar	−0.07	−0.30	0.10
	Least similar	0.48	0.03	−0.01
	Most similar	−0.01	−0.01	0.00
StDev	Avg similar	0.01	0.17	−0.05
	Least similar	−0.03	0.39	−0.14
	Most similar	−0.09	−0.02	0.00
Min	Avg similar	−0.01	−0.03	0.00
	Least similar	0.05	−0.22	0.07
	Most similar	−0.15	0.29	−0.09
Max	Avg similar	−0.01	0.02	0.00
	Least similar	0.03	0.03	0.00
	Most similar	−0.45	0.26	−0.10
Med	Avg similar	−0.04	−0.39	0.13
	Least similar	0.60	0.42	−0.16
	Most similar	0.01	0.02	0.19
Slope	Avg similar	0.01	0.22	0.59
	Least similar	−0.01	0.25	0.71

It is quite interesting that the first PC is composed primarily of the average and median price lags corresponding to the most and least similar items. In that sense, the first PC captures primarily information related to the previous day's "most typical" price. In contrast, the second PC, while also capturing some of the "typical" price information, is composed largely of price variability information from the previous day (especially price variability corresponding to the least similar items). Finally, the third PC's most noteworthy feature is its capturing of price trend information (i.e., the slope), and especially price trends corresponding to the average and least similar items. In summary, the first three PCs capture the typical price, its variability, and its trend over the previous day. In that sense, these three PCs appear to capture exactly what an eBay savvy user would be expected to gather from a day's worth of auction activity.

In the following, we operate on the reduced space by using the scores on the first three principal components rather than the refined lag variables $\tilde{\mathbf{y}}^{\text{ms}}_{t-1}$, $\tilde{\mathbf{y}}^{\text{as}}_{t-1}$, and $\tilde{\mathbf{y}}^{\text{ls}}_{t-1}$ directly. That is, let $\tilde{\mathbf{y}}^{\text{sc}}_{t-1} = (\tilde{y}^{\text{sc}}_{t-1,1}, \tilde{y}^{\text{sc}}_{t-1,2}, \tilde{y}^{\text{sc}}_{t-1,3})$ denote the scores on the first three principal components. Our temporal concurrency component captures the relationship between $\tilde{\mathbf{y}}^{\text{sc}}_{t-1}$ and y_t. A model for this relationship is described next.

Temporal Concurrency Component We now model the relationship between $\tilde{\mathbf{y}}^{\text{sc}}_{t-1}$ and y_t. Figure 4.43 shows the scatter of the components of $\tilde{\mathbf{y}}^{\text{sc}}_{t-1}$ versus y_t. We can

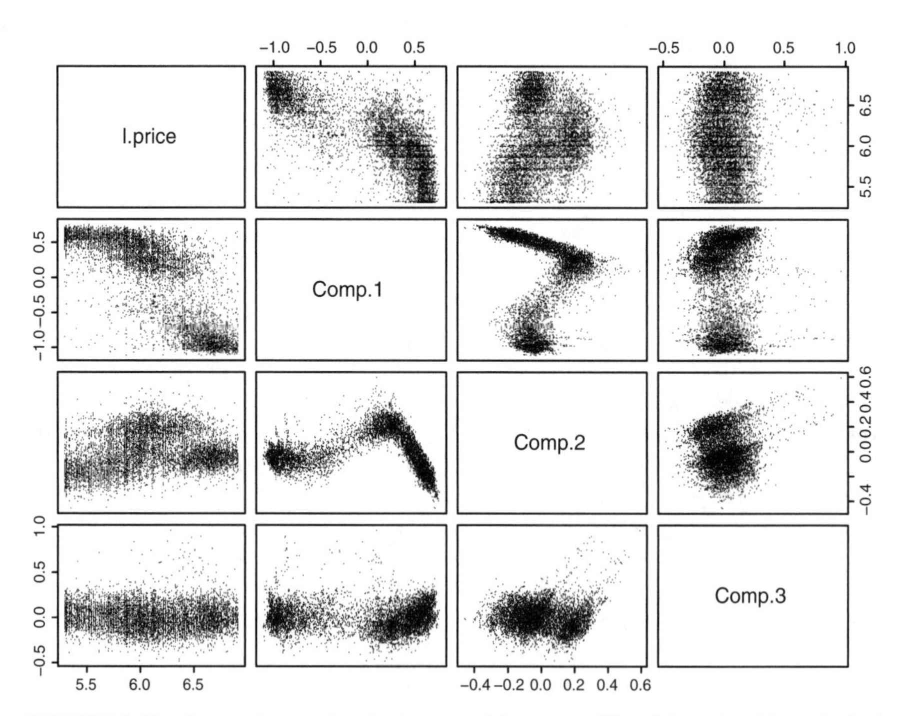

FIGURE 4.43 Scatterplot matrix of price y_t and the scores $\tilde{\mathbf{y}}^{\text{sc}}_{t-1}$ of the refined lag principal components.

see a strong relationship between $\tilde{\mathbf{y}}_{t-1}^{\text{sc}}$ and y_t. However, we can also see that this relationship cannot be described easily by any functional form. For this reason, we resort, once again, to a nonparametric approach.

Let the relationship between y_t and $\tilde{\mathbf{y}}_{t-1}^{\text{sc}}$ be given by

$$y_t = f_1(\tilde{y}_{t-1,1}^{\text{sc}}) + f_2(\tilde{y}_{t-1,2}^{\text{sc}}) + f_3(\tilde{y}_{t-1,3}^{\text{sc}}) + \epsilon, \tag{4.57}$$

where f_1, f_2, and f_3 correspond to three unspecified functions, to be estimated from the data. Note that we do not have to worry about possible interactions between the components of $\tilde{\mathbf{y}}_{t-1}^{\text{sc}}$ since, by nature of principal components, the components of $\tilde{\mathbf{y}}_{t-1}^{\text{sc}}$ are orthogonal. This makes the additive model in (4.57) plausible and also eliminates the need for multivariate smoothers.

Following the steps outlined earlier, we can write model (4.57) once again in a mixed model notation. This leads to the temporal concurrency component

$$g_{\text{TC}}(\mathbf{x}) = \mathbf{x}_{\text{TC}}^T \boldsymbol{\beta}_{\text{TC}} + \mathbf{z}_{\text{TC}}^T \mathbf{u}_{\text{TC}}, \tag{4.58}$$

with $\mathbf{x}_{\text{TC}}$, $\mathbf{z}_{\text{TC}}$, $\boldsymbol{\beta}_{\text{TC}}$, and $\mathbf{u}_{\text{TC}}$ defined similar to (4.40) and (4.41).

4.3.6 Grand Concurrency Model and Competing Models

Collecting all individual model components, the grand concurrency model can now be written as

$$y = g_{\text{AC}}(\mathbf{x}) + g_{SC}(\mathbf{x}) + g_{\text{TC}}(\mathbf{x}) + \epsilon, \tag{4.59}$$

where g_{AC}, g_{SC}, and g_{TC} denote the auction and, the spatial and temporal concurrency components. Note that g_{AC} is entirely parametric, while g_{SC} and g_{TC} are nonparametric. In that sense, (4.59) provides a semiparametric approach to understanding concurrency in online auctions.

In the next section, we investigate the usefulness of this model by measuring its predictive accuracy on a holdout sample. To this end, we compare our model to a few competing models. The choice of these competitors is driven by the question, "What do I gain by using the more complex modeling approaches via g_{AC}, g_{SC}, and g_{TC} as opposed to a simpler approach?" In this context, to investigate the impact of all auction related variables in g_{AC}, our first competitor model (C1) uses only the brand and feature related variables from Tables 4.4 and 4.5 (i.e., it ignores all information related to the auction transaction in Table 4.6). The second competitor model (C2) uses brand, feature, *and* auction related information. This is comparable to concurrency components g_{AC} and g_{SC}, however, *without* the spatial approach to modeling the relation between the features. This type of model is also the one most commonly used in the literature on online auctions (e.g., Lucking-Reiley et al., 2007). The third competitor model (C3) uses the traditional definition of a time lag (i.e., y_{t-1} corresponds to the observation prior to y_t) and thus investigates the impact

TABLE 4.9 Competing Concurrency Models

	Explanatory Variables		
Model	Brand and features (Processor speed, Screen/memory size, Chip, DVD drive, Brand, used item)	Auction transaction (opening bid, auction length, number of bids, reserve price, buy-it-now)	Traditional lag (y_{t-1} corresponds to observation prior to y_t)
C1	✓	–	–
C2	✓	✓	–
C3	✓	✓	✓

of the temporal concurrency component g_{TC}. Table 4.9 summarizes all competing concurrency models.

Model Predictive Performance We investigate the performance of the grand concurrency model (4.59) on a holdout sample. That is, we use 70% of the data to train the model and we measure its performance on the remaining 30%. We investigate the model in equation (4.59) in several steps: we first investigate the impact of its three components separately (models M1, M2, M3), then we consider combinations of pairs of two out of the three components (models M4, M5, M6), and finally we consider all three components in the model (M7). This results in seven different models, summarized in Table 4.10. We then compare the performance of these seven models with the three competing models (C1, C2, C3) from Table 4.9. We compare predictive performance using two different performance measures: root mean squared error (RMSE) between the predicted and the true price and the MAPE between the two. Table 4.11 shows the results.

Table 4.11 allows for interesting insight, not only into the performance of the concurrency model, but also into the factors that drive price in online auctions. Consider model M1 that includes only auction related information; that is, information related to the product transaction. This model results in a MAPE of about 36%. The MAPE reduces to 14% or 18% when including, in addition to g_{AC}, either the feature space

TABLE 4.10 Grand Concurrency Model and Its Components

Model	g_{AC}	g_{SC}	g_{TC}
M1	✓	–	–
M2	–	✓	–
M3	–	–	✓
M4	✓	✓	–
M5	✓	–	✓
M6	–	✓	✓
M7	✓	✓	✓

TABLE 4.11 Prediction Accuracy of Different Concurrency Models

Model	RMSE	MAPE
M1	195.23	0.3559
M2	97.31	0.1502
M3	114.15	0.1915
M4	93.37	0.1431
M5	110.15	0.1821
M6	97.37	0.1502
M7	88.53	0.1298
C1	129.29	0.1918
C2	129.11	0.1902
C3	292.54	0.5257

The second column reports the RMSE, and the third column reports the MAPE.

information (M4) or the temporal price information (M5). When including both feature space and temporal information into the grand concurrency model (M7), the MAPE reduces to less than 13%. It is quite remarkable that, using only information related to the auction transaction (M1), one can predict eBay's price with about 36% accuracy. Not surprisingly, among the single-component models the feature space (M2) has the highest predictive performance with a MAPE of only 15%. It is interesting to note the strong impact of time on eBay's prices: using only price information from the previous day (M3), the forecasting accuracy is about 19%. However, note that the refined lag variables in Section 4.3 use price information from *similar* auctions, and therefore not all previous auctions affect the current price.

It is also noteworthy to compare the performance of the grand concurrency model (M7) with that of simpler, competing models (C1–C3). Modeling the effect of features in the traditional, linear way (C1) results in only 19% predictive accuracy, compared to 15% for the spatial approach (M2). Also, adding auction related information, such as auction design or competition (but omitting the interaction term between opening bid and number of bids), barely improves the predictive performance (C2). This speaks further to the important relationship between the opening bid and the number of bidders captured well by the interaction term. But most interesting is the performance of a model that also includes price information from previous auctions, albeit in the form of the more traditional definition of a lag variable (C3). We can see that for this model, the predictive performance deteriorates horrendously, despite the inclusion of auction and feature information, and worsens below that of our worst performing individual concurrency component M1. This again underlines the importance of using different statistical techniques for modeling very unevenly spaced event data. Our approach, while not being the only option, proves to be a very viable solution to that problem.

4.3.7 Conclusion

This section proposes an innovative solution to modeling competition in online auctions. Taking advantage of the flexibility of semiparametric models, we propose an approach that consists of three main model components: one component that captures all information related to the auction transaction, another component that captures feature combinations and their effect on the response, and a third component that captures the effect of time.

Modeling and understanding auction competition is important, and it is quite surprising that, with the exception of Jank and Shmueli (2007) and Jank and Zhang (2009a,b), only little statistical work has been done in this area to date. Competition is a fact of today's life. Whether it is stocks, sales promotions of competing products, or mortgage rates offered by lenders in the same market that influence one another, action often results in re-action. Competition is particularly prevalent in the online environment since the openness of the Internet allows individuals to observe each other's moves in real time. In this section, we consider competition of online auctions on eBay, but our approach can be generalized to other applications. Modeling competition is important to understand the factors that drive auctions and also to predict the outcome of future actions.

Our modeling approach presents several novelties. The translation of item features and similarities among items into the concept of "space" allows us to use a large array of existing modeling tools that have been developed and used primarily for geographically referenced data. Although measuring distance in a product feature space is not new, the novelty here is in combining different distance measures, explaining the distance between two items in a spatial model, and using flexible semiparametric models for detecting complicated but important relationships among the features and their effect on the response. The flexibility of semiparametric models allows us not only to discover patterns and relationships that standard parametric models miss, but also to improve upon the predictability of new auctions using input from concurrent auctions.

A second innovation is in handling and summarizing unevenly spaced time-series data in a meaningful way that allows for integration into a semiparametric model. Our examples show that treating such series as evenly spaced leads to unacceptable distortions in analysis results. Our proposed process starts from defining a meaningful "time lag," through selecting summary statistics that capture the most important information, to finally reducing this space to achieve a parsimonious model. We believe that this general process can be used in many applications outside of online auctions, since unevenly spaced time series are becoming increasingly commonplace.

One caveat is that we consider competition in eBay's auctions only from the point of view of the *final* auction price. However, eBay's concurrency is even further complicated if one considers the *current* auction price at any given moment during the ongoing auction. The current price in an ongoing auction is a time series on a closed time interval, and this time series is related to other price time series from other auctions. One first step to capture this more complex type of concurrency is the work by Hyde et al. (2006), who take a functional data analytic approach for the purpose of visualizing the degree of concurrency that arises between current prices (and their dynamics)—see Chapter 3.

4.4 MODELING BID AND BIDDER ARRIVALS

We now turn our attention to the bidders and the choices that they make. In particular, we consider two of the most important aspects of any auction: the arrival process of bidders (Russo et al., 2008, 2010) and—related to this—the arrival process of bids (Shmueli et al., 2004, 2007). In fact, auctions depend on the existence of interested bidders—without any bidders, there is no transaction. Moreover, since auctions do not transact at a fixed price, their outcome strongly relies on the total *number* of bidders participating. From the point of view of the seller, more bidders typically translate into a higher selling price. In contrast, the winning bidder would probably prefer as few competing co-bidders as possible to keep the price low. While the number of bidders is an important component of an auction, the *timing* of bids is equally important. For instance, while earlier bids are typically lower, online auctions often experience the so-called bid sniping (Roth and Ockenfels, 2002) and bidding frenzy (Ariely and Simonson, 2003) in which bidders try to outbid one another in the final moments of an auction, and the price often makes significant jumps during this period. In the following, we describe both the arrival process of bidders (i.e., the number of bidders and the time when they arrive) and the associated arrival process of bids. More specifically, we investigate a nonhomogeneous Poisson process for modeling the bid arrivals and show that this process adequately captures the different phases of an online auction. We will then generalize this idea and show that the process can also be used to characterize bidder activity. We start with the bid arrival process.

4.4.1 Modeling Bid Arrivals

Many of the theoretical results derived for traditional (offline) auctions have been shown to fail in the online setting for reasons such as globalism, computerized bidding and the recording of complex bids, longer auction durations, more flexibility in design choice by the seller, and issues of trust. A central factor underlying many important results is the *number of bidders* participating in an auction. Typically, it is assumed that this number is fixed (Pinker et al., 2003) or fixed but unknown (McAfee and McMillan, 1987). In online auctions, the number of bidders and bids is not predetermined, and it is known to be affected by the auction design and its dynamics. Thus, in both the theoretical and empirical domains the number of bidders and bids plays an important role.

Shmueli et al. (2007) proposed a new and flexible model for the bid arrival process. Having a model for bid arrivals has several important implications. First, many researchers in the online auction arena use simulated bid arrival data to validate their results. Bapna et al. (2002a), for example, used simulated bid arrival data to validate their model on a bidder's willingness to pay. Gwebu et al. (2005) designed a complex simulation study to analyze bidders' strategies using assumptions about bidder as well as bid arrival rates. It has also been noted that the placement of bids influences the bidder arrival process (Beam et al., 1996). Hlasny (2006) reviewed several econometric procedures for testing the presence of latent shill bidding (where sellers fraudulently bid on their own item) based on the arrival rate of bids. While a clear understanding of the process of bidding can have an impact on the theoretical literature, it can also be useful in applications. These range from automated electronic negotiation through monitoring auction server performance to designing automated bidding agents. For instance, Menasce and Akula (2004) studied the connection between bid arrivals and auction server performance. They found that the commonly encountered "last-minute bidding" creates a huge surge in the workload of auction servers and degrades their performance. They then proposed a model to improve a server's performance through auction rescheduling using simulated bid arrival data.

Modeling the bid arrival process (rather than that of the bidders) also promises to produce more reliable results, because bid placements are typically completely observable from the auction's bid history, whereas bidder arrivals are not. eBay, for instance, posts the temporal sequence of all the bids placed over the course of the auction. In particular, every time a bid is placed, its exact time stamp is posted. In contrast, the time when bidders first arrive at an auction is unobservable from the bid history. Bidders can browse an auction without placing a bid, thereby not leaving a trace or revealing their interest in that auction. That is, they can look at a particular auction, inform themselves about current bid and competition levels in that auction, and make decisions about their bidding strategies. All these activities can take place without leaving an observable trace in the bid history that the auction site makes public. In fact, it is likely that bidders first browse an auction and only later place their bid. The gap between the bidder arrival time and the bid placement also means that the bidder arrival is not identical to the bid arrival and can therefore not be inferred directly from the observed bid times. Another issue is that most online auctions allow bid revision, and therefore many bidders place multiple bids. This further adds to the obscurity of defining the bidder arrival–departure process. Our approach is therefore to model the bid arrival process based on empirical evidence.

Features of Bid Arrivals We start by describing two of the most prominent features of bid arrivals, the multistage arrival intensities and the self-similarity of the process.

Multistage Arrival Intensities: Time-limited tasks are omnipresent in the offline world: voting for a new president, purchasing tickets for a popular movie or sporting event, filing one's federal taxes, and so on. In many of these cases, arrivals are especially intense as the deadline approaches. For instance, during the 2001 political elections in Italy, more than 20 million voters cast their ballots between 13:00 and 22:00 (Bruschi et al., 2002) when ballots were scheduled to close at 22:00. Similarly, a high proportion of U.S. tax returns are filed near the April 15 deadline. For instance, about one-thirds of all returns are not filed until the last 2 weeks of tax season. According to Ariely et al. (2005), deadline effects have been noted in studies of bargaining, where agreements are reached in the final moments before the deadline (Roth et al., 1998). Such effects have been shown among animals, which respond more vigorously toward the expected end of a reinforcement schedule, and in human task completion where individuals become increasingly impatient toward the task's end. Furthermore, people use different strategies when games are framed as getting close to the end (even when these are arbitrary breakpoints; Croson, 1996). In addition to the deadline effect, there is an effect of earliness where the strategic use of time moves transactions earlier than later, for example, in the labor market (Roth and Xing, 1994; Avery et al., 2001).

Such deadline and earliness effects have also been observed in the online environment. Several researchers have noted deadline effects in internet auctions (Bajari and Hortacsu, 2003; Borle et al., 2006; Ku et al., 2004; Roth and Ockenfels, 2002; Wilcox, 2000; Shmueli et al., 2007). In many of these studies, it was observed that a nonnegligible percent of bids arrive at the very last minute of the auction. This phenomenon, called "bid sniping," has received much attention, and numerous explanations have been suggested to explain its existence. Empirical studies of online auctions have also reported an unusual amount of bidding activity at the auction start followed by a longer period of little or no activity (Borle et al., 2006; Jank and Shmueli, 2008b). Bapna et al. (2003) referred to bidders who place a single early bid as "evaluators." Finally, "bid shilling," a fraudulent act where the seller places dummy bids to drive up the price, is associated with early and high bidding (Kauffman and Wood, 2000). The existence of these bid timing phenomena are important factors in determining outcomes at both the auction level and the market level. They have therefore received much attention from the research community.

Self-Similarity (and Its Breakdown): While both the offline and online environments share the deadline and earliness effects, the online environment appears to possess the additional property of *self-similarity* in the bid arrival process.[13] Self-similarity refers to the "striking regularity" of shape that can be found among the distribution of bid arrivals over the intervals $[t, T]$, as t approaches the auction deadline T. Self-similarity is central in applications such as web, network, and Ethernet traffic. Huberman and Adamic (1999) found that the number of visitors to websites follows a universal power law. Liebovitch and Schwartz (2003) reported that the arrival process of email viruses is self-similar. However, this has also been reported in other online environments. For instance, Aurell and Hemmingsson (1997) showed that times between bids in the interbank foreign exchange market follow a power law distribution.

Several authors reported results that indicate the presence of self-similarity in the bidding frequency in online auctions. Roth and Ockenfels (2002) found that the arrival of last bids by bidders during an online auction is closely related to a self-similar process. They approximated the CDF of bid arrivals in "reverse time" (i.e., the CDF of the elapsed times between the bid arrivals and the auction deadline) by the power functional form $F_T(t) = (t/T)^{\alpha}$ $(\alpha > 0)$, over the interval $[0, T]$, and estimated α from the data using OLS. This approximates the distribution of bids over intervals that range from the last 12 h to the last 10 min, but accounts for neither the final minutes of the auction nor the auction start and middle. Yang et al. (2003) found that the number of bids and the number of bidders in auctions on eBay and on its Korean partner (auction.co.kr) follows a power law distribution. This was found for

[13] This property was also found in the offline process of bargaining agreements, as described in Roth and Ockenfels (2002).

auctions across multiple categories. The importance of this finding, which is closely related to the self-similarity property, is that the more bidding one observes up to a fixed time point, the higher the likelihood of seeing another bid before the next time point. According to Yang et al. (2003), such power law behaviors imply that the online auction system is driven by self-organized processes, involving all bidders who participate in a given auction activity.

The implications of bid arrivals following a self-similar process instead of an ordinary Poisson model are significant: The levels of activity throughout an auction with self-similar bid arrivals would increase at a much faster rate than that expected under a Poisson model. It would be especially meaningful toward the end of the auction, which has a large impact on the bid amount process and the final price. The self-similar property suggests that the rate of incoming bids increases steadily as the auction approaches its end. Indeed, empirical investigations have found that many bidders wait until the very last possible moment to submit their final bid. By doing so, they hope to increase their chance of winning the auction since the probability that another competitor successfully places an even higher bid before closing is diminishing. This common bidding strategy of "bid sniping" (or "last-minute bidding") would suggest a steadily increasing flow of bid arrivals toward the auction end. However, empirical evidence from online auction data indicates that bid times over the last minute or so of hard-close auctions tend to follow a uniform distribution (Roth et al., 1998). This has not been found in soft-close, or "going-going-gone" auctions, such as those on uBid.com (or the auctions previously held on Amazon.com and Yahoo!), where the auction continues several minutes after the last bid is placed.

Thus, in addition to the evidence for self-similarity in online auctions, there is also evidence of its breakdown during the very last moments of a hard-close auction. Roth and Ockenfels (2002) note that the empirical CDF plots for intervals that range between the last 12 h of the auction and the last 1 min all look very similar except for the last 1 min plot. Being able to model this breakdown is essential, since the last moments of the auction (when sniping takes place) are known to be crucial to the final price. In the absence of such a model, we introduce a bid arrival process that describes the frequency throughout the *entire* auction. Rather than focusing on the last several hours and excluding the last moments, our model accommodates the changes in bidding intensity from the very start to the very end of a hard-close auction.

To illustrate the self-similarity in the bid arrival process in online auctions, we collected data on 189 seven-day auctions (with a total of 3651 bid times) on eBay for new Palm M515 personal digital assistants (PDA). Figure 4.44 displays the empirical CDFs for the 3651 bid arrivals for the purposes of examining the self-similarity property. The CDF is plotted at several resolutions, "zooming in" from the entire auction duration (of 7 days) to the last day, last 12 h, 6 h, 3 h, 5 min, 2 min, and the very last minute. We see that (1) the last day curve (thick black) is different from the other curves in that it starts out concave, (2) the last day through last 3 h curves (the four lowermost curves) are all very similar to each other, and (3) the last minutes curves (gray) gradually approach the 1 min curve, which is nearly uniform. These visual similarities are confirmed by the results of two-sample Kolmogorov–Smirnoff

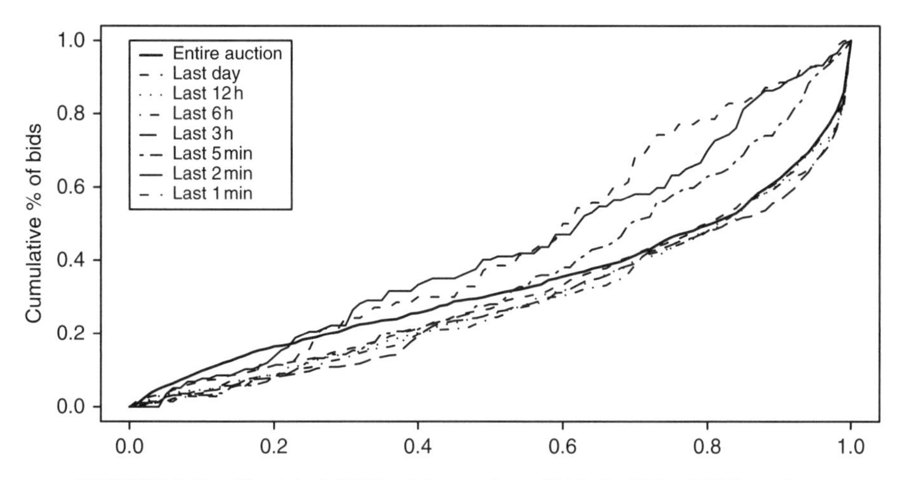

FIGURE 4.44 Empirical CDFs of the number of bids for Palm M515 auctions.

tests where we compared all the pairs of distributions and found similarities only among the curves within each of the three groups.

This replicates the results in Roth and Ockenfels (2002), where self-similarity was observed in the bid time distributions of the last 12 h, 6 h, 3 h, 1 h, 30 min, and 10 min periods of the auction and where this self-similarity breaks down in the last minute of the auction to become a uniform distribution. However, we examine a few additional time resolutions that give further insight: First, by looking at the last 5 min and last 2 min bid distributions, we see that the self-similarity gradually transitions into the 1 min uniform distribution. Second, our inspection of the entire auction duration (which was unavailable in the study by Roth and Ockenfels (2002)) reveals an additional early bidding stage. Self-similarity, it appears, is not prevalent throughout the entire auction duration! Such a phenomenon can occur if the probability of a bid not getting registered on the auction site is positive at the last moments of the auction and increases as the auction comes to a close. There are various factors that may cause a bid to not get registered. One possible reason is the time it takes to manually place a bid (Roth and Ockenfels (2002) found that most last-minute bidders tend to place their bids manually rather than through available sniping software agents). Other reasons are hardware difficulties, Internet congestion, unexpected latency, and server problems on eBay. Clearly, the closer to the end the auction gets, the higher the likelihood that a bid will not get registered successfully. This increasing likelihood of an unsuccessful bid counteracts the increasing flow of last minute bids. The result is a uniform bid arrival process that "contaminates" the self-similarity of the arrivals until that point.

In the next sections, we describe a flexible nonhomogeneous Poisson process (NHPP) that captures the empirical phenomenon described above, introduced by Shmueli et al. (2004). In addition to the self-similarity, it also accounts for the two observed phenomena of "early bidding" and "last-minute bidding" (sniping). We start by showing how a purely self-similar process can be represented as a $NHPP_1$ using a special intensity function. Next, we introduce an improved model ($NHPP_2$) that has a dual intensity function. This model is aimed at capturing the last-minute breakdown of self-similarity and naturally incorporates the phenomenon of "sniping" or last minute bidding. The beginning of the process is no longer a pure self-similar process, but rather a contaminated one. Finally, we present the most flexible model ($NHPP_3$) that accounts for two frequently observed phenomena in the online auction literature: "early bidding" and "last-minute bidding" (sniping). All three models are combined in the so-called BARISTA model of Shmueli et al. (2007).

$NHPP_1$: A Nonhomogeneous Poisson Process That Leads to a Self-Similar Process A nonhomogeneous Poisson process differs from an ordinary Poisson process in that its intensity is not a constant but rather a function of time. We suggest a particular intensity function that leads to a self-similar process: Suppose bids arrive during $[0, T]$ in accordance with a nonhomogeneous Poisson process $N(t)$, $0 \leq t \leq T$, with intensity function

$$\lambda(s) = c\left(1 - \frac{s}{T}\right)^{\alpha - 1}, \text{ some } 0 < \alpha < 1 \text{ and } c > 0, \tag{4.60}$$

so $\lambda(s) \to \infty$ as $s \to T$. That is, the bidding becomes increasingly intense as the auction deadline approaches. The r.v. $N(t)$ has a $Poisson(m(t))$ distribution, where

$$m(t) = \int_0^t \lambda(s)ds = \frac{Tc}{\alpha}\left[1 - \left(1 - \frac{t}{T}\right)^{\alpha}\right]. \tag{4.61}$$

Given that $N(T) = n$, the joint distribution of the arrival times $X_1, ..., X_n$ is equivalent to that of the order statistics associated with a random sample of size n from a distribution whose pdf is shaped like λ on the interval $[0, T]$, namely,

$$f(s) = \frac{\lambda(s)}{m(T)} = \frac{\alpha}{T}\left(1 - \frac{s}{T}\right)^{\alpha - 1} \quad 0 < s < T. \tag{4.62}$$

$f(s)$ is a proper density for any $\alpha > 0$. Some special cases include uniform arrivals ($\alpha = 1$) and the triangular density ($\alpha = 2$).

The CDF and the survival function corresponding to (4.62) are

$$F(s) = \frac{m(s)}{m(T)} = 1 - \left(1 - \frac{s}{T}\right)^{\alpha} \quad 0 \leq s \leq T \tag{4.63}$$

and

$$R(s) = \left(1 - \frac{s}{T}\right)^{\alpha} \quad 0 \leq s \leq T \tag{4.64}$$

with the hazard function

$$h(s) = \frac{f(s)}{R(s)} = \frac{\alpha}{T - s} \quad \text{for } 0 \leq s < T. \tag{4.65}$$

Observe that for $0 < t \leq T$ and $0 \leq \theta \leq 1$, we have

$$\frac{R(T - \theta t)}{R(T - t)} = \theta^{\alpha}, \tag{4.66}$$

which is not dependent on t. This is equivalent to the formulation in Roth and Ockenfels (2002), who model the number of bids in the *last* minutes and obtain the CDF

$$F^{*}(t) = \left(\frac{t}{T}\right)^{\alpha}. \tag{4.67}$$

The difference between the CDF in (4.63) and that in (4.67) results from time reversal. The former looks at the number of bids from the *beginning* of the auction until time s, while the latter looks at the number of bids in the *last* s time units of the auction. By setting $t = T - s$ in (4.67) and subtracting from 1, we obtain the same expression for the CDF as in (4.63). Let F_e denote the empirical CDF corresponding to the $N(t)$ bid arrival times on $[0, T]$ and let $R_e = 1 - F_e$ denote the associated survival function.

Intensity Function Structure and Self-Similarity One of the main features of a self-similar process is that the distribution (the CDF, autocorrelation function, etc.) remains the same at any level of aggregation. Using the intensity function in (4.60), we show that this is also a property of NHPP_1: If we have m independent processes $N_j(t)$, $1 \leq j \leq m$, with intensity functions

$$\lambda_j(s) = c_j \left(1 - \frac{s}{T}\right)^{\alpha - 1}, \quad 1 \leq j \leq m \tag{4.68}$$

(i.e., different c_j's, but the same α), then all have the right form to possess the self-similar property (4.66). The aggregated process $N(t) = \sum_{j=1}^{m} N_j(t)$ is a NHPP with intensity function

$$\lambda(s) = \sum_{j=1}^{m} \lambda_j(s) = c\left(1 - \frac{s}{T}\right)^{\alpha - 1}, \quad c = \sum_{j=1}^{m} c_j \tag{4.69}$$

and also has property as (4.66). Conversely, if we start with a NHPP $N(t)$ having intensity function $\lambda(s)$ given by (4.69) and randomly categorize each arrival into one of m "types" with respective probabilities $c_1/c, \ldots, c_m/c$, then the resulting processes $N_1(t), \ldots, N_m(t)$ are independent NHPPs with intensity functions as in (4.68). If we aggregate two processes with $\alpha_1 \neq \alpha_2$, the resulting process is a NHPP, but with an intensity function of the wrong form (even when $c_1 = c_2$) to satisfy (4.66).

Another self-similarity property that emerges from the intensity function is that when "zooming in" into smaller and smaller intervals, the intensity function has the same form, but on a different timescale: Fix a time point $\beta T \in [0, T]$. The process $N_\beta(s) := N(s)$ on the shortened interval $[\beta T, T]$ is a NHPP with intensity function $\lambda_\beta(s) = \lambda(s)$ for $x \leq s \leq T$. This can be written as

$$\begin{aligned}\lambda_\beta(s) &= c(1 - \frac{s}{T})^{\alpha-1} \\ &= [c(1-\beta)^{\alpha-1}]\left(1 - \frac{s - \beta T}{T(1-\beta)}\right)^{\alpha-1} \\ &= \lambda(\beta)\left(1 - \frac{s - \beta T}{T(1-\beta)}\right)^{\alpha-1},\end{aligned}$$

where $\lambda(\beta) \to \infty$ as $\beta \to 1$. Note that originally

$$\lambda_0(s) = \lambda(0)\left(1 - \frac{s}{T}\right)^{\alpha-1}.$$

If we think of the fixed time point βT as the new *zero* and we change time units so that $T(1-\beta)$ minutes on the old scale $= T$ shminutes[14] on the new scale, then the N_β process is defined on the interval $[0, T]$ and has intensity function

$$\lambda_\beta(s) = \lambda(\beta)\left(1 - \frac{s}{T}\right)^{\alpha-1}, \quad 0 \leq t \leq T, \text{ time now measured in shminutes.}$$

Thus, at any given point βT in $[0, T]$, the process of remaining bids is similar to the original process on $[0, T]$, but is on a reset and faster ($1/(1-\beta)$ faster) clock and is more intense by a factor of $\lambda(\beta)/\lambda(0)$.

Probabilities and Ratios of Arrival Let X be a random variable with CDF as in (4.63). The mean arrival time over the entire period $[0, T]$ is

$$E(X) = \int_0^T R(s)ds = T/(1+\alpha) \tag{4.70}$$

and the variance is

$$\text{Var}(X) = \frac{\alpha T^2}{(\alpha+2)(\alpha+1)^2}. \tag{4.71}$$

[14] We use the term "shminute" to signify a new unit of time.

The probability that a bid will be placed within a time interval of length t depends on the location of the interval through the following function: For $0 < x < x + t < T$,

$$\begin{aligned} p(x,t) :&= P(\text{receive a bid during } [x, x+t]) \\ &= 1 - \exp[m(x) - m(x+t)] \\ &= 1 - \exp\left\{\frac{cT}{\alpha}\left(\left(1 - \frac{x+t}{T}\right)^{\alpha} - \left(1 - \frac{x}{T}\right)^{\alpha}\right)\right\}. \end{aligned}$$

In the context of *sniping*, or last-minute bidding, a useful probability is that of the event where there are no bids after time x. This is given by

$$1 - p(x, T - x) = \exp\left\{-\frac{cT}{\alpha}\left(1 - \frac{x}{T}\right)^{\alpha}\right\} \to 1 \quad \text{as } x \to T. \tag{4.72}$$

Another quantity of interest is the conditional distribution of the next bid after time x, given that there *is* a next bid:

$$\begin{aligned} p^*(x,t) :&= P(\text{bid arrives during } [x, x+t] \mid \text{there is a bid after time } x) \\ &= \frac{p(x,t)}{p(x, T-x)} \\ &= \frac{\exp[\frac{cT}{\alpha}(1 - \frac{x}{T})^{\alpha}] - \exp[\frac{cT}{\alpha}\left((1 - \frac{x+t}{T})^{\alpha}\right)]}{\exp[-\frac{cT}{\alpha}(1 - \frac{x}{T})^{\alpha}] - 1}. \end{aligned} \tag{4.73}$$

Finally, a variable of special interest (Roth and Ockenfels, 2002) is the ratio of the number of arrivals within a fraction θ of the last t moments of the auction and the number of arrivals within the last t moments. For $0 < t \le T$ and $0 \le \theta \le 1$, define

$$\pi(t,\theta) := \frac{N(T) - N(T - \theta t)}{N(T) - N(T - t)} = \frac{R_e(T - \theta t)}{R_e(T - t)}. \tag{4.74}$$

We set $\pi(t,\theta) = 0$ if $R_e(T - t) = 0$. It can be shown (Shmueli et al., 2004) that $\pi(t,\theta)$ converges uniformly in probability as $c \to \infty$. This means that if the bidding is reasonably intense over the interval $[0, T]$, then there is a high probability—the higher the value of c, the greater the probability—that all the π functions ($\pi(t,\theta)$, $0 < t \le T$) will be close to θ^{α}, for all t that are not too close to 0. Thus, the model does not guarantee convergence for small t. This accommodates, or at least does not contradict, the empirical evidence that self-similarity of bid arrival processes breaks down at the very last minute (Roth and Ockenfels, 2002). In their graphs of the empirical CDF, the empirical $\pi(t,\theta)$ functions are displayed as functions of θ, for various choices of t. If t is not too small, these graphs all seem similar to $g(\theta) = \theta^{\alpha}$.

Regarding the behavior of $\pi(t, \theta)$ for fixed $t \in (0, T)$ and $\theta \in (0, 1)$, it can be shown that as $c \to \infty$

$$P\left(|\pi(t, \theta) - \theta^{\alpha}| > \frac{x}{\sqrt{c}}\right) \to P\left(|Z| > x\sqrt{\frac{t^{\alpha}T^{1-\alpha}}{\alpha\theta^{\alpha}(1-\theta^{\alpha})}}\right), \tag{4.75}$$

where Z is the unit normal variable (see Shmueli et al. (2007) for detailed proof).

Model Estimation Under the NHPP_1 model, conditional on the event $N(T) = n$, the bid arrival times $X_1, \ldots, X_n$ are distributed as the order statistics of a random sample of size n from the distribution in (4.63). These variables, when randomly ordered, are equivalent to a random sample of size n from that distribution.

Given a set of n arrival times on the interval $[0, T]$, we want to assess whether the NHPP_1 model adequately describes the data, and if so, to estimate the parameter α that influences the intensity of the arrivals and controls the shape of the distribution of bid times. Due to the special form of the CDF in (4.63), which can be approximated by the empirical CDF $F_e(t)$, a log–log graph of $1 - F_e(t)$ versus $(1 - t/T)$ should reveal the adequacy of the fit. If the model is reasonable, we expect the points to fall on a line that has a slope of α. Then α can be estimated as the slope of the least squares line fitted to the n points. This type of plot is widely used for assessing self-similarity in general, the only difference being that we have a finite interval $[0, T]$ whereas usually the arrival interval is infinite $[0, \infty]$.

Here, we present two alternative estimators of α: one based on the moment method and the other on maximum likelihood.

Moment Estimator: Using (4.70), we can derive the moment estimator (conditional on $N(t) = n$):

$$\hat{\alpha} = \frac{T}{\bar{\bar{X}}} - 1. \tag{4.76}$$

This is quick and easy to compute. Although the estimator is biased for estimating α, it is consistent since it is a continuous function of $\bar{X}$. Thus, for a very large sample, we expect to get an estimate close to α.

Maximum Likelihood Estimator: Using the density in (4.62), the log-likelihood function of α, given n observations from NHPP_1, is

$$L(\alpha | x_1, \ldots, x_n) = n \log \alpha / T + (\alpha - 1) \sum_{i=1}^{n} \log\left(1 - \frac{x_i}{T}\right). \tag{4.77}$$

This means that this distribution is a member of the exponential family. An important and useful implication is the existence of a sufficient statistic, namely, $\sum_{i=1}^{n} \log\left(1 - \frac{X_i}{T}\right)$. This enables a great reduction in storage for the purpose of estimation. All that is required is the sufficient statistic and not the n arrival times.

The maximum likelihood estimate for α is then given by

$$\hat{\alpha} = -n \left[\sum_{i=1}^{n} \log \left(1 - \frac{X_i}{T} \right) \right]^{-1}. \tag{4.78}$$

This is similar to the Hill estimator that is used for estimating a heavy tailed distribution. The two differences between this estimator and the Hill estimator are that in our case the arrival interval is finite ($[0, T]$), whereas in the general heavy tailed case the arrival interval is infinite ($[1, \infty]$). Second, the Hill estimator is used when the assumed process is close to a Pareto distribution only for the upper part of the distribution. Therefore, the average in the Hill estimator is taken over only the k latest arrivals, while in our case we assume that the entire bid arrival process comes from the same distribution and thus all arrival times are included in the estimator.

Note that the above estimator is conditional on $N(T) = n$. The unconditional likelihood is given by

$$L(c, \alpha) = P(N(T) = n) f(x_1, \ldots, x_n) = \frac{(\exp(-Tc/\alpha))c^n}{n!} \prod_{k=1}^{n} \left(1 - \frac{x_k}{T} \right)^{\alpha - 1}, \tag{4.79}$$

which leads to the maximum likelihood estimates

$$\hat{\alpha} = -N(t) \left[\sum_{i=1}^{N(t)} \log \left(1 - \frac{X_i}{T} \right) \right]^{-1} \tag{4.80}$$

and

$$\hat{c} = \frac{N(T)}{T} \hat{\alpha}. \tag{4.81}$$

The equivalent form of (4.78) and (4.80) stems from the fact that the parameter c affects the size $N(T)$ of the sample, but not the shape of the conditional distribution of the arrival times (see Shmueli et al. (2007) for a general version of this result). Asymptotics for $\widehat{\alpha}$ as defined in (4.78) as $n \to \infty$, and (4.80) as $c \to \infty$, are also given in Shmueli et al. (2007).

Simulation of NHPP_1 Although simulating self-similar processes on an infinite interval appears to be difficult from a practical point of view, simulation on a finite interval is simple and efficient. We show a simple algorithm for simulating arrivals from NHPP_1 that create a self-similar process. Since the CDF is easily invertible

$$F^{-1}(s) = T - T(1 - s)^{1/\alpha}, \tag{4.82}$$

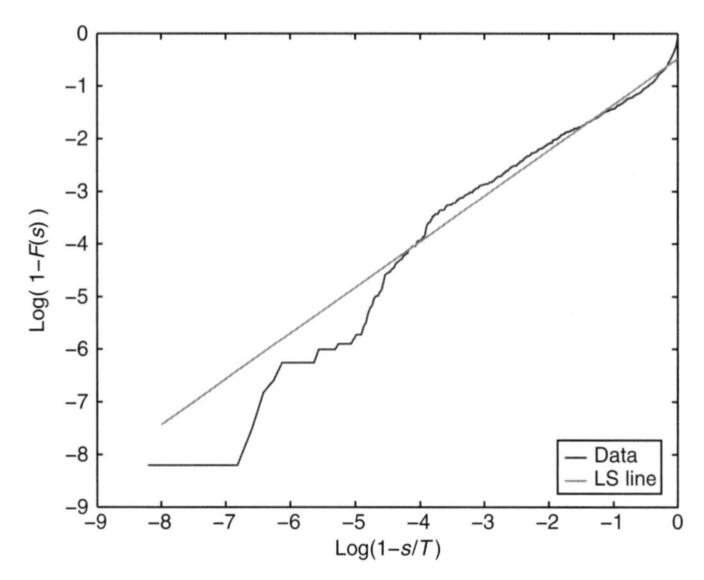

FIGURE 4.45 Log–log plot of $1 - F_e(s)$ versus $1 - s/T$ for Palm bid times.

we can use it to simulate a specified number of NHPP$_1$ arrivals using the inversion method. To generate n arrivals, generate n (0, 1) uniform variates $u_k, k = 1, \ldots n$, and transform each by plugging it into (4.82) in place of s:

$$x_k = T - T(1 - u_k)^{1/\alpha}. \tag{4.83}$$

Empirical and Simulated Results For this study, we collected bidding data for 7-day auctions of Palm M515 personal digital assistant units from mid-March to June 2003 on eBay.com, resulting in 3561 bid times. Figure 4.45 displays $\log(1 - F_e(t))$ versus $\log(1 - t/T)$ for these 3561 Palm M515 bid times. Had the data originated from a pure NHPP$_1$-type self-similar process, we would expect a straight line. From the figure, it appears that the data follow a line for values of $\log(1 - t/T)$ in the approximate range $(-3, -0.25)$, which corresponds to the period of the auction between days 1.55 and 6.66. This includes about 42% of the arrivals. Figure 4.46 displays the same type of log–log plot, but it uses only bid arrivals inside the interval [1.55, 6.66]. It can be seen that the bid arrivals in this interval are more consistent with the NHPP$_1$ model than are the bids from the entire 7-day auctions. The straight line is the least squares line fitted to the data. If we focus on bids that arrived only during this period (and scale them to a [0, 5.11] interval), we obtain the following estimates for α: $\hat{\alpha}_{\text{LS}} = 0.91$, $\hat{\alpha}_{\text{moments}} = 1.03$, and $\hat{\alpha}_{\text{ML}} = 0.95$.

NHPP$_2$: Accounting for Last-Moment Breakdown of Self-Similarity The NHPP$_1$ model suggests that the rate of the incoming bids increases steadily as the auction approaches its end. Indeed, empirical investigations have found that many

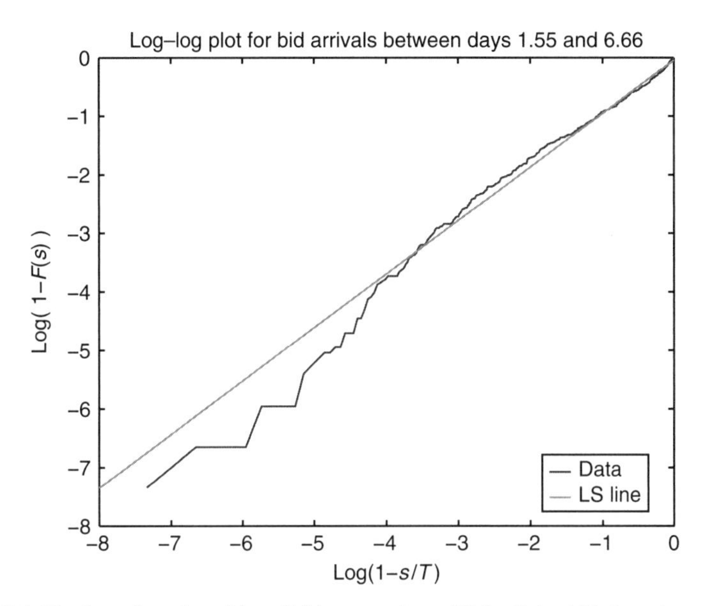

FIGURE 4.46 Log–log plot of $1 - F_e(s)$ versus $1 - s/T$ for Palm bid times between days 1.55 and 6.66.

bidders wait until the very last possible moment to submit their final bid. By doing so, they hope to increase their chance of winning the auction since the probability that another competitor successfully places an even higher bid before closing is diminishing. This common bidding strategy, often referred to as "last-minute bidding" or "bid sniping," would suggest a steadily increasing flow of bid arrivals toward the auction end. However, we pointed out earlier that empirical evidence from online auction data indicates that bid times over the last minute or so of closed-ended auctions tend to follow a uniform distribution. Such a phenomenon can occur if the probability of a bid not getting registered on the auction site is positive at the last moments of the auction and increases as the auction comes to a close. This increasing likelihood of an unsuccessful bid counteracts the increasing flow of last-minute bids. The result is a uniform bid arrival process that "contaminates" the self-similarity of the arrivals until that point. In addition, there is no clear-cut line between the self-similar process at the beginning and the uniform process at the end. Rather, self-similarity transitions gradually into a uniform process—we will now describe a model for such a phenomenon.

As before, we assume a nonhomogeneous Poisson process, except that now the parameter α of the intensity function turns from α_1 to α_2 at the last d moments of the auction:

$$\lambda_2(s) = \begin{cases} c\left(1 - \dfrac{s}{T}\right)^{\alpha_1 - 1}, & \text{for } 0 \le s \le T - d, \\ c\left(\dfrac{d}{T}\right)^{\alpha_1 - \alpha_2}\left(1 - \dfrac{s}{T}\right)^{\alpha_2 - 1}, & \text{for } T - d \le s \le T. \end{cases} \tag{4.84}$$

Note that this intensity function is continuous, so there is no jump at time $T - d$. Also, note that in this formulation the self-similar process transitions gradually into a uniform process if $\alpha_2 = 1$. In such a case, the intensity function flattens out during the final d moments, conveying a uniform arrival process during $[T - d, T]$.

In general, the beginning of the process is a contaminated self-similar process, and the closer it gets to the transition point (time $T - d$), the more contaminated it becomes. The random variable $N(t)$ that counts the number of arrivals until time t follows a Poisson distribution with mean value function

$$m_2(s) = \begin{cases} \dfrac{Tc}{\alpha_1}\left(1 - (1 - \frac{s}{T})^{\alpha_1}\right), \text{ for } 0 \leq s \leq T - d, \\ \dfrac{Tc}{\alpha_1}\left\{\left[1 - \left(\dfrac{d}{T}\right)^{\alpha_1}\right] + \dfrac{\alpha_1}{\alpha_2}\left(\dfrac{d}{T}\right)^{\alpha_1}\left[1 - \left(\frac{d}{T}\right)^{-\alpha_2}\left(1 - \frac{s}{T}\right)^{\alpha_2}\right]\right\}, \\ \qquad \text{for } T - d \leq s \leq T. \end{cases} \tag{4.85}$$

Note that

$$m_2(T) = \frac{Tc}{\alpha_1}\left[1 - (1 - \frac{\alpha_1}{\alpha_2})\left(\frac{d}{T}\right)^{\alpha_1}\right]. \tag{4.86}$$

The CDF of this process is then given by

$$F_2(t) = \frac{m_2(t)}{m_2(T)} = \begin{cases} \dfrac{1 - (1 - \frac{t}{T})^{\alpha_1}}{1 - \left(1 - \frac{\alpha_1}{\alpha_2}\right)\left(\frac{d}{T}\right)^{\alpha_1}}, & \text{for } 0 \leq t \leq T - d, \\ 1 - \dfrac{\frac{\alpha_1}{\alpha_2}\left(\frac{T-t}{d}\right)^{\alpha_2}\left(\frac{d}{T}\right)^{\alpha_1}}{1 - \left(1 - \frac{\alpha_1}{\alpha_2}\right)\left(\frac{d}{T}\right)^{\alpha_1}}, & \text{for } T - d \leq t \leq T, \end{cases} \tag{4.87}$$

and the density by

$$f_2(t) = \begin{cases} \dfrac{\frac{\alpha_1}{T}\left(1 - \frac{t}{T}\right)^{\alpha_1 - 1}}{1 - \left(1 - \frac{\alpha_1}{\alpha_2}\right)\left(\frac{d}{T}\right)^{\alpha_1}}, & \text{for } 0 \leq t \leq T - d, \\ \\ \dfrac{\frac{\alpha_1}{T}\left(\frac{d}{T}\right)^{\alpha_1 - \alpha_2}\left(1 - \frac{t}{T}\right)^{\alpha_2 - 1}}{1 - \left(1 - \frac{\alpha_1}{\alpha_2}\right)\left(\frac{d}{T}\right)^{\alpha_1}}, & \text{for } T - d \leq t \leq T. \end{cases} \tag{4.88}$$

This is a proper density for any $\alpha_1 \neq 0, \alpha_2 > 0$, and $0 < d < T$. When $\alpha_1 = 0$, the form of the density is different from (4.88). This process describes a contaminated self-similar process. Of course, if $\alpha_1 = \alpha_2$, then NHPP$_2$ reduces to NHPP$_1$, which is a pure self-similar process. The contamination of the self-similarity means that as we get closer to $[T-d, T]$, the arrival process becomes less and less self-similar in a way that gradually changes into uniform arrivals in $[T-d, T]$. In Figure 4.44, it can be seen that the distribution during the last 2 min is somewhere between the uniform last-minute distribution and the earlier (almost) self-similar distribution. This transition can also be seen through the function $\pi(\theta, t)$, which changes form over three regions in the θ, t plane:

$$\pi_2(t,\theta) = \frac{1-F_2(T-t\theta)}{1-F_2(T-t)} = \begin{cases} \dfrac{(t\theta)^\alpha + (\alpha-1)d^\alpha}{t^\alpha + (\alpha-1)d^\alpha}, & \text{for } \ t \geq \frac{d}{\theta}, \\ \dfrac{\alpha t\theta d^{\alpha-1}}{t^\alpha + (\alpha-1)d^\alpha}, & \text{for } \ d < t < \frac{d}{\theta}, \\ \theta, & \text{for } \ t \leq d. \end{cases} \tag{4.89}$$

Simulation of NHPP$_2$ To simulate the two-stage NHPP on the interval $[0, T]$, we can use the inversion method. The inverse CDF can be written in the following form:

$$F_2^{-1}(s) = \begin{cases} T - T\left\{1 - s\left[1-(1-\dfrac{\alpha_1}{\alpha_2})\left(\dfrac{d}{T}\right)^{\alpha_1}\right]\right\}^{1/\alpha_1}, & \text{for } \ 0 \leq s \leq T-d, \\ T - d\left\{\dfrac{1-s}{1-F_2(T-d)}\right\}^{1/\alpha_2}, & \text{for } \ T-d \leq s \leq T. \end{cases} \tag{4.90}$$

The algorithm for generating n arrivals is then the following:

1. Generate n uniform variates $u_1, \ldots, u_n$.
2. For $k = 1, \ldots, n$ set

$$x_k = \begin{cases} T - T\left\{1 - u_k\left[1-\left(\dfrac{d}{T}\right)^{\alpha_1}\right]/F_2(T-d)\right\}^{1/\alpha_1}, & \text{if } \ u_k < F_2(T-d), \\ T - d\left\{(1-u_k)/\left[1-F_2(T-d)\right]\right\}^{1/\alpha_2}, & \text{if } \ u_k > F_2(T-d). \end{cases} \tag{4.91}$$

Note that

1. when $u_k = F_2(T-d)$, we get $x_k = T-d$, and
2. when $\alpha_2 = 1$, F_2 is linear on the interval $[T-d, T]$.

Quick and Crude (CDF-Based) Parameter Estimation For estimating α_2, we can use the exact relation

$$\alpha_2 = \frac{\log\ R(t_2)/R(t_2')}{\log(T - t_2)/(T - t_2')}, \tag{4.92}$$

where $R(t) = 1 - F_2(t)$ and t_2, and t_2' are within $[T - d, T]$. To estimate α_2, we pick reasonable values of t_2, and t_2' and use the empirical CDF. For the special case where the end of the auction is characterized by uniform arrivals, we have $R_e(t) \approx$ const$(T - t)$ for $t \approx T$. Thus,

$$\frac{\log(R_e(t_2)/R_e(t_2'))}{\log((T - t_2)/(T - t_2'))} \approx 1. \tag{4.93}$$

For estimating α_1, we cannot write an equation such as (4.92). However, it is possible to use an approximation that can be easily computed from the data. The approximation works for both intervals (i.e., for estimating α_1 and α_2), but we use it only for the first interval $[0, T - d]$. The calculations for the other period $[T - d, T]$ are the same. The idea is to choose an interval $[T - s, T - t]$ that we are confident lies within the period of interest. For example, in the Palm bid arrivals, we are confident that the first 5 days of the auction occur within the first period. Thus, we choose an interval (or try several) that is contained in $[0, 5]$.

The mean value function for each of the two intervals $0 \le y \le T - d$ and $T - d \le y \le T$ is of the form

$$m_2(y) = \beta_j - \theta_j \left(1 - \frac{y}{T}\right)^{\alpha_j} \tag{4.94}$$

for $j = 1, 2$. For the first interval, fix $0 \le T - s < T - t \le T - d$. Writing α for α_1, we have

$$\begin{aligned}
\frac{N(T-t) - N(T-\sqrt{st})}{N(T-\sqrt{st}) - N(T-s)} &\approx \frac{E\left[N(T-t) - N(T-\sqrt{st})\right]}{E\left[N(T-\sqrt{st}) - N(T-s)\right]} \\
&= \frac{m(T-t) - m(T-\sqrt{st})}{m(T-\sqrt{st}) - m(T-s)} \\
&= \frac{\theta_1\left[1 - (1 - \frac{T-t}{T})^\alpha\right] - \theta_1\left[1 - (1 - \frac{T-\sqrt{st}}{T})^\alpha\right]}{\theta_1\left[1 - (1 - \frac{T-\sqrt{st}}{T})^\alpha\right] - \theta_1\left[1 - (1 - \frac{T-s}{T})^\alpha\right]} \\
&= \frac{(ts)^{\alpha/2} - t^\alpha}{s^\alpha - (st)^{\alpha/2}}
\end{aligned}$$

$$= \frac{(s^{\alpha/2} - t^{\alpha/2})t^{\alpha/2}}{(s^{\alpha/2} - t^{\alpha/2})s^{\alpha/2}}$$

$$= \left(\frac{t}{s}\right)^{\alpha/2}. \tag{4.95}$$

Taking logs in (4.95), we get

$$\alpha \approx 2\frac{\log\left[N(T-t) - N(T-\sqrt{st})\right] - \log\left[N(T-\sqrt{st}) - N(T-s)\right]}{\log t - \log s}. \tag{4.96}$$

This can be written in terms of $F_e(s)$, the empirical CDF:

$$\alpha \approx 2\frac{\log[F_e(T-t) - F_e(T-\sqrt{st})] - \log[F_e(T-\sqrt{st}) - F_e(T-s)]}{\log t - \log s}. \tag{4.97}$$

The same approximation works on the interval $[T-d, T]$, but it is preferable to use the exact relation given in (4.92).

To learn more about the quick and crude estimates, we generated 5000 random arrival times from NHPP_2 on the interval $[0, 7]$, with $\alpha_1 = 0.4$, $\alpha_2 = 1$, and $d = 5/10,080$ (5 min). The top panel of Figure 4.47 shows that $\hat{\alpha_1}$ is close to 0.4 if a reasonable interval is selected. For an interval that excludes the last 20 min, the estimate is 0.4. Using (4.92), we estimated α_2. The bottom panel of Figure 4.47 shows $\hat{\alpha}_2$ as a function of the number of minutes before the auction close. The estimate seems to overestimate the generating $\alpha_2 = 1$ value. Note that values of $T - t > 5$ are beyond the "legal" interval of the last 5 min. To refine the estimate, a value such as $\alpha_2 = 1.2$ can be used as an initial value in a maximum likelihood procedure.

For estimating d, the relation (4.87) between F_2 and d for $0 < t < T - d$ can be written in the form

$$d = T\left\{\frac{1 - \frac{1-(1-\frac{t}{T})^{\alpha_1}}{F_2(t)}}{1 - \frac{\alpha_1}{\alpha_2}}\right\}^{1/\alpha_1}. \tag{4.98}$$

Thus, using a "safe" initial value of t that we are confident is contained in the interval $[0, T-d]$, we can use (4.98), with F_e in place of F_2 to estimate d. Figure 4.48 illustrates the values of $\hat{d}$ as a function of t. It can be seen that the estimate moves between 2 and 10 min, depending on the choice of t.

Maximum Likelihood Parameter Estimation Based on the density in (4.88), we obtain the likelihood function, conditional on $N(t) = n$ (see Shmueli et al. (2007) for

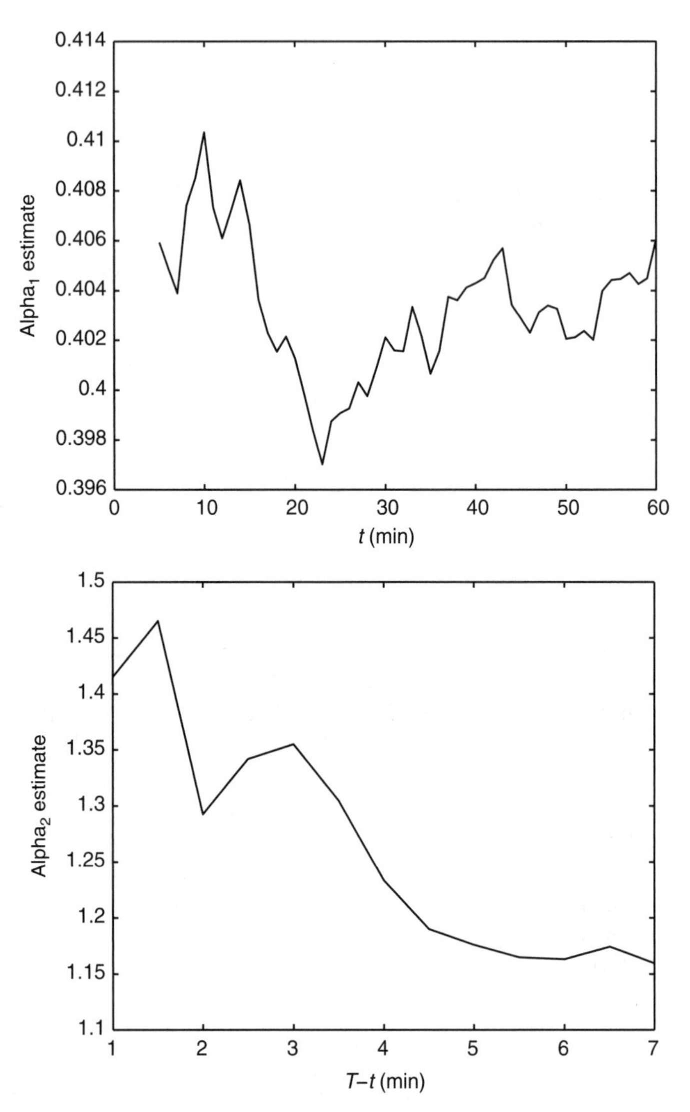

FIGURE 4.47 Quick and crude estimates of α_1 as a function of t (with $s = 6.99$) (top), and of α_2 as a function of t (bottom), for the NHPP_2 simulated data with $\alpha_1 = 0.4$, and $\alpha_2 = 1$.

a note on unconditional estimation):

$$\begin{aligned} L(x_1, \ldots, x_n | \alpha_1, \alpha_2) & \\ &= n \log \alpha_1 - n \log T \\ &\quad + (\alpha_1 - 1) \sum_{i: x_i \leq T-d} \log \left(1 - \frac{x_i}{T}\right) \end{aligned} \tag{4.99}$$

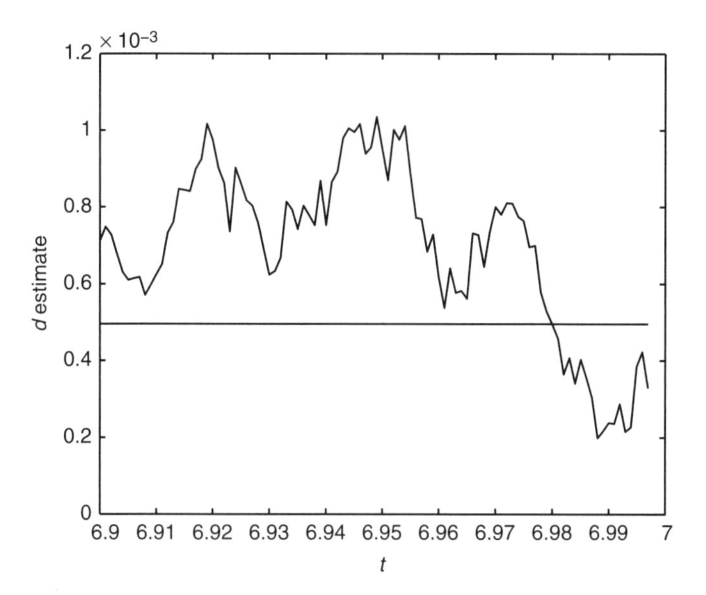

FIGURE 4.48 Quick and crude estimate of d as a function of t.

$$+(\alpha_2 - 1) \sum_{i:x_i > T-d} \log\left(1 - \frac{x_i}{T}\right)$$

$$+n_2(\alpha_1 - \alpha_2)\log\frac{d}{T} - n\,\log\left(1 - (1 - \frac{\alpha_1}{\alpha_2})\left(\frac{d}{T}\right)^{\alpha_1}\right).$$

The two equations that need to be solved to obtain ML estimates for α_1 and α_2 are

$$\sum_{i:x_i \le T-d} \log\left(1 - \frac{x_i}{T}\right) + n_2 \log\frac{d}{T} = n\frac{1 - (\alpha_2 - \alpha_1)\log\alpha_1}{\alpha_2(1 - (d/T)^{-\alpha_1}) - \alpha_1} - \frac{n}{\alpha_1},$$

$$\sum_{i:x_i > T-d} \log\left(1 - \frac{x_i}{T}\right) - n_2 \log\frac{d}{T} = n\frac{-\alpha_1/\alpha_2}{\alpha_2(1 - (d/T)^{-\alpha_1}) - \alpha_1},$$

where n_2 is the number of arrivals after $T - d$. In the uniform case ($\alpha_2 = 1$) where d is known (e.g., the "last-minute bidding" phenomenon in online auctions), the maximum likelihood estimator of α_1 is the solution of the equation

$$\sum_{i:x_i \le T-d} \log\left(1 - \frac{x_i}{T}\right) + n_2 \log\frac{d}{T} = n\frac{1 - (1 - \alpha)\log\,\alpha}{1 - \alpha - (d/T)^{-\alpha}} - \frac{n}{\alpha}. \qquad (4.100)$$

This can be solved using an iterative gradient-based method such as Newton Raphson or the Broyden–Fletcher–Goldfarb–Powell (BFGP) method, which is a more stable quasi-Newton method that does not require the computation and inversion of the Hessian matrix. A good starting value would be the estimate obtained from the probability plot or the quick and crude method.

If d is unknown and we want to estimate it from the data, then a gradient approach can no longer be used. One option is to combine a gradient approach for α_1 and α_2 with an exhaustive search over a logical interval of d values. The size of the step in d should be relevant to the application at hand, and we expect that a small change should not lead to dramatic changes in $\hat{\alpha}_1$, and $\hat{\alpha}_2$.

Empirical and Simulated Results To check for the presence of the "last-minute bidding" phenomenon in our Palm data, we started by fitting a probability plot. The top panel of Figure 4.49 shows a probability plot for the Palm data with $T = 7$ (days) and $d = 1/10,080$ (the changepoint occurs 1 min before the auction close). The plot contains several curves that correspond to different values of α. For these data, it appears that the two-phase NHPP does not capture the relatively fast beginning! For example, with $\alpha = 0.5$ that yields the closest fit, the arrivals in the data seem to occur faster than expected under the NHPP_2 model during the first 2.5 days, then slow down more than expected until day 6, and finally arrive at the expected rate until the end of the auction. To check that the first 2.5 days are indeed the cause of this misfit, we drew only the last 4.5 days (Figure 4.49, bottom panel). Here, the NHPP_2 model seems more appropriate and α can be estimated as 0.35.

Following these results, we used only bids that were placed after day 2.5 (and shifted the data to the interval [0, 4.5]). We then obtained the quick and crude estimates $\hat{\alpha_1} \approx 0.35$ and $\hat{\alpha_2} \approx 1$.

For estimating d, we used the above estimates of α_1, α_2, and $t = 5/10,080$ (5 min before the auction end), which is most likely contained within the first period $[0, 4.5 - d]$. This yields $\hat{d} = 2.8$ min. Figure 4.50 shows $\hat{d}$ as a function of t. It appears that d is between approximately 1 and 3 min.

We also used a genetic algorithm (GA) to search for values of the parameters that maximize the likelihood. This yielded the estimates $\hat{\alpha}_1 = 0.37$, $\hat{\alpha}_2 = 1.1$, and $\hat{d} = 2.1$. A combination of a gradient method for estimating α_1 and α_2 with an exhaustive search over d within a reasonable interval yielded the same estimates.

Finally, we compare our data to simulated data from an NHPP_2 with $d = 2.1$ min, $\alpha_1 = 0.37$, and $\alpha_2 = 1$. Figure 4.51 is a QQ plot of the sorted Palm bid times (only those later than 2.5 days) and the sorted simulated data. It is clear that the fit is very good.

NHPP_3: Accounting for Early and Last-Moment Bidding While our efforts so far have focused predominately on the auction end and its last-minute bidding activity, the beginning of the auction also features some interesting characteristics: early bidding! In fact, many empirical investigations of the online auctions have reported an unusual amount of bidding activity at the beginning of the auction followed by a longer period of little or no activity. The reasons for early bidding are not at all that clear. Bapna et al. (2003) refer to bidders who place a single early bid as evaluators, but there may be other reasons why people place bids early in the auction. The next model

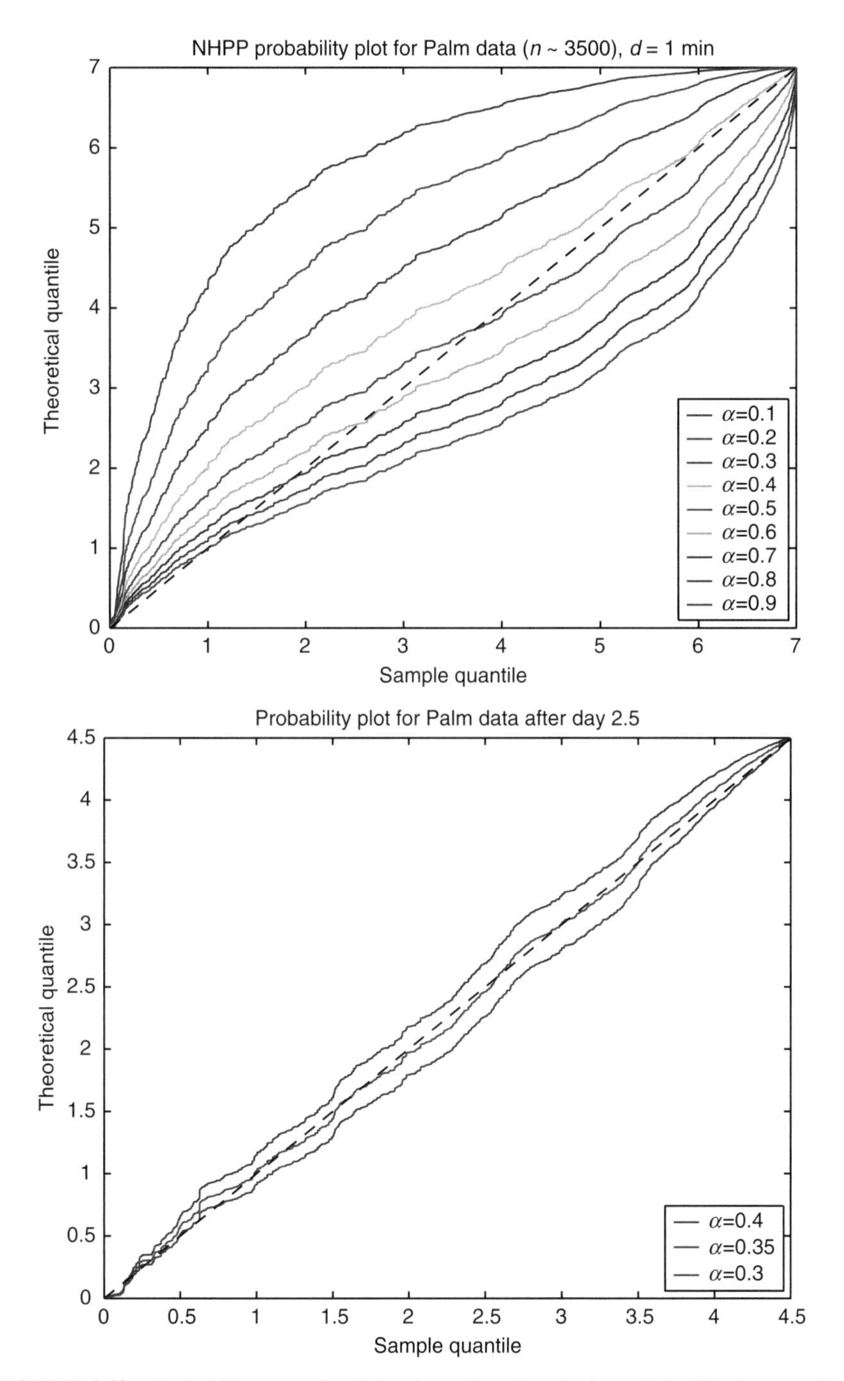

FIGURE 4.49 Probability plots for Palm data. *Top*: $T = 7, d = 1/10,080$. *Bottom*: $T = 4.5, d = 1/10,080$. Legend is ordered according to the line order.

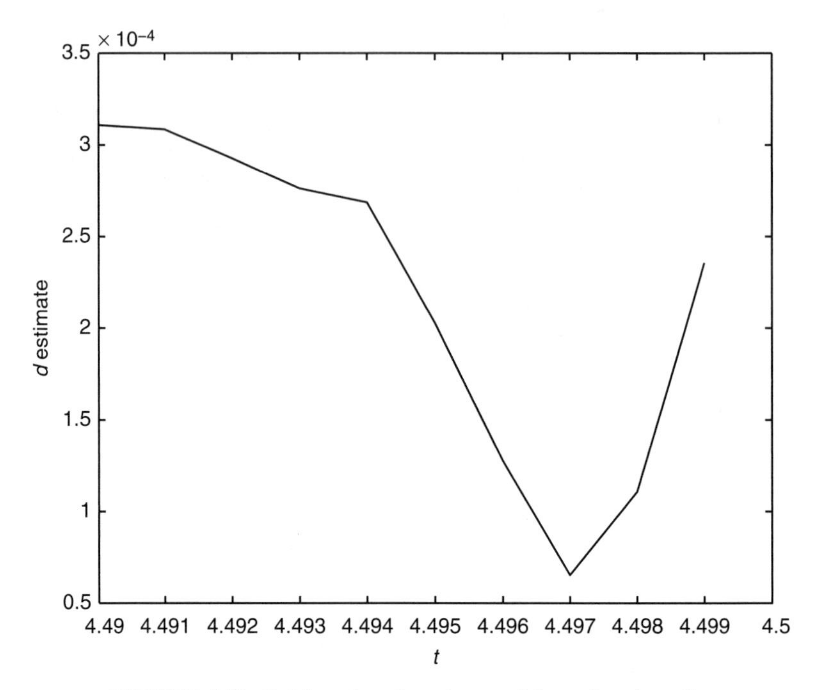

FIGURE 4.50 Quick and crude estimate of d as a function of t.

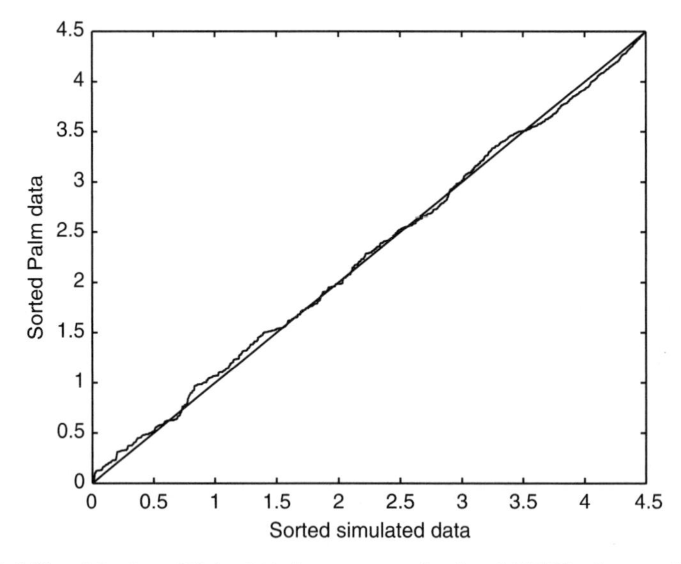

FIGURE 4.51 QQ plot of Palm bid times versus simulated $NHPP_2$ data on the interval $[0, 4.5]$ and parameters $\alpha_1 = 0.37$, $\alpha_2 = 1$, and $d = 2.1$ min.

incorporates early bidding and last-minute bidding as a generalization of the $NHPP_2$ model. To model this early activity phase, we formulate a three-stage NHPP having a continuous intensity function that switches the parameter α from stage to stage:

$$\lambda_3(s) = \begin{cases} c\left(1-\frac{d_1}{T}\right)^{\alpha_2-\alpha_1}\left(1-\frac{s}{T}\right)^{\alpha_1-1}, & \text{for } 0 \le s \le d_1, \\ c\left(1-\frac{s}{T}\right)^{\alpha_2-1}, & \text{for } d_1 \le s \le T-d_2, \\ c\left(\frac{d_2}{T}\right)^{\alpha_2-\alpha_3}\left(1-\frac{s}{T}\right)^{\alpha_3-1}, & \text{for } T-d_2 \le s \le T. \end{cases} \quad (4.101)$$

We expect α_3 to be close to 1 (uniform arrival of bids at the end of the auction) and $\alpha_1 > 1$ to represent the early surge in bidding. The random variable $N(t)$ that counts the number of arrivals until time t follows a Poisson distribution with mean

$$m_3(s) = \begin{cases} K\left(1-\left(1-\frac{s}{T}\right)^{\alpha_1}\right), \text{ for } 0 \le s \le d_1, \\ K\left(1-\left(1-\frac{d_1}{T}\right)^{\alpha_1}\right)+\frac{Tc}{\alpha_2}\left(1-\left(1-\frac{s}{T}\right)^{\alpha_1}\right), \text{ for } d_1 \le s \le T-d_2, \\ K\left(1-\left(1-\frac{d_1}{T}\right)^{\alpha_1}\right)+\frac{Tc}{\alpha_2}\left(1-\frac{d_2}{T}\right)^{\alpha_1}+\frac{Tc}{\alpha_3}\left(\frac{d_2}{T}\right)^{\alpha_2-\alpha_3}\left(1-\left(1-\frac{s}{T}\right)^{\alpha_3}\right), \\ \qquad \text{for } T-d_2 \le s \le T, \end{cases} \quad (4.102)$$

where $K = \frac{Tc}{\alpha_1}(1-\frac{d_1}{T})^{\alpha_2-\alpha_1}$.

The density function corresponding to this process is given by

$$f_3(t) = \begin{cases} C\left(1-\frac{d_1}{T}\right)^{\alpha_2-\alpha_1}\left(1-\frac{t}{T}\right)^{\alpha_1-1}, & \text{for } 0 \le t \le d_1, \\ C\left(1-\frac{t}{T}\right)^{\alpha_2-1}, & \text{for } d_1 \le t \le T-d_2, \\ C\left(\frac{d_2}{T}\right)^{\alpha_2-\alpha_3}\left(1-\frac{t}{T}\right)^{\alpha_3-1}, & \text{for } T-d_2 \le t \le T, \end{cases} \quad (4.103)$$

where

$$\begin{aligned} C &= c/m(T) \\ &= \frac{\alpha_1\alpha_2\alpha_3/T}{\left(1-\frac{d_1}{T}\right)^{\alpha_2}\alpha_3(\alpha_1-\alpha_2)+\alpha_3\alpha_2\left(1-\frac{d_1}{T}\right)^{\alpha_2-\alpha_1}+\left(\frac{d_2}{T}\right)^{\alpha_2}\alpha_1(\alpha_2-\alpha_3)}. \end{aligned}$$

The CDF is given by

$$F_3(t) = \begin{cases} \dfrac{CT}{\alpha_1}\left(1-\dfrac{d_1}{T}\right)^{\alpha_2-\alpha_1}\left[1-\left(1-\dfrac{t}{T}\right)^{\alpha_1}\right], \text{ for } 0 \le t \le d_1, \\ \dfrac{CT}{\alpha_1\alpha_2}\left[(\alpha_1-\alpha_2)\left(1-\dfrac{d_1}{T}\right)^{\alpha_2} + \alpha_2\left(1-\dfrac{d_1}{T}\right)^{\alpha_2-\alpha_1} - \alpha_1\left(1-\dfrac{t}{T}\right)^{\alpha_2}\right], \\ \qquad \text{for } d_1 \le t \le T-d_2, \\ 1-\dfrac{CT}{\alpha_3}\left(\dfrac{d_2}{T}\right)^{\alpha_2-\alpha_3}\left(1-\dfrac{t}{T}\right)^{\alpha_3}, \text{ for } T-d_2 \le t \le T. \end{cases} \tag{4.104}$$

Note that for the interval $d_1 \le t \le T - d_2$, we can write the CDF as

$$F_3(t) = F_3(d_1) + \frac{CT}{\alpha_2}\left[\left(1-\frac{d_1}{T}\right)^{\alpha_2} - \left(1-\frac{t}{T}\right)^{\alpha_2}\right]. \tag{4.105}$$

Simulation of NHPP$_3$ To simulate a three-stage NHPP on the interval $[0, T]$, we use the inversion method and follow the same logic as for NHPP$_2$. The inverse CDF can be written as

$$F_3^{-1}(s) = \begin{cases} T - T\left\{1-\dfrac{s\alpha_1}{CT}\left(1-\dfrac{d_1}{T}\right)^{\alpha_1-\alpha_2}\right\}^{1/\alpha_1}, & \text{for } 0 \le s \le d_1, \\ T - T\left\{\left(1-\dfrac{d_1}{T}\right)^{\alpha_2} - \dfrac{\alpha_2}{CT}(s-F_3(d_1))\right\}^{1/\alpha_2}, & \text{for } d_1 \le s \le T-d_2, \\ T - T\left\{\dfrac{\alpha_3}{CT}(1-s)\left(\dfrac{d_2}{T}\right)^{\alpha_3-\alpha_2}\right\}^{1/\alpha_3}, & \text{for } T-d_2 \le s \le T. \end{cases} \tag{4.106}$$

The algorithm for generating n arrivals is then the following:

1. Generate n uniform variates $u_1, \ldots, u_n$.
2. For $k = 1, \ldots, n$ set

$$x_k = \begin{cases} T - T\left\{1-\dfrac{u_k\alpha_1}{CT}\left(1-\dfrac{d_1}{T}\right)^{\alpha_1-\alpha_2}\right\}^{1/\alpha_1}, & \text{if } u_k < F_3(d_1), \\ T - T\left\{\dfrac{\alpha_2}{CT}(F_3(d_1)-u_k) + \left(1-\dfrac{d_1}{T}\right)^{\alpha_2}\right\}^{1/\alpha_2}, & \\ \qquad \text{if } F_3(d_1) \le u_k < F_3(T-d_2), & \\ T - T\left\{\dfrac{\alpha_3}{CT}u_k\left(\dfrac{d_2}{T}\right)^{\alpha_3-\alpha_2}\right\}^{1/\alpha_3}, & \text{if } u_k \ge F_3(T-d_2). \end{cases} \tag{4.107}$$

Quick and Crude (CDF-Based) Parameter Estimation The quick and crude method described for estimating the α parameters in $NHPP_2$ works also for the $NHPP_3$. In each interval, the mean of the Poisson process is in the form $m_3(y) = \beta_j - \theta_j \left(1 - \frac{y}{T}\right)^{\alpha_j}$, $j = 1, 2, 3$, and therefore the same approximation works on each of the three intervals $[0, d_1]$, $[d_1, T - d_2]$, and $[T - d_2, T]$. The idea, once again, is to pick intervals $[T - t, T - s]$ that we are confident lie in the first, second, or third phases. Then, based on the bid times in each interval, the relevant α is

$$\alpha = 2\frac{\log[F(T-t) - F(T-\sqrt{st})] - \log[F(T-\sqrt{st}) - F(T-s)]}{\log t - \log s} \tag{4.108}$$

and is estimated by plugging the empirical CDF F_e for F in the approximation.

For α_3, we can use the exact relation

$$\alpha_3 = \frac{\log R(t_3)/R(t_3')}{\log(T - t_3)/(T - t_3')} \tag{4.109}$$

where $R(t) = 1 - F_3(t)$ and t_3, and t_3' are within $[T - d_2, T]$. To estimate α_3, we pick reasonable values of t_3, and t_3' and use the empirical survival function R_e.

To assess this method, we simulated 5000 random observations from an $NHPP_3$ with parameters $\alpha_1 = 3, \alpha_2 = 0.4, \alpha_3 = 1$ and the changepoints $d_1 = 2.5$ (defining the first 2.5 days as the first phase) and $d_2 = 5/10,080$ (defining the last 5 min as the third phase). We computed the quick and crude estimate for α_1 on a range of intervals of the form $[0.001, t_1]$, where $0.5 \leq t_1 \leq 5$. Note that this interval includes values that are outside the range $[0, d_1 = 2.5]$. The top panel in Figure 4.52 illustrates the estimates obtained for these intervals. For values of t_1 between 1.5 and 3.5 days, the estimate for α_1 is relatively stable and close to 3. Similarly, the bottom panel in Figure 4.52 describes the estimates of α_3, using (4.109), as a function of the choice of t_3 with $t_3' = 7 - 1/10,080$. The estimate is relatively stable and close to 1.

For estimating α_2, an interval such as $[3, 6.9]$ is reasonable. Figure 4.53 shows the estimate as a function of the interval choice. It is clear that the estimate is relatively insensitive to the exact interval choice, as long as it is reasonable.

Using functions of the CDF, we can also obtain expressions for d_1 and d_2. Let t_1, t_2, t_2', and t_3 be such that $0 \leq t_1 \leq d_1$, $d_1 \leq t_2' < t_2 \leq T - d_2$, and $T - d_2 \leq t_3 \leq T$. For d_1, we use the ratio $\frac{F_3(t_2) - F_3(t_2')}{F_3(t_1)}$ and for d_2 we use the ratio $\frac{F_3(t_2) - F_3(t_2')}{1 - F_3(t_3)}$. These lead to the following expressions:

$$d_1 = T - T\left\{\frac{\alpha_1}{\alpha_2} \cdot \frac{F_3(t_1)}{F_3(t_2) - F_3(t_2')} \cdot \frac{(1 - t_2'/T)^{\alpha_2} - (1 - t_2/T)^{\alpha_2}}{1 - (1 - t_1/T)^{\alpha_1}}\right\}^{1/(\alpha_2 - \alpha_1)},$$

$$d_2 = T\left\{\frac{\alpha_3}{\alpha_2} \cdot \frac{1 - F_3(t_3)}{F_3(t_2) - F_3(t_2')} \cdot \frac{(1 - t_2'/T)^{\alpha_2} - (1 - t_2/T)^{\alpha_2}}{(1 - t_3/T)^{\alpha_3}}\right\}^{1/(\alpha_2 - \alpha_3)}.$$

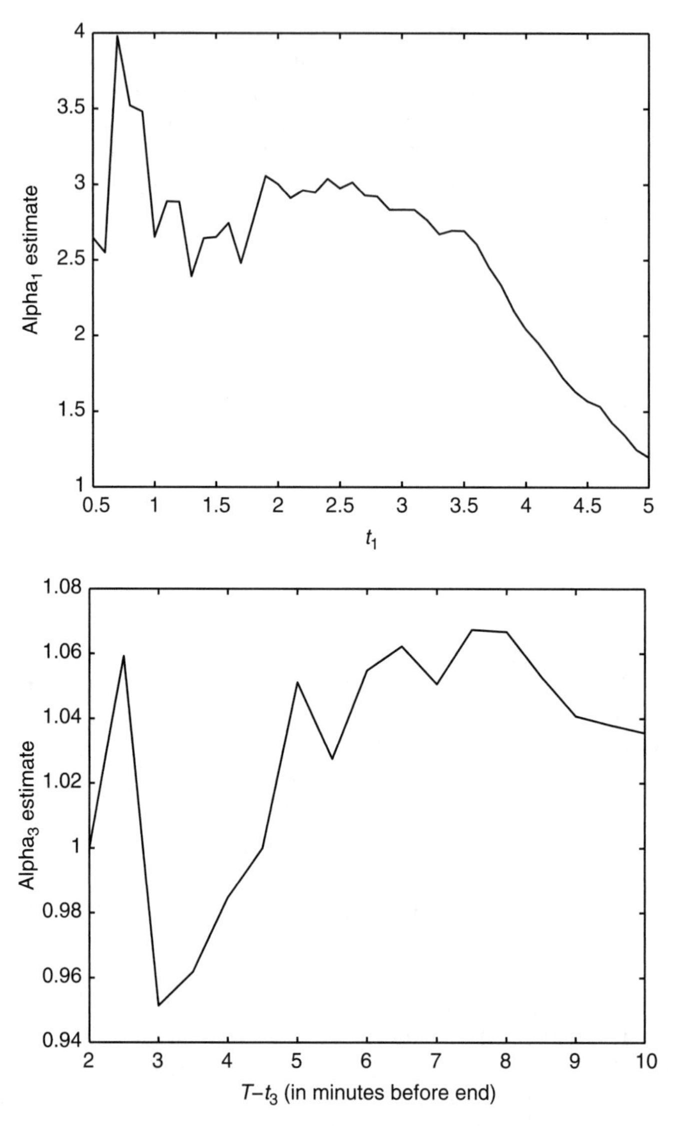

FIGURE 4.52 Quick estimates of α_1, α_2, and α_3 as a function of the input intervals for simulated $NHPP_3$ data.

Thus, we can estimate d_1 and d_2 by selecting "safe" values for t_1, t_2', t_2, and t_3 (which are confidently within the relevant interval) and using the empirical CDF at those points.

Using this method, we estimated d_1 and d_2 for the simulated data, setting the α parameters to their true values. Using $t_1 = 1, t_2' = 3, t_2 = 6$, and $t_3 = 7 - 2/10080$ yields $\hat{d}_1 = 2.5$ and $\hat{d}_2 = 4.73$ min (Figure 4.54).

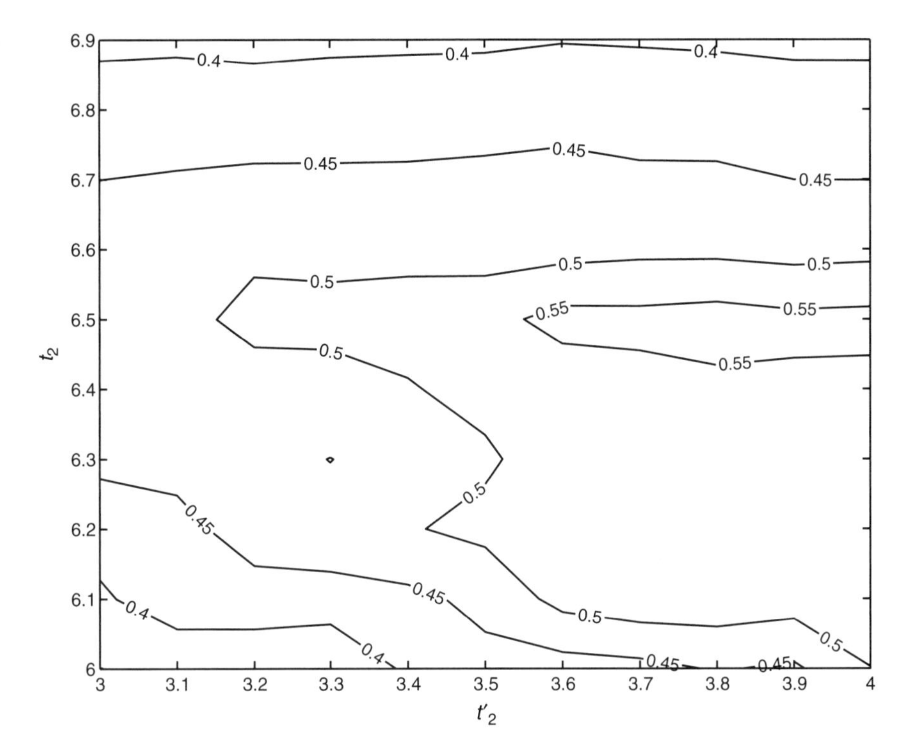

FIGURE 4.53 Quick and crude estimate of α_2 as a function of $[t'_2, t_2]$ choice. $\hat{\alpha}_2$ is between 0.4 and 0.55 in the entire range of intervals. The more extreme intervals ($t'_2 < 3.4$ or $t_2 > 6.8$) yield $\hat{\alpha}_2 = 0.4$.

Maximum Likelihood Estimation Conditional on $N(T) = n$, the NHPP$_3$ likelihood function is given by

$$\begin{aligned} &L(x_1, \ldots, x_n | \alpha_1, \alpha_2, \alpha_3, d_1, d_2) \\ &= n \log C + n_1(\alpha_2 - \alpha_1) \log\left(1 - \frac{d_1}{T}\right) + n_3(\alpha_2 - \alpha_3) \log \frac{d_2}{T} \\ &+(\alpha_1 - 1)S_1 + (\alpha_2 - 1)S_2 + (\alpha_3 - 1)S_3, \end{aligned} \tag{4.110}$$

where n_1 is the number of arrivals before time d_1, n_3 is the number of arrivals after $T - d_2$, $S_1 = \sum_{i:x_i \le d_1} \log\left(1 - \frac{x_i}{T}\right)$, $S_2 = \sum_{i:d_1 < x_i < T - d_2} \log\left(1 - \frac{x_i}{T}\right)$, and $S_3 = \sum_{i:x_i > T - d_2} \log\left(1 - \frac{x_i}{T}\right)$.

To estimate α_1, α_2, and α_3 for given values of d_1, and d_2, the following three equations must be solved (equating the first derivatives in α_1, α_2, and α_3 to zero).

$$S_1 = n_1 \log\left(1 - \frac{d_1}{T}\right) - \frac{n}{C}\frac{\partial C}{\partial \alpha_1}, \tag{4.111}$$

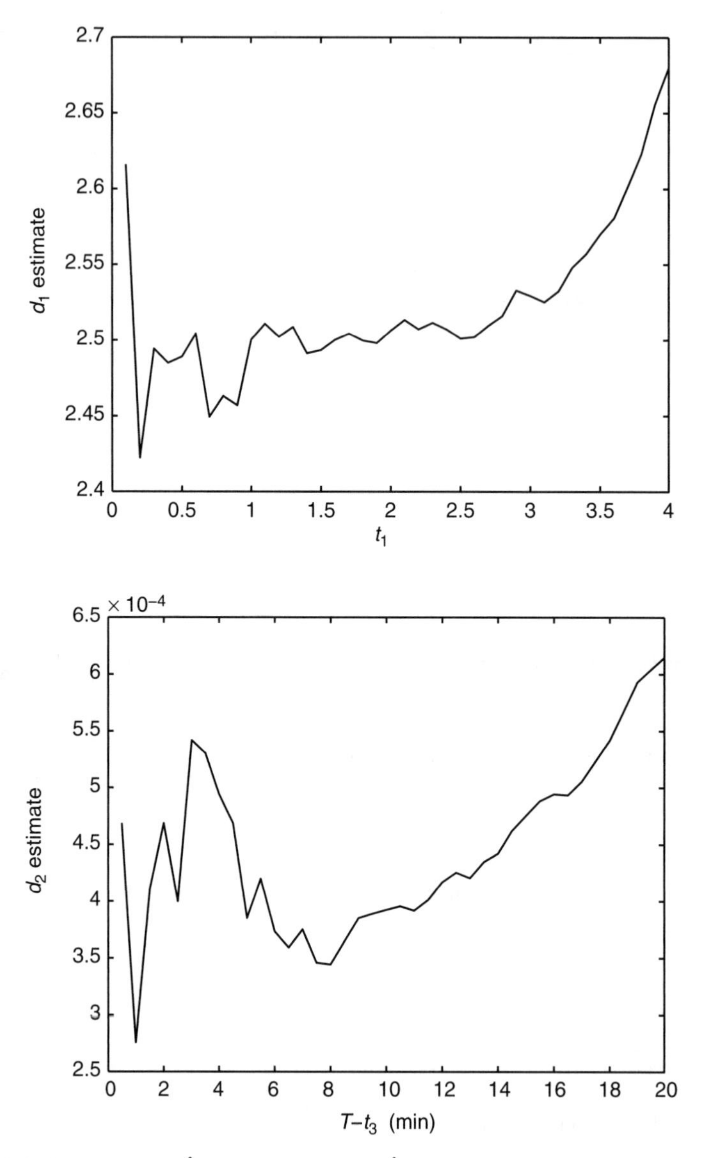

FIGURE 4.54 Graphs of $\hat{d}_1$ versus t_1 (left) and $\hat{d}_2$ versus initial values of $T - t_3$ (right) for simulated data. The estimate for d_1 is stable at ≈ 2.5. $\hat{d}_2$ using the last 2–5 min interval is in the range of 4–5 min.

$$S_2 = -n_1 \log\left(1 - \frac{d_1}{T}\right) - n_3 \log\frac{d_2}{T} - \frac{n}{C}\frac{\partial C}{\partial \alpha_2}, \tag{4.112}$$

$$S_3 = n_3 \log\frac{d_2}{T} - \frac{n}{C}\frac{\partial C}{\partial \alpha_3}, \tag{4.113}$$

where

$$\frac{\partial C}{\partial \alpha_1} = \frac{C^2 T}{\alpha_1^2}\left(1-\frac{d_1}{T}\right)^{\alpha_2} \times \left[\left(1-\frac{d_1}{T}\right)^{-\alpha_1}\left(1+\alpha_1 \log(1-\frac{d_1}{T})\right)-1\right], \tag{4.114}$$

$$\begin{aligned} \frac{\partial C}{\partial \alpha_2} &= \frac{C^2 T}{\alpha_1 \alpha_3 \alpha_2^2}\left\{\alpha_3\left(1-\frac{d_1}{T}\right)^{\alpha_2}\right. \\ &\quad \times \left[\alpha_2 \log\left(1-\frac{d_1}{T}\right)\left(\alpha_2-\alpha_1+\alpha_2\left(1-\frac{d_1}{T}\right)^{-\alpha_1}\right)-\alpha_1\right] \\ &\quad \left.+\,\alpha_1\left(\frac{d_2}{T}\right)^{\alpha_2}\left[\alpha_3+\alpha_2\log\frac{d_2}{T}(\alpha_2-\alpha_3)\right]\right\} \\ &= \frac{C^2 T}{\alpha_2^2}\left[\left(\frac{d_2}{T}\right)^{\alpha_2} - (1-\frac{d_1}{T})^{\alpha_2} - \frac{\alpha_2^2}{\alpha_1}(1-\frac{d_1}{T})^{\alpha_2}\log(1-\frac{d_1}{T})\right. \\ &\quad \times \left(1-(1-\frac{d_1}{T})^{-\alpha_1}\right) \\ &\quad - \alpha_2\left(\frac{d_2}{T}\right)^{\alpha_2}\log\frac{d_2}{T} + \alpha_2\left(1-\frac{d_1}{T}\right)^{\alpha_2} \\ &\quad \left.\times \log\left(1-\frac{d_1}{T}\right) + \frac{\alpha_2^2}{\alpha_3}\left(\frac{d_2}{T}\right)^{\alpha_2}\log\frac{d_2}{T}\right], \end{aligned} \tag{4.115}$$

$$\frac{\partial C}{\partial \alpha_3} = \frac{C^2 T}{\alpha_3^2}\left(\frac{d_2}{T}\right)^{\alpha_2}. \tag{4.116}$$

Since the equations are nonlinear in the parameters, an iterative gradient method can be used. If d_1 and d_2 are unknown and we want to estimate them from the data, then search algorithms such as genetic algorithms can be more efficient, more stable, and more easily programmable for finding a solution. Otherwise, the likelihood needs to be computed for a grid of $d_1 \times d_2$ values. In addition, empirical evidence suggests that gradient methods tend to be unstable for solving this maximization problem. Therefore, an exhaustive search over a reasonable grid of the parameter space and a stochastic search algorithm are good practical solutions.

Genetic Algorithm Search: An alternative to an exhaustive search is the genetic algorithm (GA). The genetic algorithm belongs to a general class of stochastic, global optimization procedures that imitate the evolutionary process of nature. The basic building blocks of GA are *crossover*, *mutation*, and *selection*, similar to their biological counterparts found in the evolution of genes. GA is an iterative process and each iteration is called a new generation. Starting from a parent population, two parents create offspring via crossover and mutation. The crossover operator imitates the mixing of genetic information during sexual reproduction. The mutation operator imitates the occasional changes of genetic information due to external influences. An offspring's fitness is evaluated relative to an objective function. The offspring with the highest fitness is then selected for further reproduction. Although GAs' operations appear heuristic, Holland (1975) provides theoretical arguments for convergence to a high-quality optimum.

The GA begins with a population of parents. One first evaluates the fitness of the parents and then allocates reproductive opportunities in such a way that those parents with the highest fitness are given a chance to produce offspring. This results in a new population. The expectation is that some members of this new population acquire those characteristics from the parents that enable them to better adapt to the environmental conditions, thus providing an improved solution to the problem. Let $\theta^{(t,1)}, \theta^{(t,2)}, \ldots, \theta^{(t,S)}$ denote the parent population in the tth generation, where S is the size of this population. In our application, we set $S = 100$.

Using this parent population, we create offspring via crossover and mutation. Crossover can be thought of as a swapping of genes. In the context of parameter estimation, we swap the components of two parameter vectors.

We use a GA to estimate the parameters of the NHPP_3 model as follows. After creating the parent population of size $S = 100$, we select the top 10% of parents with the highest fitness and perform crossover and mutation on a randomly chosen pair, thereby creating a pair of new offspring. We repeat this 50 times to obtain an offspring population of the same size S as the parent population. After creating a set of suitable offspring, the next step is to evaluate an offspring's fitness according to its likelihood value. Let θ denote an offspring and let $L(\theta) = L(x_1, \ldots, x_n|\theta)$ denote the corresponding likelihood value. For two offsprings θ_1 and θ_2, θ_1 has higher fitness if $L(\theta_1) > L(\theta_2)$. We restricted the range of possible solutions to the hypercube $(\alpha_1, \alpha_2, \alpha_3, d_1, d_2) \in [0, 10] \times [0, 1] \times [0, 5] \times [0, 5] \times [0, 0.1]$. This yielded the estimates $\hat{\alpha}_1 = 3.06$, $\hat{\alpha}_2 = 0.39$, $\hat{\alpha}_3 = 1.01$, $\hat{d}_1 = 2.51$, and $\hat{d}_2 = 4.68/10,080$ for the simulated data. All these estimates are completely in line with the quick and crude estimates and very close to the values that were used to generate the data. The runtime for this procedure was only a few minutes. The combined numerical maximization and grid search procedure did not converge.

Empirical and Simulated Results We use the quick and crude method to estimate the parameters for the 3651 Palm bid arrival times. From domain knowledge, we chose the first day for estimating α_1; that is, we believe that bids placed during the first day are contained within the first "early bidding" phase. Looking at the estimate as a function of the interval chosen (Figure 4.55, top panel), we see that the estimate is between 4 and 5 if we use the first 1 and 2 days. It is interesting to note that after the first 2 days, the estimate decreases progressively reaching $\hat{\alpha_1} = 2.5$ on the interval $[0.01, 3]$, indicating that the changepoint d_1 is around 2.

The parameter α_3 was estimated using (4.109) with $t_3' = 7 - 0.1/10,080$ and a range of values for t_3. From these, α_3 appears to be approximately 1. It can be seen in the bottom panel of Figure 4.55 that this estimate is relatively stable within the last

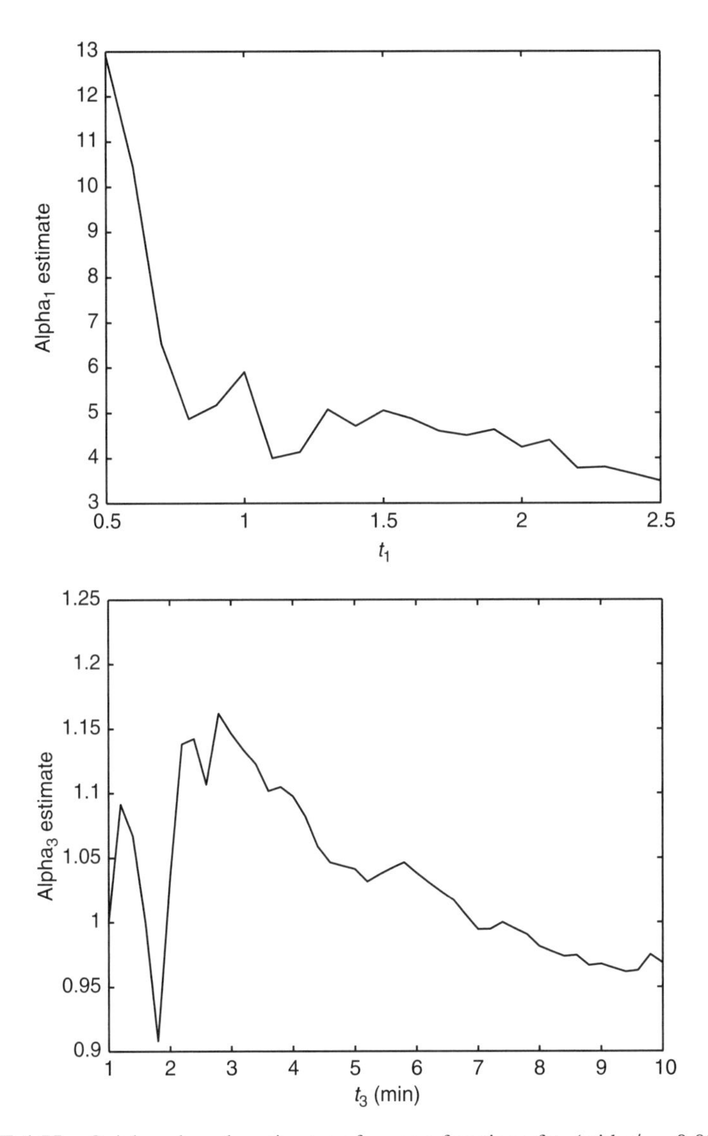

FIGURE 4.55 Quick and crude estimates of α_1 as a function of t_1 (with $t_1' = 0.001$) (top) and of α_3 as a function of t_3 (with $t_3' = 0.5/10,080$) (bottom). $\hat{\alpha}_1$ is stable around 5 for t_1 in the range 0.75–1.75 days. A shorter interval does not contain enough data. A longer interval leads to a drop in the estimate, indicating that $d_1 < 2$. $\hat{\alpha}_3$ is around 1.1 when t_3 is within the last 2–4 min.

10 min. Also, note that selecting t_3 too close to t_3' results in unreliable estimates (due to a small number of observations between the two values).

Finally, we chose the interval [3, 6.9] for estimating α_2. This yielded the estimate $\hat{\alpha_2} = 0.36$. Figure 4.56 shows the estimate as a function of the interval choice. Note

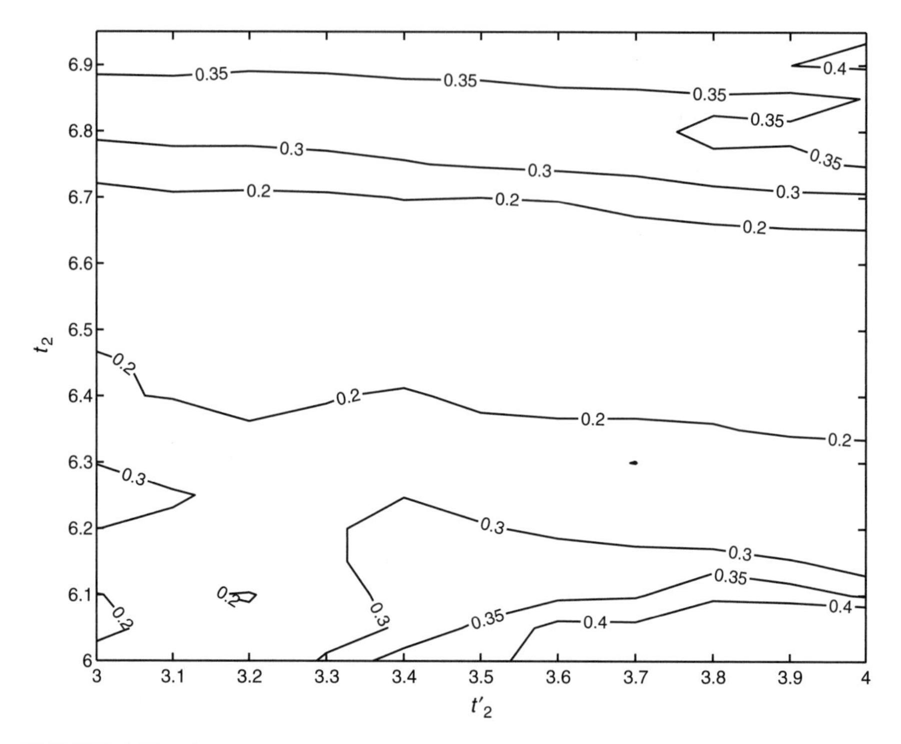

FIGURE 4.56 Quick and crude estimate of α_2 as a function of $[t'_2, t_2]$. Shorter, "safer" intervals are at the lower right. Longer intervals, containing more data, are at the upper left. $\hat{\alpha}_2$ is between 0.2 and 0.4 for all intervals. For $t_2 > 6.9$, the estimate is approximately 0.35.

that the estimate is stable between 0.2 and 0.4 for the different intervals chosen. It is more sensitive to the choice of t_2, the upper bound of the interval, and thus an overly conservative interval could yield to large inaccuracies.

Using these estimates ($\hat{\alpha}_1 = 5, \hat{\alpha}_2 = 0.36, \hat{\alpha}_3 = 1.1$), we estimated d_1 and d_2. Figure 4.57 shows graphs of the estimates as a function of the intervals selected. The estimate for d_1 (top panel) appears to be stable at approximately $\hat{d}_1 = 1.75$. The estimate for d_2 (bottom panel) appears to be around 2 min. From the increasing values obtained for $T - t_3 > 3$ min, we also learn that $d_2 < 3$.

Table 4.12 displays the above estimates and compares them with the two other estimation methods: An exhaustive search over a reasonable range of the parameter space (around the quick and crude estimates) and the much quicker genetic algorithm. It can be seen that all methods yielded estimates in the same vicinity.

Finally, to further validate this estimated model, we simulated data from an NHPP_3 with the above ML estimates as parameters. Figure 4.58 shows a QQ plot of the Palm data versus the simulated data. The points appear to fall on the line $x = y$, thus supporting the adequacy of the estimated model for the Palm bid times.

The estimated model for the Palm data reveals the auction dynamics over time. We can learn that the "average" auction has three phases: the beginning takes place during

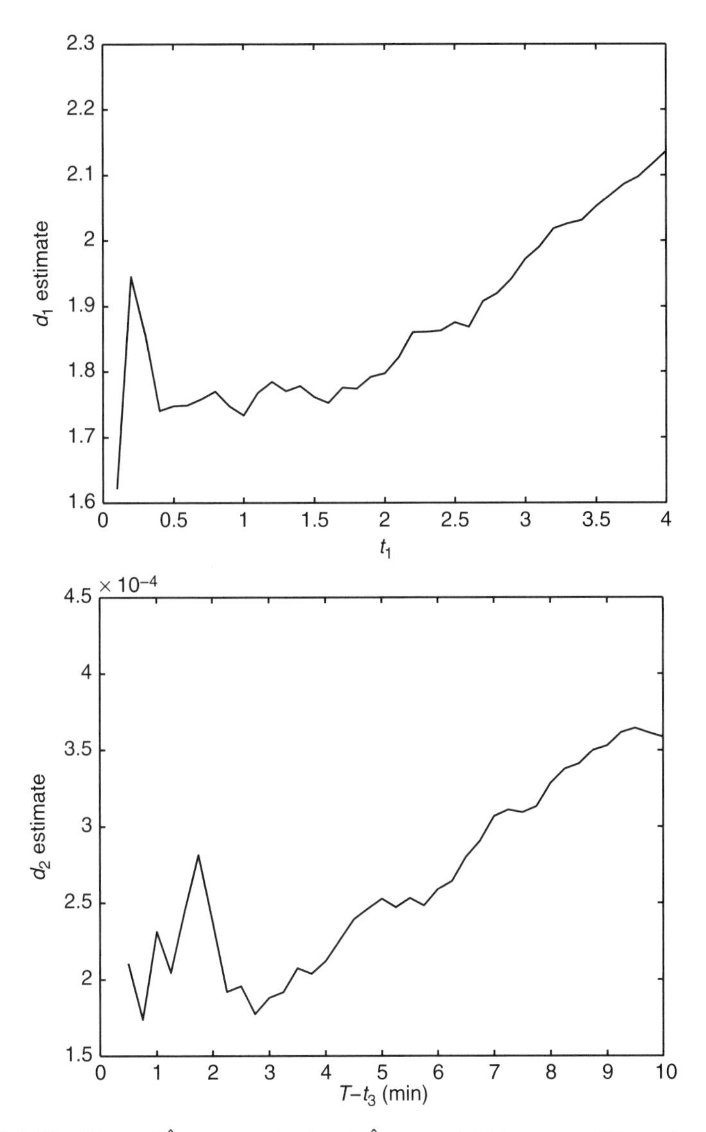

FIGURE 4.57 Plots of $\hat{d}_1$ versus t_1 (top) and $\hat{d}_2$ versus initial values of $T - t_3$ (bottom) for Palm data. The estimate for d_1 seems stable at ≈ 1.75. $\hat{d}_2$ is approximately 2 min.

TABLE 4.12 Estimates for Five NHPP$_3$ Parameters Using the Three Estimation Methods

	$\hat{\alpha}_1$	$\hat{\alpha}_2$	$\hat{\alpha}_3$	$\hat{d}_1$	$\hat{d}_2$ (min)
CDF-based Q&C	5	0.36	1.1	1.75	2
Exhaustive search	4.9	0.37	1.13	1.7	2
Genetic algorithm	5.56	0.37	1.1	1.54	2.11

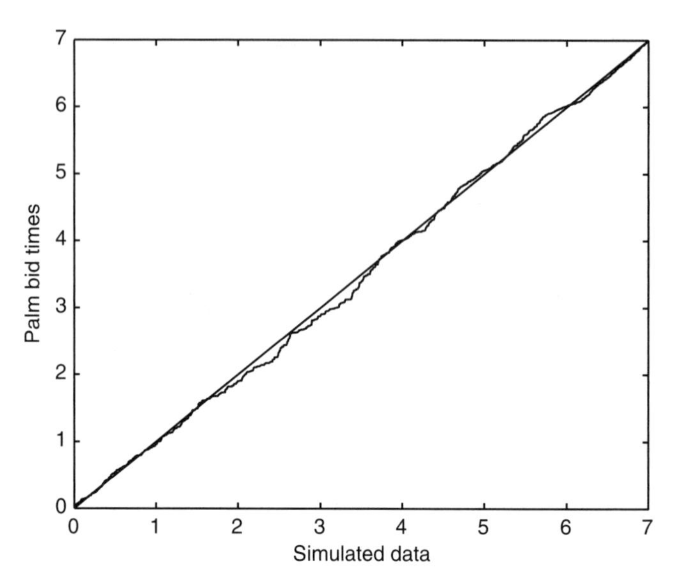

FIGURE 4.58 QQ plot of Palm bid times versus simulated data from an $NHPP_3$ with parameters $\alpha_1 = 4.9, \alpha_2 = 0.37, \alpha_3 = 1.13, d_1 = 1.7$, and $d_2 = 2/10080$.

the first 1.7 days, the middle continues until the last 2 min, and then the third phase kicks in. The bid arrivals in each of the three phases can be described by an NHPP process, but they each have different intensity functions. The auction beginning is characterized by an early surge of interest, with more intense bidding than during the start of the second phase. Then, the increase in bid arrival rate slows down during the middle of the auction. The bids do tend to arrive faster as the auction progresses, but at the very end during the last 2 min of the auction, we observe a uniform bid arrival process. Finally, it is interesting to note that in these data the third phase of bidding seems to take place within the last 2–3 min rather than the last 1 min. Thus, we use the term "last-moment bidding" rather than "last-minute bidding".

4.4.2 Modeling Bidder Arrivals

In the previous section, we described models that capture the main features of the bid arrival process in online auctions. Now, we take the discussion from bids to bidders, introducing models that describe bidder behavior in online auctions.

The online auction literature is rich with studies that assume an ordinary homogeneous Poisson bidder arrival process. This assumption underlies various theoretical derivations, is the basis for the simulation of bid data, and is often used to design field experiments. Bajari and Hortacsu (2003) specified and estimated a structural econometric model of bidding on eBay, assuming a Poisson bidder arrival process. Etzion et al. (2003) suggested a model for segmenting consumers at dual channel online merchants. On the basis of the assumption of Poisson arrivals to the website, they model consumer choice of channel, simulate consumer arrivals and actions, and compute relationships between auction duration, lot size, and the constant Poisson arrival rate λ. Zhang et al. (2002) modeled the demand curve for consumer products in online auctions based on Poisson bidder arrivals and fitted the model to bid data. Pinker et al. (2003) and Vackrat and Seidmann (2000) used a Poisson process for modeling the arrival of bidders in going-going-gone auctions. They used the intensity function $\lambda(t) = \lambda_\alpha e^{-t/T}$, $0 \leq t \leq T$, where T is the auction duration and λ_α is the intensity of website traffic into the auction. Their model describes the decline in the number of new bidders as the auction progresses. Haubl and Popkowski Leszczyc (2003) designed and carried out an experiment for studying the effect of fixed-price charges (e.g., shipping costs) and reserve prices on consumer's product valuation. The experiment used simulated data based on Poisson arrivals of bidders. These studies are among the many that rely on a Poisson arrival process assumption.

The difficulty in creating bidder models that match empirical behavior stems from the fact that several aspects of bidder behavior (namely, bidder arrivals, departures, and strategies) are largely unobservable. Bid placements, on the other hand, are fully observable. For instance, the NHPP_1–NHPP_3 models described earlier are based on observed bid placements. This is different from models of bidder behavior. Russo et al. (2008, 2010) developed models of *bidder activity* (where "activity" is defined as bidder arrival, bidder departure, and bid placement) that are consistent with observable online auction phenomena. In the following, we describe these bidder behavior models and their relationship to bid arrival models.

Let $N(s)$, $0 \leq s \leq T$, denote the bid arrival process associated with an online auction having a start time 0 and a deadline (hard close) T. As discussed earlier, Roth and Ockenfels (2002) noted two interesting characteristics common to the aggregations of such processes (see also Figure 4.44):

(P1) An *increasing intensity* of bid arrivals as the auction deadline approaches (i.e., bid sniping)

(P2) A *striking similarity* in shape among the left- truncated bid time distributions on $[s, T]$, as s approaches T (i.e., self-similarity)

For $0 \leq s < t \leq T$, define $N(s, t) = N(t) - N(s)$. *Increasing bid intensity* refers to the stochastic monotonicity of $N(t - \delta, t)$ as a function of t for any fixed $\delta > 0$, while *self-similarity* refers to the regularity in shape of the distribution functions

$$F_s(\eta) := \frac{N(T - \eta s, T)}{N(T - s, T)}, \quad 0 \leq \eta \leq 1$$

for all s sufficiently close to T (for $s \in [0, T - b]$, some b). Since $N(t - \delta, t)$ and $F_s(\eta)$ are determined empirically, we need to precisely define these properties in terms of the expected bid counts: we say that the N-*process* has *increasing bid intensity* if $\mathbb{E}(N(t-\delta, t))$ is increasing, and is *self-similar* over the interval $[b, T]$ if $E_s(\eta)$ defined as

$$E_s(\eta) := \frac{\mathbb{E}(N(T-\eta s, T))}{\mathbb{E}(N(T-s, T))}, \quad \text{for } (s, \eta) \in [0, T-b] \times [0, 1] \tag{4.117}$$

is independent of the value of s in $[0, T-b]$. In other words, $E_s(\eta)$ is the same function of $\eta \in [0, 1]$ for all $s \in [0, T-b]$. From the Cauchy equation (see p. 41 of Aczel (1966)), for (4.117) to hold we must have for some $\gamma > 0$,

$$\frac{\mathbb{E}N(s)}{\mathbb{E}N(T)} = 1 - \left(1 - \frac{s}{T}\right)^{\gamma}, b \leq s \leq T, \tag{4.118}$$

in which case $E_s(\eta) = \eta^{\gamma}$ (see footnote 29 of Roth and Ockenfels (2002)).

In the following, we define a *general bid process* (GBP) of bidder arrivals, bidder departures, and bid placement that (under sufficient conditions) yields a bid-arrival sequence that possesses one or both of the above properties. We derive an expression for the probability that an individual bidder is *active* at time s (has not departed the auction as of that time), and an expression for $\mathbb{E}(N(s))$. Moreover, we show that $N(s)/N(T) \to \mathbb{E}(N(s))/\mathbb{E}(N(T))$ uniformly in $s \in [0, T]$ as the number of bidders increases. Under a simplifying restriction, the GBP reduces to the *general poisson bid process* (GPBP), an aggregation of nonhomogeneous Poisson processes having randomly determined start times and stopping rules. This latter process is related to the NHPP process from the previous section and is shown to possess the above property (P1) under very general conditions. A further simplification results in the *self-similar bid process* (SSBP) that, as its name suggests, also possesses the above property (P2).

The General Bid Process Auction theory focuses on bidder behavior and, in particular, on finding optimal bidding strategies for different auction formats. However, the online implementation of auctions has created a different environment where nonoptimal bidding is often observed. Although various empirical studies have documented and quantified these phenomena, there exists a gap in the development of models of bidder behavior that are consistent with them. This owes largely to the fact that one cannot directly observe bidder behavior in publicly available online auction data. Typically, bid placements are completely observable from the auction's bid history, whereas bidder arrivals and departures, and bidder strategies, are not. On eBay, for example, the temporal sequence of all bids placed over the course of the auction is publicly available. In particular, every time a bid is placed its exact time stamp is posted. In contrast, the time when bidders first arrive at an auction is unobservable from the bid history. Bidders can browse an auction without placing a bid and thereby not leave a trace or reveal their interest in that auction. That is, they can look at a particular auction, inform themselves about current bid and competition levels in that auction, and make decisions about their bidding strategies without leaving an observable trace in the bid history that the auction site makes public.

Our goal is to establish a model of bidding activity that is consistent with phenomena observable in the bid arrival process.[15] We now define bidder activity more formally.

Suppose that m bidders participate in an online auction starting at time 0 and closing at time T. The parameter m can be fixed or random (see Remark 3 below). With each bidder associate a random triple $\Theta = (X, \Pi, \mathcal{H})$, which we refer to as the bidder's *type*, that is comprised of an absolutely continuous random variable $X \in [0, T)$, a continuous function Π that maps $[0, T]$ into $(0, 1]$, and a family $\mathcal{H} = \{H_s\}$ of real-valued distribution functions indexed by a real parameter $s \in [0, T]$, with $H_s(\cdot)$ having support $[s, T]$, where it is differentiable. The variable X represents the arrival time of an individual bidder into the auction, the function Π determines upon each of his bid placements whether he will remain in the auction and make a future bid, and the family $\mathcal{H}$ determines the timing of each bid that he makes. Given that the bidder's type is $\theta = (x, \pi, \{H_s\})$, he enters the auction at time x and places an initial bid at time $Y_1 \sim H_x$. If $Y_1 = y_1$, he departs the auction with probability $\pi(y_1)$ or otherwise places a second bid at time $Y_2 \sim H_{y_1}$. If $Y_2 = y_2$, he departs the auction with probability $\pi(y_2)$ or otherwise places a third bid at time $Y_3 \sim H_{y_2}$, and so on, ultimately placing a random number of bids during $[0, T]$. All m bidders are assumed to act in a similar manner, independently of each other.

In online auctions, *bid snipers* often attempt to place their final bid close to the deadline, so as to forestall a response from competing bidders. It has been observed that these late bids often fail to transmit due to technical reasons such as network congestion. Thus, as $s \to T$, there may be a growing probability that a late placed bid fails to be recorded. This phenomenon can most efficiently be accommodated within our GBP by building the probability of a failed last bid directly into the π function as in (4.130).

For a more precise definition of the GBP, let $X_1, X_2, \ldots, X_m$ denote the arrival times of the m bidders. For $1 \leq k \leq m$, let $N_k(s)$ denote the number of bids placed by bidder k during the period $[0, s]$, $0 \leq s \leq T$, and let $Y_{k,j}$ denote the time of the jth bid placed by the kth bidder, $1 \leq j \leq N_k(T)$. For convenience, set $Y_{k,0} = X_k$ and suppose that

(A1) $\Theta_1 = (X_1, \Pi_1, \mathcal{H}_1), \ldots, \Theta_m = (X_m, \Pi_m, \mathcal{H}_m)$ are independent and identically distributed.

(A2) $\Pr\left(Y_{k,j+1} \leq t \mid Y_{k,j} = y_{k,j}\right) = H_{k,y_{k,j}}(t)$, for $1 \leq k \leq m$ and $j \geq 0$.

[15] eBay auctions use a proxy bidding mechanism, whereby bidders are advised to place the highest amount they are willing to pay for the auctioned item. The auction mechanism then automates the bidding process to ensure that the bidder with the highest proxy bid is in the lead at any given time. Thus, when bidder A places a proxy bid that is lower than the highest proxy bid of (say) bidder B, the new displayed highest bid and its time stamp will appear with bidder B's username (although bidder A is the one who placed the bid). In our discussion, we shall consider such a bid as having been placed by A, rather than B, as it is A's action that led to a change in the displayed bid.

(A3) The sequences $\{Y_{1,j}\}_{j\geq 0}, \ldots, \{Y_{m,j}\}_{j\geq 0}$ are independent.

(A4) $\Pr\left[N_k(T) = r \mid Y_{k,r} = y_{k,r} \text{ and } \Pi_k = \pi_k\right] = \pi_k(y_{k,r})$ for $r \geq 1$ and $1 \leq k \leq m$.

Remark 1. We observe that condition (A1) allows dependence among the elements of Θ_k. Thus, the model can accommodate a tendency (say) for infrequent bidders to place their bid(s) late. We observe further that the above model accounts for heterogeneity in bidder probabilities of remaining in the auction after placing a bid, both across bidders and from bid to bid. Empirical evidence (Bapna et al., 2004) suggests that a realistic distribution of Π should reflect the dichotomy of one-time bidders (*opportunists*) versus multibid bidders (*participators*).

Remark 2. Since π is a continuous function on a compact set, it achieves its minimum ($= \pi_{\min}$) on $[0, T]$. Since π maps into $(0, 1]$, we have $\pi_{\min} > 0$. Thus, with each bid, the probability of departure is bounded below by $\pi_{\min}$. The total number of bids placed by the bidder on $[0, T]$ is therefore stochastically bounded above by a geometric($\pi_{\min}$) variable and thus has finite moments of all order.

Remark 3. (Poisson Bidder Arrivals). The parameter m can be either fixed or random. If random, it is natural to assume that m is Poisson distributed, as would be the case when bidders arrive in accordance with a nonhomogeneous Poisson process having an intensity $\mu g(t), t \in [0, T]$, where $\mu > 0$ and g is a density function on $[0, T]$. We then have that $X_1, \ldots, X_m$ is a random sample of random size $m \sim$ Poisson (μ) from a fixed distribution with density function $g(t)$, $0 < t < T$.

A Single-Bidder Auction: The following two results pertain to an auction involving a single bidder ($m = 1$) whose *type* is $\theta = (x, \pi, \{H_s\})$. Let N_1 denote the resulting bid counting process. Obviously, a single bidder would not bid against himself. However, it is convenient to study the actions of a single bidder in the context of an m-bidder auction. We say that the bidder is *active* at time s if $N_1(T) > N_1(s)$ (he does not depart the auction during $[0, s]$). In particular, a bidder is active during the period prior to his arrival. Let $p(s \mid \theta)$ denote the probability that this event occurs, and for $0 \leq s \leq t \leq T$ define

$$G_s(t \mid \theta) = \Pr\left(Y_{N_{(s)+1}} \leq t \mid \theta\right), \quad \text{on } \{N_1(T) > N_1(s);\ X < s\}.$$

Given that the bidder arrived at time x ($\leq s$), and is still active at time s, his next bid time $Y_{N_1(s)+1}$ has the distribution above. Let $g_s(\theta)$ denote the right derivative of $G_s(t \mid \theta)$ evaluated at s.

To derive the form of $g_s(\theta)$, we define the functional sequence

$$\phi_0(s;t) = h_s(t),$$

$$\phi_1(s;t) = \int_s^t (1 - \pi(u))\,\phi_0(s;u)\phi_0(u;t)du,$$

$$\vdots$$

$$\phi_n(s;t) = \int_s^t (1 - \pi(u))\,\phi_0(s;u)\phi_{n-1}(u;t)du.$$

We now have

$$g_s(\theta) = \frac{\sum_{n=0}^{\infty} \phi_n(x;s)}{\sum_{n=0}^{\infty} \int_s^T \phi_n(x;u)du}, \qquad x < s.$$

Proposition 1. *Suppose that*

$$\sup_{s \le v \le s+\delta} H_v(s+\delta) := \omega(s,\delta) \to 0 \text{ as } \delta \to 0, \text{ all } s \in [0,T). \tag{4.119}$$

Then, for $0 \le s \le T$,

$$p(s \mid \theta) = 1_{x<s} \exp\left(-\int_x^s \pi(t) g_t(\theta) dt\right) + 1_{x>s}.$$

Proof. If $x > s$, the result is trivial. Fix $s \in [x, T)$ and define

$$\pi_\delta = \inf\{\pi(t) : s \le t \le s+\delta\} \quad \text{and} \quad \pi^\delta = \sup\{\pi(t) : s \le t \le s+\delta\}.$$

Writing $p(s)$ for $p(s \mid \theta)$, we have

$$p(s+\delta) \ge p(s)\left[1 - G_s(s+\delta \mid \theta) + G_s(s+\delta \mid \theta)(1-\pi^\delta)(1-\omega(s,\delta))\right]$$

since an active bidder at time s remains active at time $s+\delta$ if either (1) $Y_{N_1(s)+1} \in (s+\delta, T)$ or (2) $Y_{N_1(s)+1} < s+\delta$, the bidder stays active, and $Y_{N_1(s)+2} > s+\delta$. It follows that

$$\liminf_{\delta \to 0} \frac{p(s+\delta) - p(s)}{\delta} \ge -\pi(s)p(s)g_s(\theta).$$

Moreover,

$$p(s+\delta) \leq p(s)\,[1 - G_s(s+\delta \mid \theta) + G_s(s+\delta \mid \theta)(1-\pi_\delta)]$$

since an active bidder at time s is active at time $s+\delta$ only if (1) $Y_{N_1(s)+1} > s+\delta$ or (2) $Y_{N_1(s)+1} < s+\delta$ and the bidder stays active. It follows that

$$\limsup_{\delta \to 0} \frac{p(s+\delta) - p(s)}{\delta} \leq -\pi(s)p(s)g_s(\theta).$$

Thus, $p'(s) = -\pi(s)p(s)g_s(\theta)$. Hence the proof.

Remark 4. Condition (4.119) is a mild condition, which will be assumed from here onward. For the condition to fail would require a pathological $\mathcal{H}$.

The following is an easy consequence of the above and is hence stated without proof:

Proposition 2. *For* $0 \leq s \leq T$,

$$\mathbb{E}\,[N_1(s) \mid \theta] = 1_{x<s} \int_x^s p(t \mid \theta) g_t(\theta) dt.$$

A Multibidder Auction: Returning to the case of m bidders, we state a uniform limit result for $N(s)/N(T)$. It should be noted that many online auctions involve a small number of bidders. *Self-similarity*, as studied in Roth and Ockenfels (2002), is not a phenomenon that can be easily observed from an individual auction where the number of bids are few. To observe the *self-similarity* property, one usually must aggregate many *equivalent* auctions (i.e., auctions for the same item, over the same duration, etc.). Such aggregations often involve hundreds of bidders, and hence our interest in $m \to \infty$.

Proposition 3. *If* $\mathbb{E}\,(N_1(T)) < \infty$, *then as* $m \to \infty$,

$$\sup_{0 \leq s \leq T} \left| \frac{N(s)}{N(T)} - \frac{\mathbb{E}\,(N_1(s))}{\mathbb{E}\,(N_1(T))} \right| \to 0 \ \textit{almost surely}.$$

Proof. For fixed $s \in [0, T]$, $N(s)$ is the sum of m independent and identically distributed random variables. By the *strong law of large numbers*, we have almost sure pointwise (in s) convergence:

$$\frac{N(s)}{N(T)} \to \frac{\mathbb{E}\,(N_1(s))}{\mathbb{E}\,(N_1(T))} \quad \textit{almost surely}.$$

By the continuity of the limit as a function of s, and by Polya's theorem, the above convergence is uniform in s. Hence the proof.

Remark 5. By Remark 2, we have $\mathbb{E}\,(N_1(T)|\theta) \leq 1/\pi_{\min}$, so that by the double expectation formula, $\mathbb{E}\,(N_1(T)) \leq \mathbb{E}\,(1/\Pi_{\min})$, where $\Pi_{\min} = \min\{\Pi(s) : s \in [0, T]\}$. Thus, the finiteness of $\mathbb{E}\,(1/\Pi_{\min})$ is sufficient for the convergence in Proposition 3.

Remark 6. **(Poisson Bidder Arrivals).** In the case where bidders arrive in accordance with a nonhomogeneous Poisson process with intensity $\mu g(t)$ (as in Remark 3), Proposition 3 and a standard coupling argument yields the following result:

$$\varepsilon > 0 \text{ and } \mathbb{E}N_1(T) < \infty \Rightarrow \lim_{\mu \to \infty} \Pr\left(\sup_{0<s<T} \left| \frac{N(s)}{N(T)} - \frac{\mathbb{E}N_1(s)}{\mathbb{E}N_1(T)} \right| > \varepsilon \right) = 0. \tag{4.120}$$

The practical significance of (4.120) is that in an auction with a high number of participants, or in aggregations of many equivalent auctions, the observed distribution of bid times on $[0, T]$ will be uniformly close to the deterministic function $\mathbb{E}N_1(s)/\mathbb{E}N_1(T)$ with high probability.

The General Poisson Bid Process We consider now a simplifying restriction on the form of $\mathcal{H}$. Given $\theta = (x, \pi, \{H_s\})$, suppose that all the H_s functions are driven by the single (randomly determined) function H_0 as follows:

$$H_s(t) = \frac{H_0(t) - H_0(s)}{1 - H_0(s)} \quad \text{for } 0 \leq s \leq t < T. \tag{4.121}$$

Under condition (4.121), the bidder's *jth* bid time ($j \geq 1$), conditional on $Y_{j-1} = y_{j-1}$, is distributed as H_0 restricted to $[y_{j-1}, T]$. We note that a bid time chosen from H_s, that is not realized by time $t > s$, is probabilistically equivalent to one chosen from H_t.

In this section, N_1 denotes the single-bidder process of bid arrivals under condition (4.121). We define an auxiliary bid counting process $M(s)$, $0 \leq s \leq T$, under the additional conditions that $\Pr(X = 0) = 1$ (the bidder enters the auction at time $s = 0$) and $\Pr(\Pi(s) \equiv 0) = 1$ (the bidder never departs the auction). We observe that $M(0) = 0$ and that the M-process possesses independent increments. Moreover, writing H for H_0, we have

$$\begin{aligned}
\limsup_{\delta \to 0} & \frac{\Pr(M(s+\delta) - M(s) \geq 2)}{\delta} \\
&\leq \limsup_{\delta \to 0} \left[\frac{H(s+\delta) - H(s)}{1 - H(s)} \right]^2 \frac{1}{\delta} \\
&= \left(\frac{1}{1 - H(s)} \right)^2 h(s) \limsup_{\delta \to 0} [H(s+\delta) - H(s)] = 0
\end{aligned} \tag{4.122}$$

and by (4.122)

$$\lim_{\delta\to 0}\frac{\Pr\left(M(s+\delta)-M(s)=1\right)}{\delta}=\lim_{\delta\to 0}\left[\frac{\Pr\left(M(s+\delta)-M(s)\geq 1\right)}{\delta}\right]$$
$$=\lim_{\delta\to 0}\frac{H(s+\delta)-H(s)}{(1-H(s))\,\delta}$$
$$=\frac{h(s)}{1-H(s)}.$$

The M-process is thus a nonhomogeneous Poisson process with intensity function

$$\lambda(s)=\frac{h(s)}{1-H(s)}.$$

Intuitively, when the auction clock reaches time s, no matter how many bids have been placed, and no matter when they have been placed, the probability that a bid will be placed during $(s, s+\delta)$ is approximately $\delta h(s)/(1-H(s))$. Given θ (associated with the N_1-process), we may use the function $\pi(\cdot)$ to randomly label an *M-process* arrival at time s as A or B with respective probabilities $1-\pi(s)$ and $\pi(s)$. The resulting offspring processes M_A and M_B are independent nonhomogeneous Poisson processes with respective intensity functions

$$\lambda_A(s)=\frac{(1-\pi(s))h(s)}{1-H(s)}\quad\text{and}\quad\lambda_B(s)=\frac{\pi(s)h(s)}{1-H(s)}.$$

Note that arrivals from the N_1-*process* are the M arrivals that occur after the bidder arrival time X, up to and including the first arrival from M_B. That is, N_1 is a nonhomogeneous Poisson process with intensity function λ, restricted to the random interval $[X,\ T]$, and stopped upon the first arrival from M_B. The N-process (involving all m bidders) is an aggregation of m such independent processes.

Given $\theta=(x,\pi,\{H_s\})$, a bidder is active at time s if and only if either $x>s$ or $x<s$ and there are no arrivals from M_B during the period $[x,s]$. Accordingly,

$$p(s\mid\theta)=1_{x<s}\Pr\left[M_B(s)-M_B(x)=0\right]+1_{x\geq s}$$

$$=1_{x<s}\Pr\left[\text{Poisson}\left(\int_x^s\lambda_B(t)dt\right)=0\right]+1_{x\geq s}\tag{4.123}$$

$$=1_{x<s}\exp\left[-\int_x^s\frac{\pi(t)h(t)}{1-H(t)}dt\right]+1_{x\geq s}.$$

From the above, we obtain the conditional bid intensity $\lambda(\cdot|\theta)$ of an individual bidder of type θ:

$$\lambda(s|\theta) = \begin{cases} \exp\left[-\int_x^s \frac{\pi(t)h(t)}{1-H(t)}dt\right] \frac{h(s)}{1-H(s)}, & s > x, \\ 0, & s \leq x. \end{cases}$$

In the case where $\limsup_{s\to T} \pi(s) < 1$ and $\lim_{s\to T} h(s) > 0$, the conditional intensity explodes as s approaches T, that is, $\lim_{s\to T} \lambda(s|\theta) = \infty$. The condition on $\pi(\cdot)$ can be dropped if $h(s)$ increases to infinity sufficiently fast as s approaches T (e.g., at any polynomial rate).

A Constant Probability of Departure: We now suppose that (upon each bid placement) the bidder has a randomly determined time invariant probability of departure:

$$\Pr\left(\Pi(s) = \Pi(0),\ 0 \leq s \leq T\right) = 1. \tag{4.124}$$

Under the above assumption, the number of bids placed by an individual bidder is geometrically distributed with a randomly determined parameter. A time invariant departure probability, while certainly not true for the entire auction duration, should be approximately true over short intervals. Under (4.121) and (4.124), statement (4.123) reduces to

$$p(s \mid \theta) = 1_{x<s} \left[\frac{1-H(s)}{1-H(x)}\right]^{\pi(0)} + 1_{x>s}. \tag{4.125}$$

By Proposition 2, we get

$$\begin{aligned} \mathbb{E}\left(N_1(s) \mid \theta\right) &= 1_{x<s} \int_x^s \left[\frac{1-H(t)}{1-H(x)}\right]^{\pi(0)} \frac{h(t)}{1-H(t)}dt \\ &= \frac{1_{x<s}}{\pi(0)} \left(1 - \left[\frac{1-H(s)}{1-H(x)}\right]^{\pi(0)}\right), \end{aligned}$$

so that by the double expectation formula, writing Π for $\Pi(0)$,

$$\mathbb{E}\left(N_1(s)\right) = \mathbb{E}\left(\frac{1_{X<s}}{\Pi}\left(1 - \frac{1-H(s)}{1-H(X)}\right)^{\Pi}\right). \tag{4.126}$$

Hence, by Proposition 3, under conditions (4.124) and (4.121), if $\mathbb{E}\Pi^{-1} < \infty$,

$$\sup_{0<s<T} \left| \frac{N(s)}{N(T)} - \mathbb{E}\frac{1_{X<s}}{\Pi}\left(1 - \frac{1-H(s)}{1-H(X)}\right)^{\Pi} \frac{1}{\mathbb{E}\Pi^{-1}} \right| \to 0 \quad a.s. \text{ as } m \to \infty$$

and for Poisson arrivals (as in Remarks 3 and 6),

$$\lim_{\mu\to\infty} \Pr\left(\sup_{0<s<T} \left| \frac{N(s)}{N(T)} - \mathbb{E}\frac{1_{X<s}}{\Pi}\left(1 - \frac{1-H(s)}{1-H(X)}\right)^{\Pi} \frac{1}{\mathbb{E}\Pi^{-1}} \right| > \varepsilon \right) = 0.$$

The Self-Similar Bid-Process: Suppose that conditions (4.121) and (4.124) hold and that for some constant $r > 0$ (the same for all bidders), we have

$$H_0(s) = 1 - \left(1 - \frac{s}{T}\right)^{r/\pi(0)}. \tag{4.127}$$

Suppose, in addition, that all bidders arrive by time $b < T$:

$$\Pr(0 \leq X < b) = 1. \tag{4.128}$$

Under (4.127), the higher a bidder's likelihood of departure upon the placement of a bid at time s, the stochastically greater the time of his next bid (i.e., the more inclined he is to choose his next bid near the deadline T). Condition (4.127) ties the selection function H_0 directly to the constant departure probability $\pi(0)$. Again, writing Π for $\Pi(0)$ and assuming that $\mathbb{E}\Pi^{-1} < \infty$, we have by (4.126)

$$\begin{aligned} \mathbb{E}N_1(T-s, T) &= \mathbb{E}\left(\Pi^{-1}\left(\frac{s}{T-X}\right)^r 1_{X<T-s} + \Pi^{-1}1_{X>T-s}\right) \\ &= \mathbb{E}\Pi^{-1}\left(\frac{1}{T-X}\right)^r s^r, \quad \text{for } s < T-b. \end{aligned}$$

Thus, for $(s, \eta) \in [0, T-b] \times [0, 1]$, we have

$$E_s(\eta) = \eta^r, \tag{4.129}$$

where $E_s(\eta)$ is defined in (4.117).

By statement (4.129) and Proposition 3, we have for large m

$$N_s(\eta) = \frac{N(T-\eta s, T)}{N(T-s, T)} \approx \eta^r, \quad \text{for } (\eta, s) \in [0, 1] \times [0, T-b].$$

The NHPP Process: The efforts of bidders to place bids late in the auction are often thwarted by transmission failures. We build this phenomenon into the GPBP by assuming (4.127) and supposing that each bidder has a randomly determined time invariant probability of departure upon the placement of all bids on $[0, T-d]$, for some small $d > 0$, and that this probability is magnified by a constant β for all bids placed after time $T-d$ (β and d being the same for all bidders):

$$\Pi(s) = \Pi(0)1_{s \le T-d} + \beta\Pi(0)1_{s > T-d}. \tag{4.130}$$

The resulting process of bid arrivals is a GPBP (but not a SSBP) and can thus also be characterized as an aggregation of independent nonhomogeneous Poisson processes with randomly determined start times and stopping rules. At time $s \in (b, T-d]$, the bid intensity associated with an individual bidder of type θ is

$$\begin{aligned}\lambda_1(s \mid \theta) &= p(s \mid \theta)\frac{h(s)}{1-H(s)} \\ &= \left(\frac{1-H(s)}{1-H(x)}\right)^{\pi(0)} \frac{h(s)}{1-H(s)} \quad \text{by (4.125)} \\ &= \left(1-\frac{s}{T}\right)^{r-1}\left(1-\frac{x}{T}\right)^{-r}\frac{r}{\pi(0)T} \quad \text{by (4.127)},\end{aligned}$$

and hence, for a given collection of m bidder types, the intensity of the bid arrival sequence at $s \in (b, T-d]$ is given by

$$\lambda(s \mid m, \theta_1, \ldots, \theta_m) = \left[\frac{r}{T}\sum_{k=1}^{m}\frac{1}{\pi_k(0)}\left(1-\frac{x_k}{T}\right)^{-r}\right]\left(1-\frac{s}{T}\right)^{r-1}.$$

For $s \in (T-d, T]$,

$$\begin{aligned}\lambda_1(s \mid \theta) &= p(T-d \mid \theta)\left(\frac{1-H(s)}{1-H(T-d)}\right)^{\beta\pi(0)}\frac{h(s)}{1-H(s)} \\ &= \left(\frac{d}{T}\right)^{r-r\beta}\left(1-\frac{x}{T}\right)^{-r}\frac{r}{\pi(0)T}\left(1-\frac{s}{T}\right)^{r\beta-1},\end{aligned}$$

and hence, for a given collection of m bidder types, the intensity of the bid arrival sequence at $s \in (b, T)$ is given by

$$\lambda(s \mid m, \theta_1, \ldots, \theta_m) = \begin{cases} c\left(1-\dfrac{s}{T}\right)^{r-1}, & s \in (b, T-d], \\ c\left(\dfrac{d}{T}\right)^{r-r\beta}\left(1-\dfrac{s}{T}\right)^{r\beta-1}, & s \in (T-d, T], \end{cases}$$

where

$$c = \left[\frac{r}{T} \sum_{k=1}^{m} \frac{1}{\pi_k(0)} \left(1 - \frac{x_k}{T}\right)^{-r} \right]$$

is the result of the realization $(m, \theta_1, \ldots, \theta_m)$. This is the form of the two-stage NHPP_2 see also Shmueli et al. (2007) with $d_1 = 0$, $d_2 = d$, $\alpha_2 = r$, and $\alpha_3 = \beta r$. It can be shown similarly that multiple shifts in the departure probability will lead to the NHPP_3 process.

Remark 7. We note that in the case of $r\beta < 1$, $\lambda(s \mid m, \theta_1, \ldots, \theta)$ as a function of s is increasing to infinity (i.e., increasing and exploding) as s approaches T.

Working with the general Poisson model under condition (4.124), we now derive the distribution of an individual bidder's final bid time Y_{final}. Again, writing Π for $\Pi(0)$, we have by (4.125),

$$P\left(Y_{\text{final}} > s\right) = \mathbb{E}\left(\frac{1 - H(s)}{1 - H(X)}\right)^{\Pi} 1_{X<s} + \Pr\left(X > s\right) \tag{4.131}$$

and thus

$$\mathbb{E}Y_{\text{final}} = \mathbb{E}X + \int_0^T \mathbb{E}\left(\frac{1 - H(s)}{1 - H(X)}\right)^{\Pi} 1_{X<s} ds.$$

For the *self-similar bid process*, the above statements simplify to

$$P\left(Y_{\text{final}} > s\right) = \mathbb{E}\left(\frac{T - s}{T - X}\right)^{r} 1_{X<s} + \Pr\left(X > s\right)$$

and

$$\mathbb{E}Y_{\text{final}} = \mathbb{E}X + \int_0^T \mathbb{E}\left[\left(\frac{T - s}{T - X}\right)^{r} 1_{X<s}\right] ds.$$

In the case of m-bidders, $Y_{m,\text{final}}$ is the sample maximum from a random sample of size m from the distribution of Y_{final}. Hence, properties of $Y_{m,\text{final}}$ are easily derived from those of Y_{final} given above. Also, by using conditioning we can extend the above results to the case of random m.

Example 1 (One-Stage BARISTA (NHPP_1), Pure Self-Similarity). Fix $r > 0$. Suppose $p_1, p_2, \ldots, p_m$ is an i.i.d. sequence of variables on $(0, 1)$ and for $1 \leq k \leq m$, we have $\pi_k(s) = p_k$ (upon each bid placement, bidder k departs the auction with invariant probability p_k), and

$$H_k(s) = 1 - \left(1 - \frac{s}{T}\right)^{r/p_k}. \tag{4.132}$$

The above equality ties the selection function H_k of bidder k to his invariant departure probability p_k. The greater this probability, the later in the auction bidder k tends to place his next bid. We have in this case

$$\begin{aligned} p(s \mid \theta_k) &= \exp\left[-p_k \int_{x_k}^{s} \frac{h_k(t)}{1 - H_k(t)} dt\right] \\ &= \left[\frac{1 - H_k(s)}{1 - H(x_k)}\right]^{p_k} \\ &= \left(1 - \frac{x_k}{T}\right)^{-r} \left(1 - \frac{s}{T}\right)^{r} \end{aligned} \tag{4.133}$$

so that

$$\lambda(s \mid m, \theta_1, \ldots, \theta_m) = c\left(1 - \frac{s}{T}\right)^{r-1}, \quad \max(x_1, \ldots, x_m) \le s \le T, \tag{4.134}$$

where

$$c = \frac{r}{T} \sum_{k=1}^{m} \frac{1}{p_k} \left(1 - \frac{x_k}{T}\right)^{-r}. \tag{4.135}$$

That is, conditional on the number of bidders m and their corresponding types $\theta_1, \ldots, \theta_m$ the intensity function λ has a pure self-similar form over the interval $[\max(x_1, \ldots, x_m), T]$.

Simulation 1. Suppose that $m = 1000$, $T = 7$, $X \sim U(0, 5.6)$, $(p_k < t) = 4t^2$ for $t \in (0, 1/2)$, and $r = 2/5$. In our simulation, 1000 bidders arrive uniformly during the first 5.6 days of a 7-day auction, and their departure probabilities are individually invariant and constitute a random sample from a triangular distribution on $(0, 1/2)$. The parameter $r = 2/5$ guarantees that h increases as t approaches $T = 7$. The top panel in Figure 4.59 displays the empirical cumulative distribution functions of several sets of normalized left-truncated bid times based on our simulation. The bottom panel displays the same curves based on the eBay auction data. Note the similarity of the curves (excluding the 5 min curve) for the simulated and real data. The top display shows (as expected when the intensity has a self-similar form) the 5 min curve to be of the same shape as the others. In the bottom panel, we see the 5 min curve breaking away from the self-similar form of the other curves.

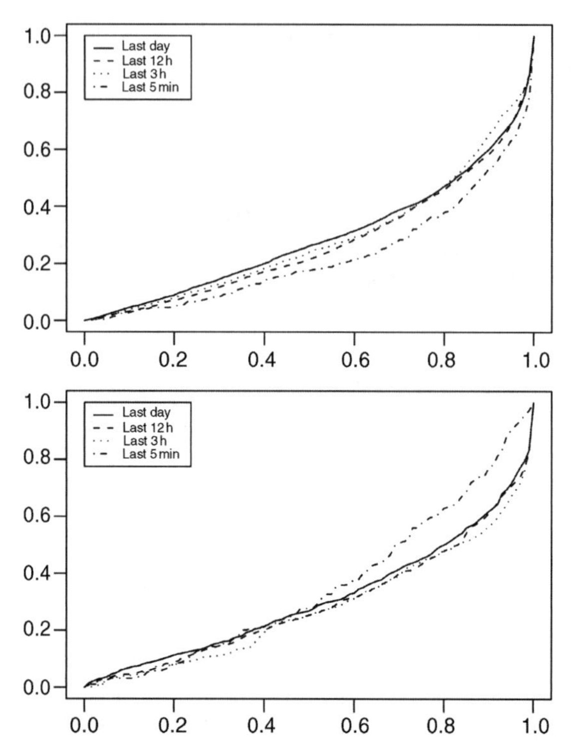

FIGURE 4.59 Empirical CDF functions of left-truncated bid times based on Simulation 1 (top) and Palm data (bottom).

Example 2 (Two-Stage BARISTA). Fix $\beta > 0$ and $d \in (0, T)$. Suppose that we have the same setup as in Example 1 with the following exception: during the interval $[T-d, T]$, the departure probabilities of all active bidders are magnified by the common factor β so that

$$\pi_k(s) = p_k 1_{0 \le s < T-d} + \beta p_k 1_{s \ge T-d}, \quad \text{for } 1 \le k \le m. \tag{4.136}$$

In this case,

$$p(s \mid \theta_k) = \begin{cases} \exp\left[-p_k \int_{x_k}^{s} \frac{h_k(t)}{1-H_k(t)} dt\right], & x_k \le s \le T-d, \\ \exp\left[-p_k \int_{x_k}^{T-d} \frac{h_k(t)}{1-H_k(t)} dt - \beta p_k \int_{T-d}^{s} \frac{h_k(t)}{1-H_k(t)} dt\right], & T-d \le s \le T \end{cases}$$

$$
= \begin{cases} \left(1 - \frac{x_k}{T}\right)^{-r} \left(1 - \frac{s}{T}\right)^{r}, & x_k \leq s \leq T - d, \\ \left(1 - \frac{x_k}{T}\right)^{-r} \left(\frac{d}{T}\right)^{r - r\beta} \left(1 - \frac{s}{T}\right)^{r\beta}, & T - d \leq s \leq T, \end{cases}
$$

so that with c as in Example 1 we have

$$
\lambda(s \mid m, \theta_1, \ldots, \theta_m) = \begin{cases} c\left(1 - \frac{s}{T}\right)^{r-1}, & \max(x_1, \ldots, x_m) \leq s \leq T - d, \\ c\left(\frac{d}{T}\right)^{r - \beta r} \left(1 - \frac{s}{T}\right)^{r\beta - 1}, & T - d \leq s \leq T. \end{cases} \tag{4.137}
$$

Note that the above λ has the form of the intensity function of a two-stage BARISTA process ($NHPP_2$) on $[\max(x_1, \ldots, x_m), T]$ with $d_1 = 0, d_2 = d, \alpha_2 = r,$ and $\alpha_3 = \beta r$.

Simulation 2. We maintain the same setup as in Simulation 1, except that we set $d = 1/10{,}080$ and $\beta = 2$, thereby doubling the departure probabilities of all active bidders during the final minute of the auction. Figure 4.60 displays the same collection of empirical cumulative distribution functions as in Simulation 1. Note the similarity of all the curves (including the 5 min curve) to those in the bottom panel (real data) display in Figure 4.59.

Example 2 demonstrates the existence of a plausible bidder arrival and behavior model that is able to replicate most of the prominent features of online auction bid data. As in the above simulation, the model can be used to generate bid arrival data for various purposes.

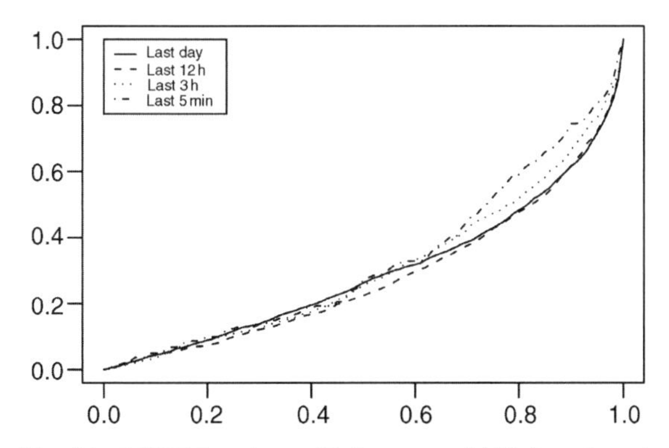

FIGURE 4.60 Empirical CDF functions of left-truncated bid times based on Simulation 2.

4.4.3 Conclusion

The ability to model the bid(-der) arrival process has many important uses: first, it enables the exploration and formalization of observed phenomena such as early and late bidding. Furthermore, such phenomena can then be explained as a function of bidder behavior. For example, it can be shown that certain bidder behavior dynamics, where each displayed bid is the minimum of a collection of (uniformly distributed) bid times contemplated by a shrinking population of bidders, lead to $NHPP_1$. Thus, if a set of auctions is shown to follow a $NHPP_1$ model, it is possible that the bidder behavior occurring behind the scenes is of this type. Another bidder behavior of interest is collusion, where a buyer is actually an agent of the seller who participates in the auction in order to "run up the bid." Kauffman and Wood (2000) hypothesize that colluders avoid bidding toward the end of the auction. In a sample of auctions infected by collusion we, would therefore expect to see less activity than usual toward the end of the auction.

A second use of a bid arrival model is in conjunction with the sequence of bid or price increments. Jank and Shmueli (2008b) explore the bidding dynamics in online auctions by fitting smoothing splines to sets of bid times and bid amounts corresponding to an auction. The knots of the splines are determined by the bidding intensity, or the intensity of the bid arrivals. An $NHPP_3$ model can be used in this application to determine favorable locations of the knots.

One of the most researched questions is what factors affect the final price obtained in an auction. Several authors have shown that the final price is higher in auctions with more activity. An open question is whether there is a function relating a particular bid(-der) arrival model to an average final price.

Knowledge of the bid arrival process is of special importance for applications that determine the frequency of page updating. For example, if eBay users are monitoring an auction from mobile devices that have costs attached to web connection, they must decide on a policy when to reconnect and update the information. In an auction that has the typical early and last-moment phases of bidding, it is better for the user to update the information more frequently during these phases and not connect as much during the middle phase.

Finally, the bid(-der) arrival model can be useful for visualization tools that display the bids over an auction or a set of auctions. To determine the scale of the time axis and avoid over- and under-crowded areas on the display, the application must know "where the action is" and to what degree. An NHPP model, even if approximate, gives a sense of the scale of interest.

4.5 MODELING AUCTION NETWORKS

In this section, we take a look at a rather novel aspect of auction data: networks generated by auction transactions. Most auction research to date assumes that auctions are independent of one another. That is, it assumes that what happens in one auction is unaffected by the outcome of another auction. We have already argued and shown in

Chapter 3 and in Section 4.3 that this independence assumption is rather unrealistic, especially for auctions that sell similar (i.e., substitute) products. We now generalize this discussion and consider auctions that are connected and thus form a type of network. Auctions (or participants of auctions) can be connected in a variety of ways: For instance, two bidders participating in the same auction are linked. While this link may exist only in the virtual world, it has an impact in that these two bidders are competing against one another. Similarly, two sellers selling the same product are linked in that they compete for the same pool of bidders. Bidders—and in particular winning bidders—are also linked to sellers and if we measure the strength of this link via the number of repeat transactions, then we could interpret the resulting network as a manifestation of the bidder's *loyalty* to a particular seller (Jank and Yahav, 2010).

Studying loyalty in auction networks is new. Much of the existing auction literature focuses on the sellers and the level of trust they signals to the bidders (e.g., Brown and Morgan, 2006). To that end, a seller's *feedback score* (i.e., the number of positive ratings minus the number of negative ratings) is often scrutinized (e.g., Li, 2006) and it has been shown that higher feedback scores can lead to price premiums for the seller (Lucking-Reiley et al., 2007; Livingston, 2005). Here, we study a complementary determinant of a bidder's decision process: *loyalty*. Loyalty is different from trust. Trust is often associated with reliability or honesty, and trust may be a necessary (but not sufficient) prerequisite for loyalty. Loyalty, however, is a stronger determinant of a bidder's decision process than trust. Loyalty refers to a state of being faithful or committed. Loyalty incorporates not only the level of confidence in the outcome of the transaction, but also satisfaction with the product, the price, and also with previous transactions by the same seller. Moreover, loyal bidders are often willing to make an emotional investment or even a small financial sacrifice to strengthen a relationship. Studying loyalty is made possible by deriving suitable measures for the link strength of the bipartite network of bidders and sellers; we discuss this next.

4.5.1 Auction Networks

Interactions in an online auction result in a network linking its participants. Bidders bidding on one auction are linked to other bidders who bid in the same auction. Sellers selling a certain product are linked to other sellers selling the same product. In this study, we focus on the network *between* buyers and sellers. Each time a bidder transacts with a particular seller, both are linked.[16] A seller can set up more than one auction, and thus repeat transactions (i.e., purchases) measure the strength of this link. For instance, a bidder transacting 10 times with the same seller has a stronger link compared to a bidder who transacts only twice. In that sense, the network strength measures an important aspect of the relationship between buyers and sellers, namely, customer loyalty.

[16] Note that in our data, bidders and sellers form disjoint groups; that is, a node is either a bidder node or a seller node, but not both. Thus, our network forms a *bipartite network*.

We emphasize that one can measure loyalty in different ways. While one could count all the repeat *bids* a bidder places on auctions hosted by the same seller, we only count the number of *winning bids* (i.e., the number of transactions). While both bids and winning bids indicate a relationship between buyers and sellers, a winning bid signals a much stronger commitment and is thus much more indicative of a buyer's loyalty. Moreover, we investigate loyalty relationships across auctions. Studying cross-auction relationships is rather rare in the literature on online auctions, and it has gained momentum only recently (Haruvy et al., 2008; Reddy and Dass, 2006; Jank and Shmueli, 2007; Jank and Zhang, 2009a,b). In this chapter, we consider network effects between auction participants and their impact on the outcome of an auction.

4.5.2 Case Study

We study the complete bidding records of *Swarowski* fine beads for every single auction that was listed on eBay between August 2007 and January 2008. Our data contain a total of 36,728 auctions out of which 25,314 are transacted. There are 365 unique sellers and 40,084 bidders out of which 19,462 made more than a single purchase. Each bidding record contains information on the auction format, the seller, the bidder, and the product details. Tables 4.13–4.15 summarize this information.

Table 4.13 shows information about the auctions and the product sold in each auction. We can see that the average auction length is over 9 days, while the median is only 3 days. One reason for this skew is a high proportion of fixed-price listings ("buy-it-now") for which the listing duration can be longer than the typical auction duration (which ranges between 1 and 10 days). The product sold in each auction (packages of beads for crafts and artisanship) is of a relatively small value, and thus both the average starting and closing prices are low. While many eBay auctions sell only one item at a time (e.g., laptop or automobile auctions), auctions in the craft category often feature multi-unit auctions; that is, the seller offers multiple counts of the same item and bidders can decide how many of these items they wish to purchase. In our data, the average item quantity per auction is 5.42. Auctions also live from the competition among bidders and while the average number of bids is slightly larger than 3, the median is only 1. As pointed out above, the items sold in these auctions are

TABLE 4.13 Auctions and Products

Attribute	Mean	Median	StDev
AuctionDuration (inDays)	9.15	3.00	37.45
StartingPrice (USD)	3.77	3.33	5.64
ClosingPrice (USD)	6.61	4.25	9.15
ItemQuantity	5.42	1.00	129.47
BidCount	3.16	1.00	4.26
Size (bead diameter)	6.41	6.00	3.35
Pieces (# of beads per item)	124.30	48.00	343.79

TABLE 4.14 Sellers

Attribute	Mean	Median	StDev
Volume	163.90	6.00	999.00
ConversionRate	0.67	0.67	0.33
SellerFeedback	2054.00	264.00	12400.00

packages of Swarowski beads. The value of a bead is, in part, defined by its size, and the average diameter of our beads equals 6.41 mm. Another measure for the value of an item is the number of pieces per package; we can see that there are on average over 124 beads in each package, but this number varies significantly from auction to auction.

We are primarily interested in the network between bidders and sellers. One main factor influencing this network is the size of the seller. We can see (Table 4.14) that the average seller volume (i.e., number of auctions per seller) is over 163. A seller's auction will only transact if (at least one) bidder places a bid. While low transaction rates (or "conversion rates") is a problem for many eBay categories (e.g., automobiles), in our data the average conversion rate is 67% per seller, which is considerably high. One factor driving conversion rates is a seller's perceived level of trust. Trust is often measured using a seller's feedback rating and it averages over 2000 in our data.

Table 4.15 shows the corresponding attributes of the bidders. Bidders bid, on average, on almost four auctions ("volume") and, in every auction, bid on average over five items. (Recall the multi-unit auctions with several items per listing.) The bidder feedback captures a bidder's experience with the auction process and its average is over 220 in our data, signaling highly experienced bidders. Our data also contain bidders' year of birth and we can see that the average bidders are over 40 years old.

Consider Figure 4.61 that shows a snapshot of our total auction network. In this figure, we display all bidders and sellers with at least two transactions (i.e., bidders who win at least two auctions and sellers who own at least two auctions). Sellers are marked by triangles, and bidders are marked by squares. A (black) line between a seller and a bidder denotes a transaction. The width of the line is proportional to the number of transactions and hence measures the strength of a link. We can see that most sellers and bidders are linked to one another (the average network degree,

TABLE 4.15 Buyers

Attribute	Mean	Median	StDev
Volume	3.62	1.00	14.29
ItemQuantity	5.05	2.00	29.25
BidderFeedback	228.10	70.00	559.53
YearOfBirth	1967.00	1965.00	16.30

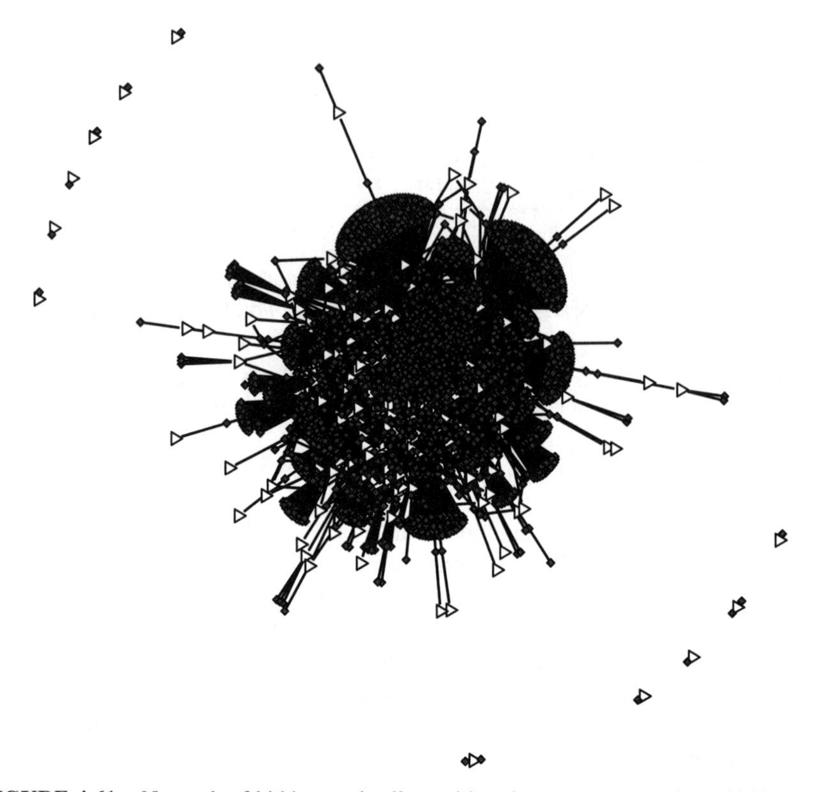

FIGURE 4.61 Network of bidders and sellers with at least two transactions: bidders who place bids in at least two auctions, and sellers who host at least two auctions.

that is, the average number of links between a bidder–seller pair, equals 1.56); we can also see a few isolated networks in which one seller is linked to only one or two bidders. Figure 4.62 shows a subset of only the top 10% of high-volume *sellers*. We can see that some sellers interact with several hundred different bidders (with 895, on average); we can also see that some sellers are "exclusive" in the sense that they are the only ones who transact with a set of bidders (e.g., at the margins of the network), while other sellers "share" a common set of bidders. Serving bidders exclusively versus sharing them with other sellers has huge implications on the outcome of the auction.

Figure 4.63 shows another subset of the data. In this figure, we display only the top 10% of all *bidders* with the highest number of transactions. We can see that many of these high-volume bidders transact with only one seller (note the many triangles that are connected with only a single arc to the network) and are hence very loyal to the same person. Figure 4.64 shows only new buyers (i.e., bidders who won an auction for the first time.) We can see that in contrast to the high-volume bidders, the network of first-time bidders is much more fragmented. We can identify five megasellers (i.e., sellers that capture most of the transactions), three high-volume sellers, and many

FIGURE 4.62 Network of the top 10th percentile of sellers: the top 10% of sellers hosting the most auctions.

medium- and low-volume sellers. Since these are only first-time buyers, loyalty does not yet play a role in bidders' decisions. However, the fact that most first-time bidders "converge" to only a few megasellers suggests that this is a very difficult market for low-volume sellers to enter.

As pointed out above, bidder–seller networks capture loyalty of participants. While most sellers and bidders are linked to one another (e.g., Figure 4.61), here we focus only on the subgraphs created by each bidder–seller pair. Next, we describe an innovative way to extract *loyalty measures* from these graphs, as proposed by Jank and Yahav (2010).

4.5.3 Extracting Loyalty from Network Information

Our loyalty measures map the entire network of bidders and sellers into a few seller-specific statistics. For each seller, these statistics capture both the *proportion of bidders* loyal to that seller and the *degree of loyalty* of each bidder. We derive the measure in two steps. First, we derive, for each seller, the *loyalty distribution*, and then we

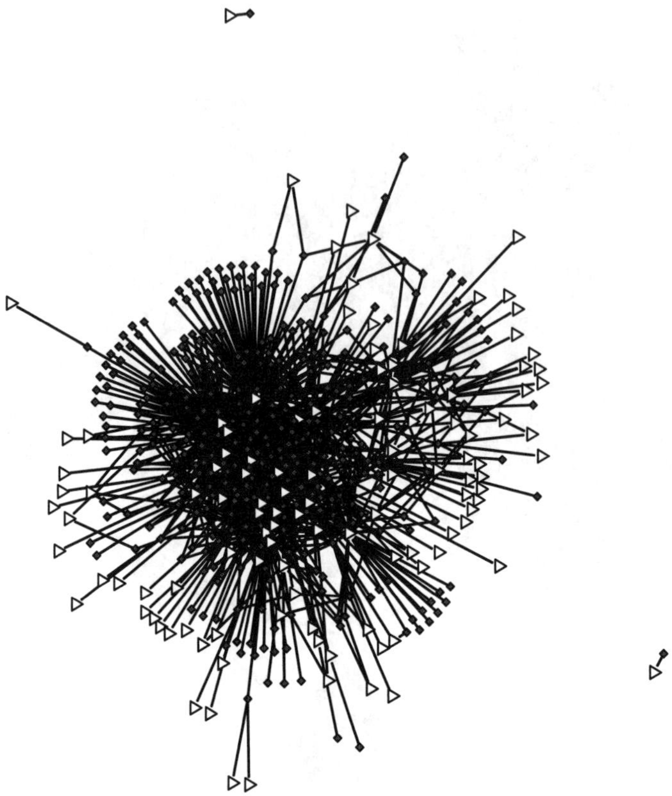

FIGURE 4.63 Network of the top 10th percentile of bidders: the top 10% of bidders bidding on most auctions.

summarize that distribution with a few statistics using functional principal component analysis. We describe each step in detail below.

Note that there exists more than one way to extract loyalty information from network data. We chose the route of loyalty distributions since they capture the two most important elements of loyalty: the proportion of customers loyal to one's business and the degree of their loyalty. Note in particular that we do not try to dichotomize loyalty (i.e., categorize it into loyal versus disloyal buyers): Since we do not believe that loyalty can be turned on or off arbitrarily, we allow it to range on a continuous scale between 0 and 1. This will allow us to quantify the impact of the *shape* of a seller's loyalty distribution on his or her bottom line. For instance, it will allow us to answer whether sellers with *pure* loyalty (i.e., all buyers 100% loyal) are better off compared to sellers with more variation among their customer base.

We would also like to caution that the resulting analysis is complex since we first have to characterize the infinite-dimensional loyalty distributions in a finite way and subsequently interpret the resulting characterizations. The resulting interpretations are more complex than, say, employing user-defined measures of loyalty (e.g., summary

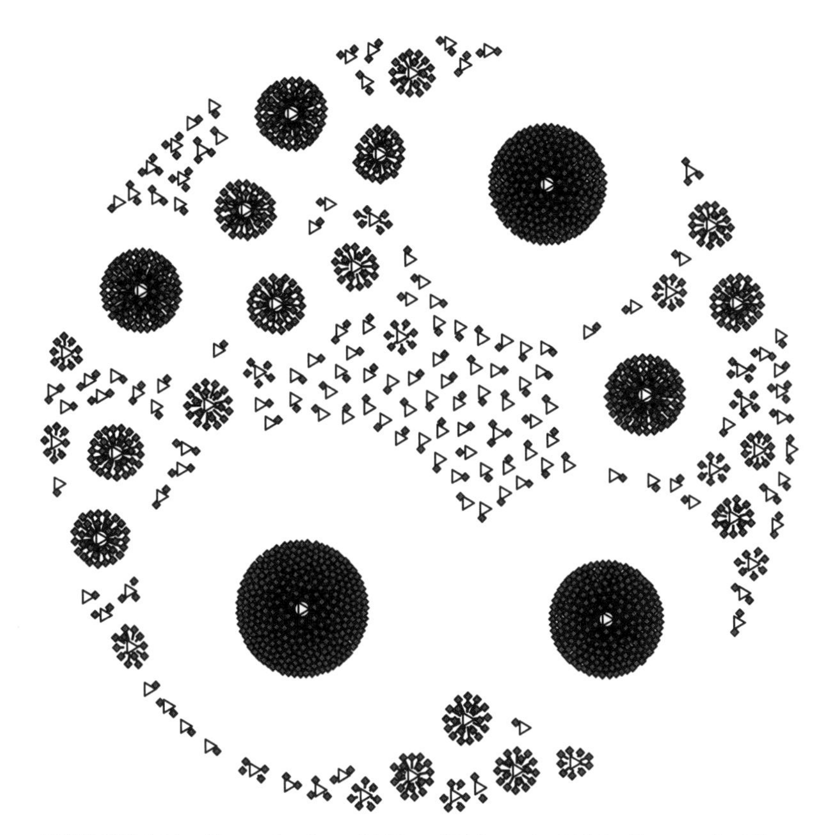

FIGURE 4.64 Network of new bidders: bidders who bid for the very first time.

statistics such as the number of loyal buyers or the proportion of at least 70% loyal buyers). While such user-defined measures are easy to interpret, there is no guarantee that they capture all the relevant information. For instance, measuring the "number of loyal bidders" would require defining a cutoff at which a buyer is considered "loyal" versus "disloyal." Any such cutoff is necessarily arbitrary and would lead to a dichotomization that we are trying to avoid. Rather than employing arbitrary, user-defined measures, we set out to let the data speak freely and first look for ways to summarize the information captured in the loyalty distributions in the most exhaustive way. This will lead us to the notion of principal component loyalty scores and their interpretations. We elaborate on both aspects below.

From Loyalty Networks to Loyalty Distributions Consider the hypothetical seller–bidder network in Figure 4.65. In this network, there are 4 sellers (labeled "A," "B,""C," and "D") and 10 bidders (labeled 1–10). An arc between a seller–bidder pair denotes an interaction, and the width of the arc is proportional to the number of repeat interactions between the pair. Consider bidder 1 who has a total of 10 interactions, all of which are with seller A; we can say that bidder 1 is 100% loyal to seller A. This is

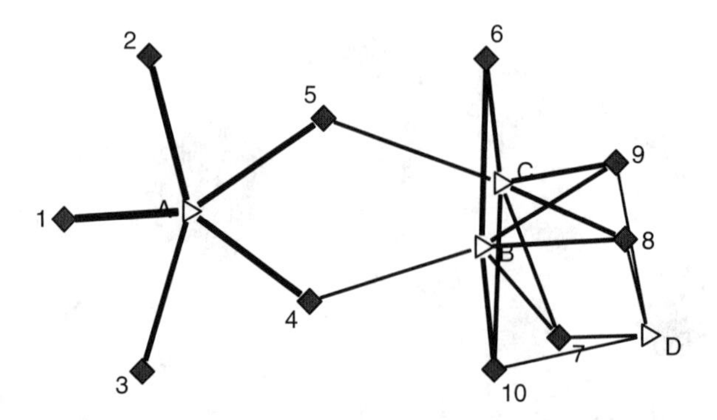

FIGURE 4.65 A hypothetical network between 4 sellers (A–D, triangles) and 10 bidders (1–10, squares). The size of the arc between a seller and a bidder corresponds to the number of interactions between the two.

similar for bidders 2 and 3, who have a total of 8 and 6 interactions, respectively, all of which are, again, with seller A. In contrast, bidders 4 and 5 are only 80% and 70% loyal to seller A since, out of their total number of interactions (both 10), they share 2 with seller B and 3 with seller C, respectively. All in all, seller A attracts mostly highly loyal bidders. This is different seller D who attracts mostly disloyal bidders, as he shares all from his bidders with either seller B or C.

For each seller, we can summarize the proportion of loyal bidders and the degree of their loyalty in the associated *loyalty distribution*. The loyalty distributions for sellers A–D are displayed in Figure 4.66. The x-axis denotes the degree of loyalty (e.g., 100% loyal, or 80% loyal) and the y-axis denotes the corresponding density. We can see that the shape of all four distributions is very different; while seller A's distribution is very left-skewed (mostly high-loyal bidders), seller D's distribution is very right-skewed (mostly disloyal bidders). The distributions of sellers B and C fall somewhat in between, yet they are still very distinct from one another.

Note that our definition of loyalty is similar to the concept of in- and out-degree analysis. More precisely, we first measure the proportion of interactions for each buyer (the normalized distribution of out-degree, that is, links that point from a particular node). Then, we measure the perceived loyalty of each seller, which can be viewed as the distribution of the weighted in-degree (i.e., the number of links that point to a particular node). This definition of loyalty is very similar to the concept of brand switching in marketing. In essence, if we have a fixed number of brands (sellers in our case) and a pool of buyers (i.e., bidders), then we measure the switching behavior from one brand to another.

While the loyalty distributions in Figure 4.66 capture all the relevant information, we cannot use them for further analysis (especially modeling). Thus, our next step is to characterize each loyalty distribution by only a few statistics. To this end, we employ a very flexible dimension reduction approach via functional data analysis.

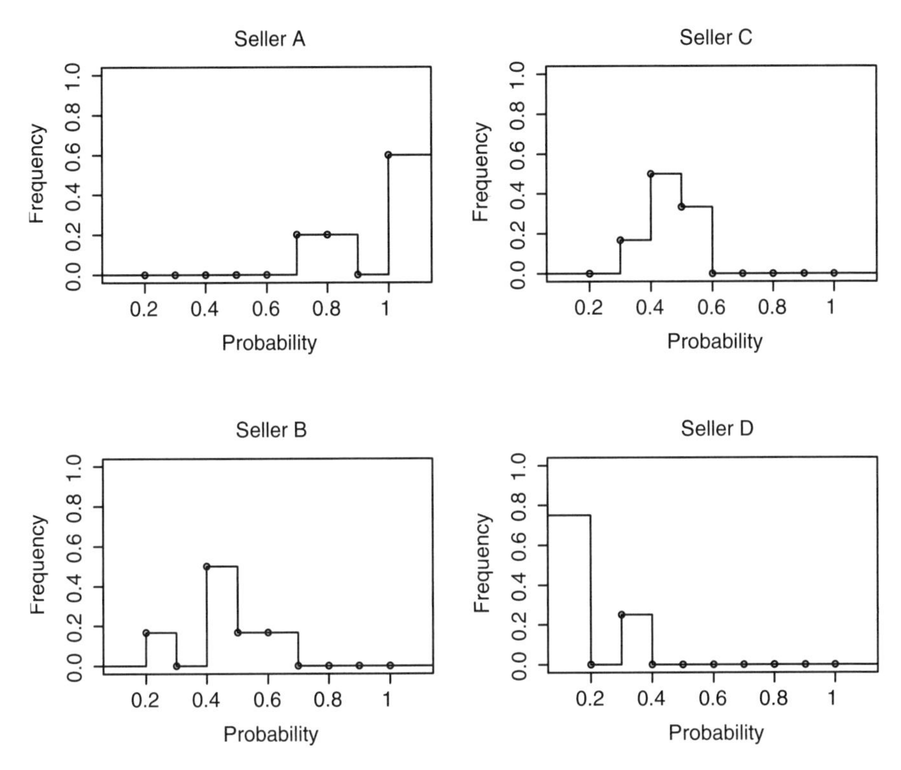

FIGURE 4.66 The resulting loyalty distributions for the hypothetical network in Figure 4.65.

From Loyalty Distributions to Loyalty Measures To investigate the effect of loyalty on the outcome of an auction, we first need to characterize a seller's loyalty distribution. While one could characterize the distributions via summary statistics (e.g., mean, median, or mode), Figure 4.66 suggests that loyalty is too heterogeneous and too dispersed. Therefore, we resort to a very flexible alternative via functional data analysis (Ramsay and Silverman, 2005). In fact, we regard each seller's loyalty distribution as a functional observation. We capture similarities (and differences) across distributions via *functional principal component analysis* (fPCA), a functional version of principal component analysis (Kneip and Utikal, 2001). In fact, while Kneip and Utikal (2001) operate on the true probability distributions, these are not known in our case; hence, we apply fPCA to the observed (empirical) distribution function that may introduce an extra level of estimation error.

Functional principal component analysis is similar in nature to ordinary PCA; however, rather than operating on data vectors, it operates on functional objects. In our context, we take the observed loyalty distributions (i.e., the histograms in Figure 4.66) as input. Although one could also first smooth the observed histograms, we decided against it since the results were not substantially different.

Ordinary PCA operates on a set of data vectors, say, $\mathbf{x}_1, \ldots, \mathbf{x}_n$, where each observation is a p-dimensional data vector $\mathbf{x}_i = (x_{i1}, \ldots, x_{ip})^T$. The goal of ordinary PCA is to find a projection of $\mathbf{x}_1, \ldots, \mathbf{x}_n$ into a new space that maximizes the variance along each component of the new space and at the same time renders the individual components of the new space orthogonal to one another. In other words, the goal of ordinary PCA is to find a PC vector $\mathbf{e}_1 = (e_{11}, \ldots, e_{1p})^T$ for which the principal component scores (PCS)

$$S_{i1} = \sum_j e_{1j} x_{ij} = \mathbf{e}_1^T \mathbf{x}_i \tag{4.138}$$

maximize $\sum_i S_{i1}^2$ subject to

$$\sum_j e_{1j}^2 = \|\mathbf{e}_1\|^2 = 1. \tag{4.139}$$

This yields the first PC, $\mathbf{e}_1$. In the next step, we compute the second PC, $\mathbf{e}_2 = (e_{21}, \ldots, e_{2p})^T$, for which, similarly to above, the principal component scores $S_{i2} = \mathbf{e}_2^T \mathbf{x}_i$ maximize $\sum_i S_{i2}^2$ subject to $\|\mathbf{e}_2\|^2 = 1$ and the *additional constraint*

$$\sum_j e_{2j} e_{1j} = \mathbf{e}_2^T \mathbf{e}_1 = 0. \tag{4.140}$$

This second constraint ensures that the resulting principal components are orthogonal. This process is repeated for the remaining PC, $\mathbf{e}_3, \ldots, \mathbf{e}_p$.

The functional version of PCA is similar in nature, except that we now operate on a set of continuous curves rather than discrete vectors. As a consequence, summation is replaced by integration. More specifically, assume that we have a set of curves $\mathbf{x}_1(s), \ldots, \mathbf{x}_n(s)$, and each measured on a continuous scale indexed by s. The goal is to find a corresponding set of PC curves, $\mathbf{e}_i(s)$, that, as previously, maximize the variance along each component and are orthogonal to one another. In other words, we first find the PC function, $\mathbf{e}_1(s)$, whose PCS

$$S_{i1} = \int e_1(s) x_i(s) ds \tag{4.141}$$

maximize $\sum_i S_{i1}^2$ subject to

$$\int e_1^2 ds = \|\mathbf{e}_1\|^2 = 1. \tag{4.142}$$

Similar to the discrete case, the next step involves finding $\mathbf{e}_2$ for which the PCS $S_{i2} = \int e_2(s)x_i(s)ds$ maximize $\sum_i S_{i2}^2$ subject to $\|\mathbf{e}_2\|^2 = 1$ and the additional constraint

$$\int e_2(s)e_1(s)ds = 0. \tag{4.143}$$

In practice, the integrals in (4.141)–(4.143) are approximated either by sampling the predictors, $\mathbf{x}_i(s)$, on a fine grid or, alternatively, by finding a lower dimensional expression for the PC functions $\mathbf{e}_i(s)$ with the help of a basis expansion. For instance, let $\boldsymbol{\phi}(s) = (\phi_1(s), \ldots, \phi_K(s))$ be a suitable basis expansion (Ramsay and Silverman, 2005). Then we can write

$$\mathbf{e}_i(s) = \sum_{k=1}^{K} b_{ik}\phi_k(s) = \boldsymbol{\phi}(s)^T\mathbf{b}_i, \tag{4.144}$$

for a set of basis coefficients $\mathbf{b} = (b_{i1}, \ldots, b_{iK})$. In this fashion, the integral in, for example, (4.143) becomes

$$\int e_2(s)e_1(s)ds = \mathbf{b}_1^T\mathbf{W}\mathbf{b}_2, \tag{4.145}$$

where $\mathbf{W} = \int \boldsymbol{\phi}(s)\boldsymbol{\phi}(s)^T ds$. For more details, see Ramsay and Silverman (2005). In this chapter, we use the grid approach.

Common practice is to choose only those eigenvectors that correspond to the largest eigenvalues; that is, those that explain most of the variations in $\mathbf{x}_1(s), \ldots, \mathbf{x}_n(s)$. By discarding those eigenvectors that explain no or only a very small proportion of the variation, we capture the most important characteristics of the observed data patterns without much loss of information. In our context, the first two eigenvectors capture over 82% of the variation in loyalty distributions.

Interpreting the Loyalty Measures Since our loyalty measures are based on their principal component representations, interpretation has to be done with care. Figure 4.67 shows the first two principal components (PCs). The first PC (top panel) shows a growing trend and, in particular, puts large negative weights on the lowest loyalty scores (between 0 and 0.2) while putting positive weight on medium to high loyalty scores (0.4 and higher). Thus, we can say that the first PC contrasts the extremely disloyal distributions from the rest. Table 4.16 (first row) confirms this notion: Note the large negative correlation with the minimum; also, the large correlation with the skewness indicates that PC1 truly captures extremes in the loyalty distributions' scores and shape. We can conclude that PC1 distinguishes distributions of "pure disloyalty" from the rest.

The second PC has a different shape. It puts most (positive) weight on the highest loyalty scores (between 0.8 and 1); it puts negative weight on scores at the medium and low scores (between 0.4 and 0.6) and thus contrasts average loyalty from extremely

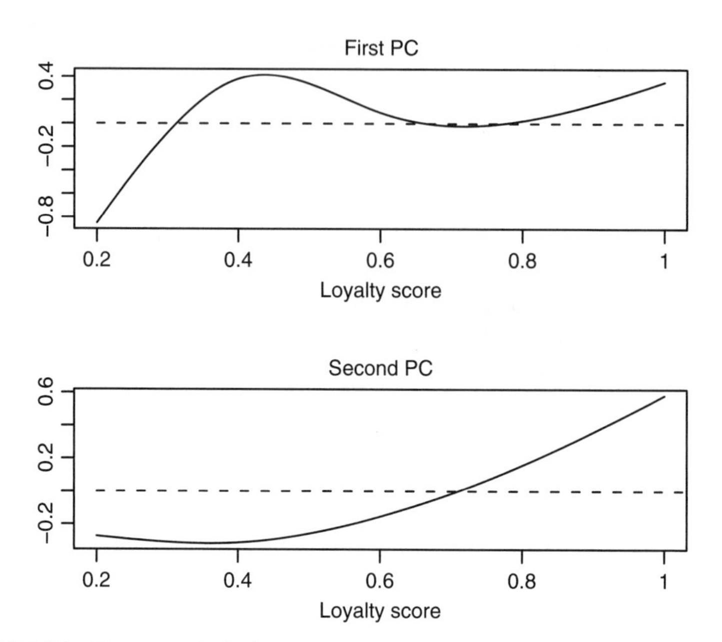

FIGURE 4.67 First two principal component curves. The dashed line indicates the x-origin.

high loyalty. Indeed, Table 4.16 (second row) shows that PC2 has a high positive correlation with the maximum and a high negative correlation with the median. In that sense, it distinguishes the mediocre loyalty from the stars.

While the above interpretations help our understanding of the loyalty components, their overall impact is still hard to grasp, especially because every individual loyalty distribution will—by nature of the principal component decomposition—comprise of a different *mix* between PC1 and PC2. Moreover, as we apply fPCA to observed densities (i.e., histograms), individual values of each density function must be heavily correlated. This adds additional constraints on the PCs and their interpretations. Hence, in the following, we discuss five *theoretical* loyalty distributions and their corresponding representation via PC1 and PC2.

Figure 4.68 shows five plausible loyalty distributions as they may develop out of a bidder–seller network. We refer to these distributions as "theoretic loyalty distributions," and we can characterize them by their specific shapes. For instance, the first distribution comprises of 100% loyal buyers, and hence we refer to it as "pure

TABLE 4.16 Correlation Between the First Two PC Scores and Summary Statistics of Sellers' Loyalty Distributions.

	Median	SDev	Max	Min	Skew
PC1	0.55	−0.2	0.52	−0.99	0.77
PC2	−0.78	−0.05	0.81	−0.02	0.63

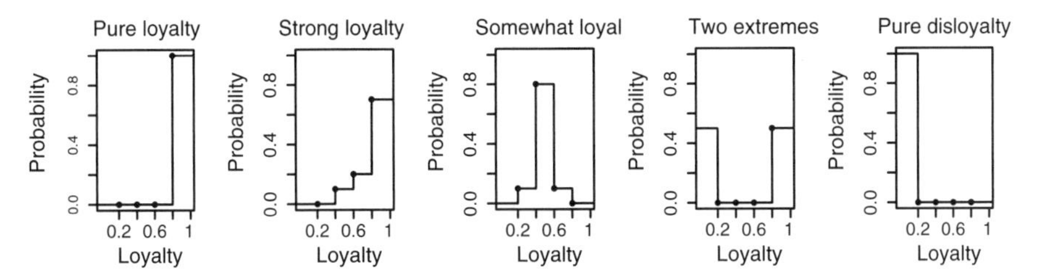

FIGURE 4.68 Five theoretic loyalty distributions.

loyalty." In contrast, the last distribution comprises of 100% disloyal buyers, and hence we name this distribution "pure disloyalty." The distribution at the center ("somewhat loyal") is interesting since it comprises mostly of buyers that exhibit some loyalty but do not purchase exclusively from only one seller.

Table 4.17 shows the corresponding PC scores. We can see that the theoretical distribution corresponding to *pure loyalty* scores very high on PC1 since it is very right-skewed and does not have any values lower than 0.9; in contrast, note the PC1 scores for *pure disloyalty*: while it is the mirror image of *pure loyalty*, it scores (in absolute terms) higher than the former because it is not only very (left-) skewed, but its extremely small values weigh heavily (and negatively) with the first part of the PC1 shape, which is in contrast to the positive values of *pure loyalty* that do not receive as much weight. As for PC2, Table 4.17 shows that *pure loyalty* scores even higher on this component as its values are extremely large, much larger than the typical (median) loyalty values. In contrast, *pure disloyalty* has very small PC2 values as low scores are given very little weight by PC2.

We can make similar observations for the remaining theoretical loyalty distributions. For instance, the distribution of *somewhat loyal* scores high on PC1 since it has not many low values. But it also receives only an average score on PC2 since it has not many high values either.

4.5.4 Conclusion

There are several statistical challenges when studying loyalty networks. First, deriving quality measures from the observed networks requires a method that can capture both

TABLE 4.17 PC1 and PC2 Scores for the Theoretical Loyalty Distributions from Figure 4.68

	Pure Lyty	Strng Lyty	Somew Loyal	Two Ex	Pure Dislyty
PC1	0.56	0.47	0.32	−0.04	−0.64
PC2	0.72	0.51	−0.08	0.35	−0.01

the intensity and the size of loyalty. We accomplish this using ideas from functional data analysis. Second, modeling the effect of loyalty is complicated by the extreme skew of loyalty networks. Jank and Yahav (2010) show that many different statistical models can lead to model misspecification and, as a consequence, to economically wrong conclusions. Similar problems likely exist in other studies on online markets (e.g., those that study seller feedback or reputation where one also records repeat observations on the same seller).

Another statistical challenge revolves around *sampling* bidder–seller networks. As pointed out earlier, we have the complete set of bidding records for a certain product (Swarovski beads, in this case) for a certain period of time (6 months). As a result, we have the complete bidder–seller network for this product, for this time frame. While sampling would be an alternative, it would result in an incomplete network (since we would no longer observe all nodes/arcs). As a result, we would no longer be able to compute loyalty without error that would bring up an interesting statistical problem. But we caution that sampling would have to be done very carefully. While one could, at least in theory, sample randomly across all different eBay categories, it would bring up several problems. The biggest problem is that we would now be attempting to compare loyalty across different product types. For instance, we would be comparing, say, a bidder's loyalty for purchasing beads (a very low price, low stake item) with that of purchasing digital cameras, computers or even automobiles (all of which are of high price and high stakes) that would be conceptually very questionable.

We also mention that we treat the bidder–seller network as *static* over time. Our data span a time frame of only 6 months, and we assume that loyalty is static over this time frame. This assumption is not too unrealistic as many marketing models consider loyalty to be static over much longer time frames (Fader and Hardie, 2006; Fader et al., 2006; Donkers et al., 2003). While incorporating a temporal dimension (e.g., by using a network with a sliding window or via down-weighting older interactions) would be an intriguing statistical challenge, it is not quite clear how to choose the width of the window or the size of the weights. Moreover, we also explicitly tested for learning effects by buyers over time and could not find any strong statistical evidence for it.

Finally, in this chapter we address one specific kind of network dependence, namely, that between buyers and sellers. However, the dependence structure may in fact be far more complex. As bidders are linked to sellers that, in turn, are linked again to other bidders, the true dependence structure among the observations may be far more complex. This may call for innovative statistical methodology and we hope to have sparked some new ideas with our work.

5

FORECASTING ONLINE AUCTIONS

In this chapter, we discuss methods and models for forecasting the outcome of an auction. By "outcome" we primarily refer to the price that an auction fetches, but similar ideas could also be applied to forecast the total number of bidders, whether or not an auction will transact, and so on. Forecasting the auction outcome (in particular its final price) is beneficial to all auction parties: Bidders can use price forecasts to make more informed bidding decisions, sellers can use similar price forecasts to determine when to post their listings, and auction houses can use forecasts for long-term budgeting and planning purposes or even for real-time adjustments such as actively inviting bidders to an auction.

Forecasting online auction outcomes is not trivial: Online auctions experience very heterogeneous arrivals of bidders and bids leading to price dynamics that change constantly. Moreover, bids are placed very unevenly throughout the auction, making traditional forecasting methods (such as moving averages or exponential smoothing) difficult to apply. In addition, auctions selling same items compete against each other for the same bidders. The competition effect may cause, for instance, the price in one auction to increase as a result of stalling prices in other auctions. Moreover, competing auctions do not start and end at the same time. In fact, auctions are often of different length and new auctions enter the market continuously.

In the following, we discuss forecasting models that can address some of these challenges. In Section 5.1, we start out with models for individual auctions that ignore the effect of competition. In that context, we describe two different approaches, one that uses only information from within the auction of interest and another that borrows strength from neighboring auctions. While the latter approach considers auctions that are similar, it does not explicitly model auction competition. We discuss forecasting competing auctions in Section 5.2. In that context, we will see that one particular challenge is the creation and selection of appropriate features that can capture and measure competition. In fact, we will see that model selection for forecasting competing auctions can be quite burdensome and we discuss several solutions. Finally, in Section 5.3, we discuss a general bidding framework for forecasting models. More specifically, we discuss the use of forecasting models to make more informed and automated bidding decisions and we compare it with well-known (manual) bidding strategies such as early bidding or last-minute sniping.

There is an important conceptual difference between this chapter and the models discussed in Chapter 4: Although both chapters discuss models for online auctions, Chapter 4 primarily focuses on *descriptive and explanatory models* while we now explicitly concentrate on forecasting. While forecasting models can share many similarities with descriptive and explanatory models , the main difference is that the goal of forecasting models is always *forward-looking*: all model components are assembled with the goal of predicting a future value. This goal leads to different ways in which we assemble, select, or evaluate models (for a discussion of further differences, see Shmueli (2009)).

5.1 FORECASTING INDIVIDUAL AUCTIONS

Online auctions are awash with choices for the bidder. In fact, there are approximately 44 million items available worldwide on eBay at any given time and approximately 4 million new items are added every day in over 50,000 categories. Oftentimes, an identical (or near-identical) product is sold in numerous, (near-) simultaneous auctions. For instance, a simple search under the keywords "iPod shuffle 512 MB MP3 player" returns over 300 hits for auctions that close within the next 7 days. A more general search under the less restrictive keywords "iPod MP3 player" returns over 3000 hits. Clearly, it would be challenging, even for a very dedicated eBay user, to inspect and simultaneously monitor all these 300 (or 3000) auctions, while being on the lookout for newly added auctions for the same product and subsequently deciding in which of these numerous auctions to participate in and to place a bid.

The decision-making process of a bidder can be supported by price forecasts. With the availability of a price forecasting system, one could, for example, create an auction ranking (from lowest predicted price to highest) and focus only on those auctions with the lowest predicted price. One of the difficulties with such an approach is that information in the online environment changes constantly: new auctions enter the market, old (i.e., closed) auctions drop out, and even within the same auction, the price changes continuously with every newly arriving bid. Thus, a well-functioning forecasting system must be adaptive to accommodate and incorporate change in this highly dynamic environment.

In the following, we discuss a dynamic forecasting model that can adapt to change. Price forecasts can generally be done in two different ways: in a *static* manner or in a *dynamic* manner. The static approach relates information that is known *before* the start of the auction to information that becomes available *after* the auction closes. This is the basic principle of several existing models (Ghani and Simmons, 2004; Ghani, 2005; Lucking-Reiley et al., 2007; Bajari and Hortacsu, 2003). For instance, one could relate the opening bid, the auction length, and a seller's reputation to the final price. Note that opening bid, auction length, and seller reputation are all known at the auction start. Training a model on a suitable set of past auctions, one can obtain static forecasts of the final price in that fashion. However, this approach does not take into account important information that arrives *during* the auction. The *current* number of competing bidders or the *current* price level are factors that are revealed only *during* the ongoing auction and that are important in determining the *future* price. Moreover, the current *change in price* also has a huge impact on the future price. If, for instance, the price had increased at an extremely fast rate over the last several hours, causing bidders to drop out of the bidding process or to revise their bidding strategies, then this could have an immense impact on the evolution of price in the next few hours and, subsequently, on the final price. We refer to models that account for newly arriving information and for the rate at which this information changes as *dynamic* models.

Dynamic price forecasting in online auctions is challenging for a variety of reasons. Traditional methods for forecasting time series, such as exponential smoothing or moving averages (MA), cannot be applied in the auction context, at least not directly, due to the special data structure. Traditional forecasting methods assume that data arrive in evenly spaced time intervals such as every quarter or every month. In such a setting, one trains the model on data up to the current time period t and then uses this model to predict at future time points $t + k$ $(k = 1, 2, \ldots)$. Implied in this process is the assumption that the distance between two adjacent time periods is equal, which is the case for quarterly or monthly data. Now consider the case of online auctions. Bids arrive in very unevenly spaced time intervals, determined by the bidders and their bidding strategies, and the number of bids within a short period of time can sometimes be very sparse, while other times extremely dense. In this setting, the distance between t and $t + 1$ can sometimes be more than a day, while at other times it may only be a few seconds. Traditional forecasting methods also assume that the time series continues, at least in theory, for an infinite amount of time and does not stop at any point in the near future. This is clearly not the case in a fixed duration online auction. The implication is a discrepancy in the estimated forecasting uncertainty. Finally, online auctions, as pointed out earlier (see Chapters 3 and 4), can experience price paths with very heterogeneous *price dynamics* (Jank and Shmueli, 2006; ?; Shmueli and Jank, 2008; Wang et al., 2008a). In the following, we therefore describe a dynamic forecasting model based on the ideas of functional data analysis.

5.1.1 A Dynamic Forecasting Model

In this section, we describe a forecasting model that incorporates the price dynamics of an online auction and adaptively adjusts these dynamics as new information arrives. One important characteristic of this model is that it uses only the information from within the auction of interest to estimate and forecast these dynamics (Wang

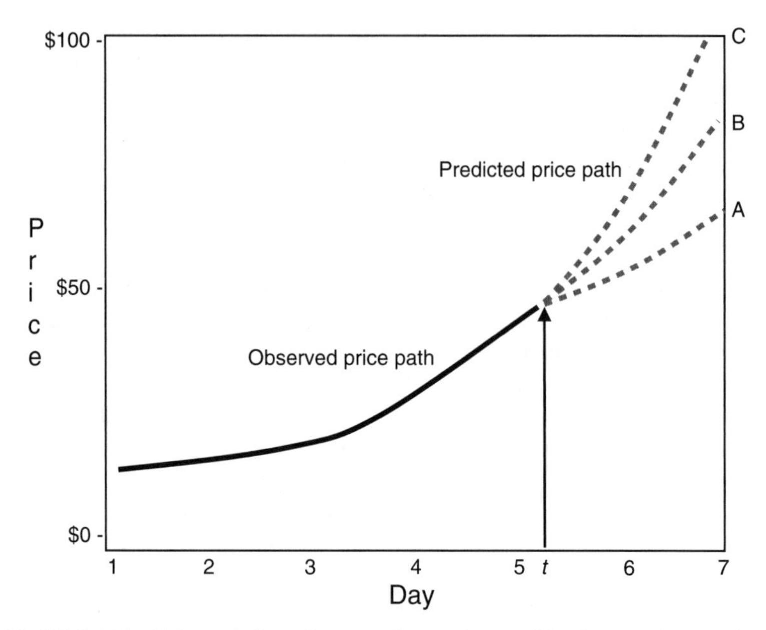

FIGURE 5.1 Schematic for a dynamic forecasting model of an ongoing auction.

et al., 2008a; Dass et al., 2009); in other words, it does not use information from other auctions. Later, we propose a forecasting model that harnesses the power of auction dynamics and also explicitly uses information from similar past auctions to estimate the price dynamics (Zhang et al., 2010). Methodologically, the main difference is that here we use an autoregressive model to estimate and predict dynamics; in contrast, later we use a k-nearest neighbor (KNN) paradigm. While the autoregressive model uses only the information from within the focal auction, the KNN model first determines a set of similar auctions (from the past), and then uses those similar auctions for prediction purposes. Note though that neither of these two models explicitly models the effect of competing auctions—we will address competing auctions in Section 5.2.

Overview Our goal is to develop a *dynamic* forecasting model, such that the model operates in the live auction and forecasts price at a future time point of the ongoing auction. This is in contrast to a *static* forecasting model that makes prediction only about the final price and takes into consideration only information available before the start of the auction. Consider Figure 5.1 for illustration. Assume that we observe the price path from the start of the auction until time t (solid black line). We now want to forecast the continuation of this price path (such as the broken gray lines, labeled "A," "B," and "C"). The difficulty in producing this forecast is the uncertainty about price dynamics into the future. If the dynamics level off, then the price increase will slow down and we might see a price path similar to the one labeled "A." On the other hand, if the dynamics remain steady, the price path might look more like the one in "B." Or, if the dynamics sharply increase, then a path like the one in "C" could be the consequence. Either way, knowledge of the future price dynamics appears to be a key factor!

Our dynamic forecasting model consequently consists of two parts: First, we develop a model for estimating and forecasting price dynamics. Then, we incorporate the estimated dynamics, together with other relevant auction covariates, into an econometric model for the price path; this price path model can be used to forecast the final price as well as the price at any time before the auction closes. One methodological challenge is the incorporation of static predictor variables, consisting of information that does not change during the auction (e.g., the opening price or the auction length). Since our model is based on the ideas of autoregressive models, such static predictors would be confounded with the model's intercept if not treated carefully. We therefore use the idea of functional regression modeling to first transform these static variables into *influence-weighted* predictors. We describe details of this approach below.

Modeling and Forecasting Dynamics One of the main characteristics of online auctions is their rapid change in dynamics. Since change in the $(p+1)$ derivative precedes change in the pth derivative (e.g., change in acceleration precedes change in velocity), we make use of derivative information for forecasting. In the following, we develop a model to estimate and forecast an auction's price dynamics.

Let $D^{(m)}y_t$ denote the mth derivative of the price y_t at time t. We model the derivative curve as a polynomial in time t with autoregressive (AR) residuals

$$D^{(m)}y_t = a_0 + a_1 t + \cdots + a_k t^k + \boldsymbol{\alpha}\mathbf{x}(t) + u_t, \tag{5.1}$$

where $\mathbf{x}(t)$ is a vector of covariates, $\boldsymbol{\alpha}$ is a corresponding vector of parameters, and u_t follows an autoregressive model of order p:

$$\begin{aligned} u_t &= \phi_1 u_{t-1} + \phi_2 u_{t-2} + \cdots + \phi_p u_{t-p} + \varepsilon_t, \\ \varepsilon_t &\sim N(0, \sigma^2). \end{aligned} \tag{5.2}$$

We allow the price dynamics to depend on the vector $\mathbf{x}(t)$, which results in a very flexible model that can accommodate different dynamics due to differences in the auction format, the product category, and so on.

We estimate model (5.1) from a training sample. Estimation is done in two steps: First, we estimate the parameters $a_1, \ldots, a_k$ and $\boldsymbol{\alpha}$. Then, using the estimated residuals $\hat{u}_t$, we estimate $\phi_1, \ldots, \phi_p$.

Forecasting is also done in two steps. Let $1 \leq t \leq T$ denote the observed time period and let $T+1$, $T+2$, $T+3$, and so on. denote time periods we wish to forecast. We first forecast the next residual via

$$\tilde{u}_{T+1|T} = \tilde{\phi}_1 u_T + \tilde{\phi}_2 u_{T-1} + \cdots + \tilde{\phi}_p u_{T-p+1}. \tag{5.3}$$

Using this forecast, we can predict the derivative at the next time point $T + 1$ via

$$D^{(m)}\tilde{y}_{T+1|T} = \hat{a}_0 + \hat{a}_1(T+1) + \cdots + \hat{a}_k(T+1)^k + \hat{\boldsymbol{\alpha}}\mathbf{x}(T+1) + \tilde{u}_{T+1|T}. \quad (5.4)$$

In a similar fashion, we can predict the derivative l steps ahead:

$$D^{(m)}\tilde{y}_{T+l|T} = \hat{a}_0 + \hat{a}_1(T+l) + \cdots + \hat{a}_k(T+l)^k + \hat{\boldsymbol{\alpha}}\mathbf{x}(T+l) + \tilde{u}_{T+l|T}. \quad (5.5)$$

Integrating Static Auction Information One structural challenge of our model is related to the incorporation of static predictors. Consider, for instance, the opening bid, which is static in the sense that its value is constant throughout the entire auction ($x(t) \equiv x, \forall t$). Ignoring all other variables, model (5.1) becomes

$$D^{(m)}y(t) = \alpha_0 + \alpha x. \quad (5.6)$$

Because the right-hand side of (5.6) does not depend on t, the least squares estimates of α_0 and α are confounded. In other words, we cannot distinguish between the intercept and the effect of the opening bid on the price velocity.

The problem outlined above is relatively uncommon in traditional time-series analysis since it is usually meaningful to include a predictor variable into an econometric model only if the predictor variable itself carries time-varying information. However, the situation is different in the context of forecasting online auctions and may merit the inclusion of certain static information. The opening bid, for instance, may in fact carry valuable information for predicting price in the ongoing auction. Economic theory suggests that sometimes bidders derive information from the opening bid about their own valuation, but the impact of this information decreases as the auction progresses (see Chapter 4 and Shmueli and Jank, 2008). What this suggests is that the opening bid can influence bidders' valuations and therefore also influence price. But what this also suggests is that the opening bid's impact on price does not remain constant but should be discounted gradually throughout the auction.

One way of discounting the impact of a static variable x is via its influence on the price evolution. That is, if x has a stronger influence on price at the beginning of the auction, then it should be discounted less during that period. On the other hand, if x only barely influences price at the auction end, then its discounting should be larger at the auction end. One way of measuring the influence of a static variable on the price curve is via functional regression analysis, as described in Chapter 4. Let $\tilde{\beta}(t)$ denote the slope coefficient from the functional regression model

$$y(t) = \alpha(t) + \beta(t)x + \varepsilon.$$

$\tilde{\beta}(t)$ thus quantifies the influence of x on $y(t)$ at any time t. We combine x and $\tilde{\beta}(t)$ and compute the *influence-weighted* version of the static variable x as

$$\tilde{x}(t) = x\tilde{\beta}(t). \quad (5.7)$$

TABLE 5.1 Categories of 768 Book Auctions

Book Category	Count	Mean	StDev
Antiquarian and collect.	84	\$89.90	\$165.82
Audiobooks	46	\$18.57	\$22.48
Children's books	102	\$12.89	\$18.03
Fiction books	162	\$7.90	\$9.34
Magazines	58	\$11.87	\$9.43
Nonfiction books	239	\$15.63	\$52.83
Textbooks and educ.	36	\$19.62	\$37.46
Other	41	\$36.48	\$80.31

The second column gives the number of auctions per category. The third and fourth columns show average and standard deviation of price per category.

$\tilde{x}(t)$ now carries time-varying information and can consequently be included as a time-varying predictor variable.

Modeling and Forecasting Price After forecasting the price dynamics, we use these forecasts to predict the auction price over the next few time periods up to the auction end. Many factors can affect the price of an auction, such as information about the auction format, the product, the bidders, and the seller. Let $\mathbf{x}(t)$ denote the vector of all such factors (possibly including influence-weighted static predictors as discussed above). Let $\mathbf{d}(t) = (D^{(1)}y_t, D^{(2)}y_t, \ldots, D^{(p)}y_t)$ denote the vector of price dynamics, that is, the vector of the first p derivatives of y at time t. The price at t can be affected by the price at $t-1$ and potentially also by its values at times $t-2, t-3$, and so on. Let $\mathbf{l}(t) = (y_{t-1}, y_{t-2}, \ldots, y_{t-q})$ denote the vector of the first q lags of y_t. We then write the general dynamic forecasting model as follows:

$$y_t = \boldsymbol{\beta}\mathbf{x}(t) + \boldsymbol{\gamma}\mathbf{d}(t) + \boldsymbol{\delta}\mathbf{l}(t) + \epsilon_t, \tag{5.8}$$

where $\boldsymbol{\beta}$, $\boldsymbol{\gamma}$, and $\boldsymbol{\delta}$ denote the parameter vectors and $\epsilon_t \sim N(0, \sigma^2)$. We use the estimated model (5.8) to predict the price at $T + l$ as

$$\tilde{y}_{T+l|T} = \hat{\boldsymbol{\beta}}\mathbf{x}(T+l) + \hat{\boldsymbol{\gamma}}\mathbf{d}(T+l) + \hat{\boldsymbol{\delta}}\mathbf{l}(T+l). \tag{5.9}$$

Case Study We investigate the performance of our forecasting model on a diverse data set containing 768 eBay book auctions, collected during October 2004. All auctions are 7 days long and span a variety of categories (Table 5.1). Prices range from \$0.10 to \$999 and are, not unexpectedly, highly skewed (see also Figure 5.2). Prices also vary significantly across the different book categories. This data set is challenging due to its diversity in products and price. We use 70% of these auctions (or 538 auctions) for training purposes. The remaining 30% (or 230 auctions) are kept in the validation sample.

Model Estimation: Our model building investigations (which are not reproduced here) suggest that among all price dynamics only velocity $f'(t)$ is significant (at least in our data). We thus estimate model $D^{(m)}y_t$ in (5.1) only for $m = 1$. We do so in the following way. Using a quadratic polynomial ($k = 2$) in time t and influence-weighted predictor variables for book-category ($\tilde{x}_1(t)$) and shipping costs ($\tilde{x}_2(t)$) results in an AR(1) process for the residuals u_t (i.e., $p = 1$ in (5.2)). The rationale behind using book-category and shipping costs in model (5.1) is that we would expect the dynamics to depend heavily on these two variables. For instance, the category of antiquarian and collectible books typically contains items that are of rare nature and that appeal to a market that is not very price sensitive and with a strong interest in obtaining the item. This is also reflected in the high average price and even higher variability for items in this category (Table 5.1). The result of these market differences may well be a different price evolution and thus different price dynamics. A similar argument applies to shipping costs, which are determined by the seller and act as "hidden" price premiums. Bidders are often deterred by excessively high shipping costs, and as a consequence auctions may experience slower price dynamics. Table 5.2 summarizes the estimated coefficients averaged across all auctions from the training set. We can see that both book-category and shipping costs result in significant price dynamics.

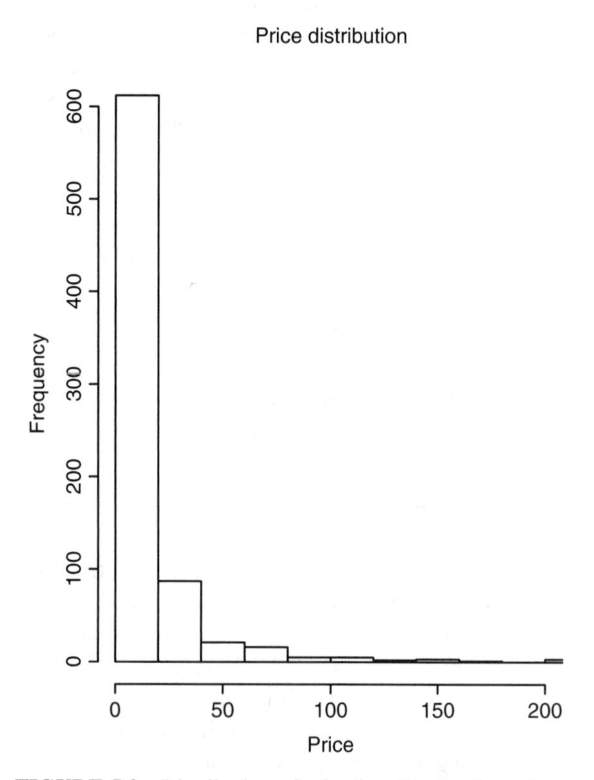

FIGURE 5.2 Distribution of price for eBay book auctions.

TABLE 5.2 Estimates for the Velocity Model $D^{(1)}\mathbf{y}_t$ in (5.1)

Predictor	Coeff	P-Value
Intercept	0.041	0.004
t	−0.012	0.055
t^2	0.004	0.041
Book category $\tilde{x}_1(t)$	1.418	0.038
Shipping costs $\tilde{x}_2(t)$	1.684	0.036
u_t	1.442	–

The second column reports the estimated parameter value, and the third column reports the associated significance levels. Values are averaged across the training set.

After modeling the price dynamics, we estimate the model for the price path. Recall that the model in (5.8) contains three components, $\mathbf{x}(t)$, $\mathbf{d}(t)$, and $\mathbf{l}(t)$. Among all reasonable price lags, we find that only the first lag is influential, so we have $\mathbf{l}(t) = y_{t-1}$. Also, as mentioned earlier, among the different price dynamics we only find the velocity to be important, so $\mathbf{d}(t) = D^{(1)}y_t$. The first two rows of Table 5.3 display the corresponding estimated coefficients.

Note that both $\mathbf{l}(t)$ and $\mathbf{d}(t)$ are predictor variables derived from price, either from its lag or from its dynamics. We also use eight nonprice related predictor variables $\mathbf{x}(t) = (x_1(t), x_2(t), x_3(t), \tilde{x}_4(t), \tilde{x}_5(t), \tilde{x}_6(t), \tilde{x}_7(t), \tilde{x}_8(t))^T$. Specifically, the eight predictor variables correspond to the average rating of all bidders until time t (which we refer to as the *current* average bidder rating at time t and denote as $x_1(t)$), the current number of bids at time t ($x_2(t)$), and the current winner rating at time t ($x_3(t)$). These first three predictor variables are time varying. We also consider five time-constant predictors: the opening bid ($\tilde{x}_4(t)$), the seller rating ($\tilde{x}_5(t)$), the seller's *positive* ratings ($\tilde{x}_6(t)$), the shipping costs ($\tilde{x}_7(t)$), and the book category ($\tilde{x}_8(t)$), where $\tilde{x}_i(t)$ again denotes the influence-weighted variables.

Table 5.3 shows the estimated parameter values for the full forecasting model. It is interesting to note that book-category and shipping costs have low statistical significance. The reason for this is that their effects have likely already been captured satisfactorily in the model for the price velocity. Also, note that the response y_t and all numeric predictors ($\tilde{x}_1(t), \ldots, \tilde{x}_7(t)$) are log-transformed. The implication of this lies in the interpretation of the coefficients. For instance, the value 0.051 implies that for every 1% increase in opening bid, the price increases by about 0.05%, on average.

Forecasting Performance: We estimate the forecasting model on the training data and use the validation data to investigate its forecasting accuracy. To that end, we assume that for the 230 auctions in the validation data we only observe the price until day 6 and we want to forecast the remainder of the auction. We forecast price over the last day in small increments of 0.1 days. That is, from day 6 we forecast day 6.1, or the price after the first 2.4 h of day 7. From day 6.1, we forecast day 6.2 and so on until the auction end, at day 7. The advantage of a sliding window approach is the possibility of feedback-based forecast improvements. That is, as the auction progresses over the

TABLE 5.3 Estimates for the Price Forecasting Model (5.8)

Des	Predictor	Coeff	P-Value
$\mathbf{d}(t)$	Price velocity $D^{(1)}y_t$	0.592	0.049
$\mathbf{l}(t)$	Price lag y_{t-1}	4.824	0.044
$\mathbf{x}(t)$	Intercept	5.909	0.110
$\mathbf{x}(t)$	Cur.Avg.Bid.Rating $x_1(t)$	0.414	0.012
$\mathbf{x}(t)$	Cur.Numb.Bids $x_2(t)$	–0.008	0.027
$\mathbf{x}(t)$	Cur.Win.Rating $x_3(t)$	0.197	0.027
$\mathbf{x}(t)$	Opening bid $\tilde{x}_4(t)$	0.051	0.031
$\mathbf{x}(t)$	Seller rating $\tilde{x}_5(t)$	−11.534	0.070
$\mathbf{x}(t)$	Pos seller rating $\tilde{x}_6(t)$	1.518	0.093
$\mathbf{x}(t)$	Shipping cost $\tilde{x}_7(t)$	0.008	0.215
$\mathbf{x}(t)$	Book category $\tilde{x}_8(t)$	3.950	0.107

The first column indicates the part of the model design that the predictor is associated with. The third column reports the estimated parameter values, and the fourth column reports the associated significance levels. Values are again averaged across the training set.

last day, the true price level can be compared with its forecasted level and deviations can be channeled back into the model for real-time forecast adjustments.

Figure 5.3 shows the forecast accuracy on the validation sample. We measure forecasting accuracy using the *mean absolute percentage error* (MAPE), given by

$$\text{MAPE}_t = \frac{1}{230}\sum_{i=1}^{230}\left|\frac{(\text{predicted price}_{t,i} - \text{true price}_{t,i})}{\text{true price}_{t,i}}\right|,$$

$$i = 1, \ldots, N;\ t = 6.1, \ldots, 7,$$

where i denotes the ith auction in the validation data. The solid line in Figure 5.3 corresponds to MAPE for our dynamic forecasting model. We benchmark the performance of our method against double exponential smoothing , which is a popular short-term forecasting method that assigns exponentially decreasing weights as the observation becomes less recent and also takes into account a possible (changing) trend in the data. The dashed line in Figure 5.3 corresponds to MAPE for double exponential smoothing.

We note that for both approaches, MAPE increases as we predict further into the future. However, while for our dynamic model MAPE increases to only about 5% at the auction end, exponential smoothing incurs an error of over 40%. This difference in performance is relatively surprising, especially given that exponential smoothing is a well-established (and powerful) tool in time-series analysis. One of the reasons for this underperformance is the rapid change in price dynamics, especially at the auction end. Exponential smoothing, despite the ability to accommodate changing trends in the data, cannot account for the price dynamics. This is in contrast to our dynamic forecasting model that explicitly models price velocity. As pointed out earlier,

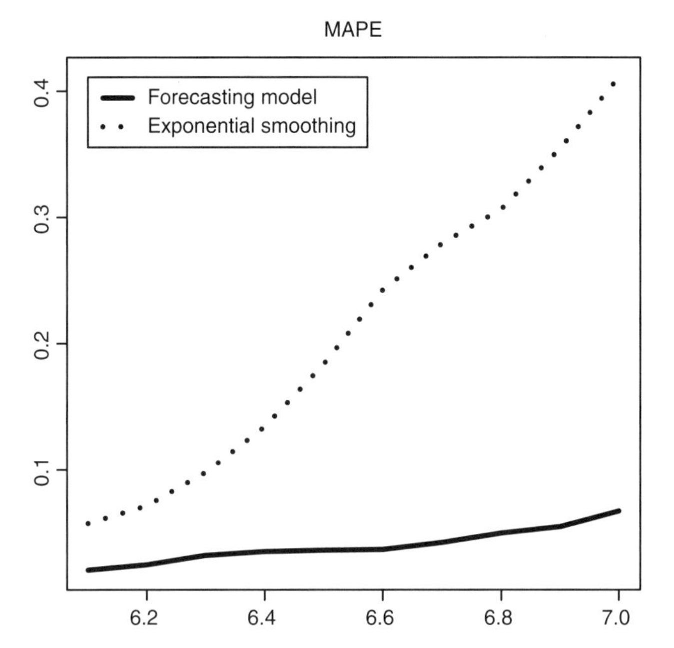

FIGURE 5.3 Mean absolute percentage error of the forecasted price over the last auction day. The solid line corresponds to our dynamic forecasting model; the dashed line correspond to double exponential smoothing. The x-axis denotes the day of the auction.

a change in a function's velocity precedes a change in the function itself. It is therefore only natural that modeling the dynamics makes a difference for forecasting the final price.

Discussion Dynamic price forecasts can be beneficial to many different auction parties. For instance, they can be used to dynamically rank auctions for the same (or similar) item by their predicted price. On any given day, there are several hundred or even thousands of auctions available online, especially for very popular items such as smart phones or game consoles. Dynamic price rankings can help bidders make decisions about which auctions are best to participate in. Auction forecasting can also be beneficial to the seller or the auction house. For instance, the auction house can use price forecasts to offer insurance to the seller. This is related to the idea by Ghani (2005), who suggested offering sellers an insurance that guarantees a minimum selling price. To do so, it is important to correctly forecast the price, at least on average. While Ghani's method is static in nature, our dynamic forecasting approach could potentially allow more flexible features like an "insure-it-now" option, enabling sellers to purchase an insurance either at the beginning of the auction or during the live auction (coupled with a time-varying premium). Price forecasts can also be used by eBay-driven businesses that provide brokerage services to buyers or sellers.

Finally, in order for dynamic forecasting to work in practice, it is important that the method is scalable and efficient. Therefore, all the components of our dynamic model are based on linear operations—estimating a smoothing spline or fitting an AR model are both done in ways very similar to least squares. In fact, the total runtime (estimation on training data plus validation on holdout data) for our data set (over 700 auctions) is less than a minute, using program code that is not even optimized for speed.

5.1.2 An Alternate Model: Functional KNN Forecasting

We now present a second model for dynamically forecasting price in online auctions introduced by Zhang et al. (2010). This model differs from the previous model in that it borrows strength from similar auctions using a k-nearest neighbors (KNN) framework; this is in contrast to the previous model that uses only information from within the auction of interest. A KNN approach is especially well suited for heterogeneous auctions such as on the eBay network since it weighs different auctions differently. Many previous approaches that aim to predict price in online auctions (e.g., Ghani and Simmons, 2004; Jap and Naik, 2008) use regression-based models. Common across regression-based models is that they use information from a set of past auctions to predict an ongoing or future auction. Moreover, for the purpose of model estimation, these models weigh the information from each past auction equally. For instance, if the goal is to predict the price of a laptop auction based on a sample of historical auctions, then estimating a regression-type model will put equal weight on the information from a *Dell* laptop and from an *IBM* laptop—which may be inappropriate if the goal is to predict an auction for a *Sony* laptop. While some of the brand and product differences can be controlled using appropriate predictor variables, there might still be intrinsic differences that are hard to measure and hence control. Caccetta et al. (2005) proposed classification and regression trees as an alternative to regression-based models. However, the authors point out that the prediction can be poor if prices in each final tree node vary significantly. Moreover, while trees, unlike regression, manage to partition the data in a very flexible way, their predictions, like those of regression, are also based on the *unweighted* information in each final node. In what follows, we describe a novel and flexible approach for forecasting online auction prices based on the ideas of *K-nearest neighbors*.

KNN is a forecasting approach that weighs the information from each record differently, depending on how similar that record is to the record of interest. For instance, if our goal is to predict the price of an auction for a *Sony* laptop, then it will put more weight on information from other *Sony* laptops and it will down-weight the information from, say, *Dell* or *IBM* laptops. More specifically, KNN predicts a record based on the weighted average of the K nearest neighbors of that record, where the weight is proportional to the proximity of the neighbor to the predicted record. KNN has been proven to converge to the true value for arbitrarily distributed samples (Stone, 1977; Devroye, 1981; Kulkarni and Posner, 1995), but studies show that its effectiveness is greatly affected by the choice of the number of neighbors (K) and the choice of distance metric (Cover and Hart, 1967; Goldstein, 1972; Short and Fukunaga, 1981; Kulkarni et al., 1998).

In the context of online auctions, the choice of the distance metric is challenging because auctions vary on many conceptually different dimensions. In particular, online auctions vary in terms of three types of information: *static, time-varying*, and *dynamic* information. Static information comprises of information that does not change throughout the auction. This includes product characteristics (e.g., brand, product condition) and auction and seller characteristics (e.g., auction length, whether there is a secret reserve price, or whether the seller is a power seller). Time-varying information gets updated during the auction (e.g., the number of bids or bidders). Static and time-varying information has been shown to be important for forecasting the auction price because differences in product or bidding characteristics all influence bidders' decisions and hence the final price. Finally, auctions also vary in terms of their *dynamic* information. Dynamic information refers to the price path and its dynamics and includes the price velocity and the rate at which this speed changes throughout the auction. As seen in the previous section, auction dynamics are important for forecasting the final price because an auction that experiences fast price movements in the earlier stage will likely see a slowdown in price in later stages; conversely, auctions whose price travels very slowly at the beginning often see price accelerations toward the end (Jank and Shmueli, 2006, 2008b; Shmueli and Jank, 2008; Wang et al., 2008a)

Price dynamics are captured via functional objects that, as discussed in Chapter 4, can be computed via a variety of (parametric or nonparametric) smoothing methods. In fact, in most of this book, we use penalized smoothing splines for computing functional objects since they are readily implemented in many different software packages and are fast to compute. However, in this section, we face a new challenge in that we want to compute *distances* between different price paths and their dynamics. Using penalized smoothing splines (or monotone splines), it is not at all obvious how to compute the distance between two functional objects. This leads us to the Beta model discussed in Chapter 4, which allows us to measure the distance between auctions' dynamics in a very parsimonious way via the Kullback–Leibler (KL) distance (Basseville, 1989). The resulting model is methodologically innovative in that it presents a *functional* KNN model, which, for short, we refer to as *fKNN.*

The fKNN forecaster, which integrates information of various types, uses different distance metrics for each data type. We first discuss the different distance measures and how they are combined into a single distance metric. We also discuss another important aspect of KNN forecasters, which is the choice of K. While choosing K too small eliminates important information, choosing K too large results in noise that deteriorates forecast accuracy. The goal is to find a value that best balances signal to noise. Stone (1977) found that K can depend on the distribution of the data and that the optimal K often grows with the sample size. In what follows, we thus investigate the optimal value of K as a function of different distance metrics as well as of data size and heterogeneity.

Overview Consider Figure 5.4, where the solid line corresponds to the price process of an auction that is observed until time T. The dotted line corresponds to

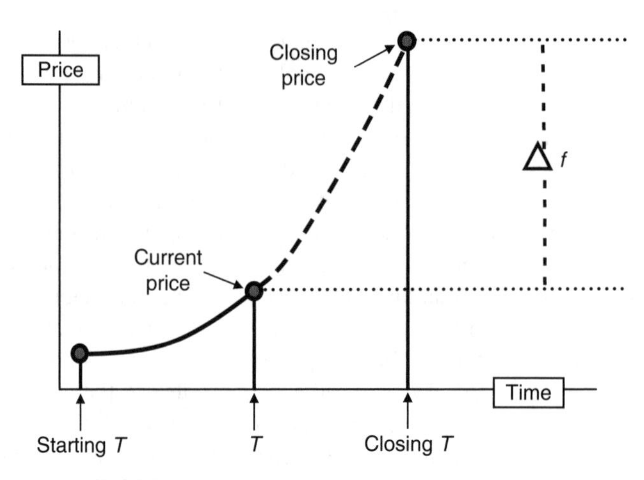

FIGURE 5.4 Illustration of the forecasting idea.

the (future) price path until the close of the auction. Our goal is to predict the closing price. As the closing price is determined by the current price plus the price increment Δ_f, our forecasting problem is equivalent to predicting Δ_f. We will therefore use fKNN to estimate Δ_f based on a training set of completed auctions.

In order to estimate Δ_f, we look for the K most similar auctions in the training set. Consider Figure 5.5 for illustration. In this scenario, the training set has six auctions, Δ_1–Δ_6. We also have associated distances, D_1–D_6, between the focal auction and each of the auctions in the training set. If K equals 3, then we will estimate Δ_f by the weighted average of the three nearest auctions, and in this case by the weighted

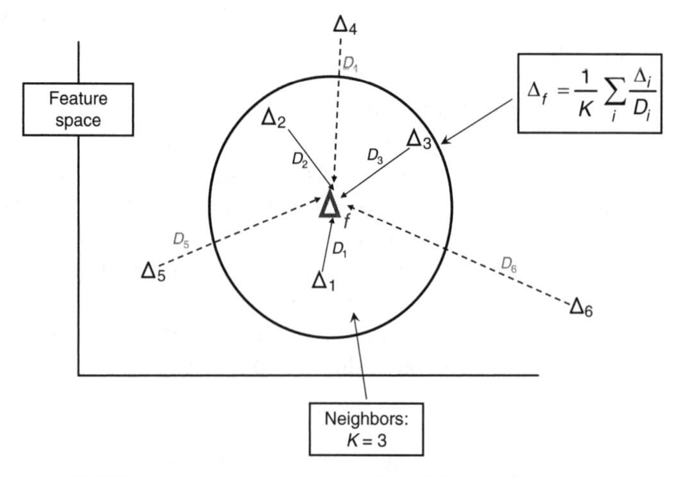

FIGURE 5.5 Illustration of the fKNN forecasting scheme.

average of $\triangle_1$—$\triangle_3$. More generally, we estimate $\triangle_f$ as

$$\triangle_f = \frac{\sum_{i=1}^{K} \frac{1}{D_i} \triangle_i}{\sum_{i=1}^{K} \frac{1}{D_i}}. \tag{5.10}$$

As we can see in equation (5.10), the two main elements of this approach are the choice of K and the choice of a distance metric D. We discuss these next.

Choice of Distance Metric As pointed out earlier, online auction data comprise of three types of information: Static information captures information that does not change during the course of the auction, time-varying information that changes during the auction, and auction dynamics that are captured and represented by functional data. Table 5.4 summarizes the three types and the specific variables for each data type. We now discuss distance metrics for both data types.

Static and Time-Varying Data: Static and time-varying information includes data measured on different scales (interval, binary, and categorical). Following Jank and Shmueli (2007), we use separate metrics for each individual scale, and then combine the individual metrics into an overall distance metric for nonfunctional data.

For binary data x_{B} and x'_{B} (e.g., an auction with the buy-it-now option versus an auction without that feature), we define the distance as

$$d^{\mathrm{B}} = \mathbf{1}(x_{\mathrm{B}} \neq x'_{\mathrm{B}}), \tag{5.11}$$

where $\mathbf{1}$ denotes the indicator function, and thus d^{B} equals 1 if and only if $x_{\mathrm{B}} \neq x'_{\mathrm{B}}$, otherwise it is 0.

We adopt a similar measure for categorical data. For instance, the categorical variable "laptop brand" could assume eight different levels (e.g., Dell, HP, IBM, Sony, Toshiba, Fujitsu, Gateway, Acer) that can be coded as a vector of seven different binary variables. Thus, each categorical variable can be represented as a set of binary variables. Let $\mathbf{x}_{\mathrm{C}}$ and $\mathbf{x}'_{\mathrm{C}}$ denote two vectors representing categorical data, and then

TABLE 5.4 Different Sources of Information Typical for Online Auctions

Data Type		Measurement Scale	Example
		Interval	Opening price, weight, size
	Static	Binary	buy-it-now, used/new
Nonfunctional		Categorical	make, model
	Time varying	Interval	Number of bids, bidders, current price
Functional		Functional	Price velocity, price acceleration

we define their distance, similar to equation (5.11), as

$$d^{\mathrm{C}} = \mathbf{1}(\mathbf{x}_{\mathrm{C}} \neq \mathbf{x}'_{\mathrm{C}}). \tag{5.12}$$

which takes the value of 1 if and only if $x_{\mathrm{C}} \neq x'_{\mathrm{C}}$ and 0 otherwise.

For interval-scaled data x_{I} and x'_{I} (e.g., two auctions with different opening prices), we use a scaled version of the Minkowski metric (Jain et al., 1999):

$$d^{\mathrm{I}} = \frac{\left|\tilde{x}_{\mathrm{I}} - \tilde{x}'_{\mathrm{I}}\right|}{\tilde{R}_{\mathrm{I}}}, \tag{5.13}$$

where $\tilde{x}$ denotes the standardized value of x and $\tilde{R}$ denotes the range of $\tilde{x}$. The advantage of the Minkowski metric is that it renders interval-scaled data onto the interval [0, 1]. Note that the maximal and minimal values of d^{I} are 1 and 0, respectively, which are also the values taken by the binary and categorical distance metrics in equations (5.11) and (5.12). Having metrics in comparable magnitudes makes it easier to *combine* individual distance metrics.

We combine individual distance metrics in the following way. Let $\mathbf{x} = \{x_1, x_2, \ldots, x_p\}$ be a vector of p nonfunctional features, including binary, categorical, and interval data. We compute the overall distance between $\mathbf{x}$ and $\mathbf{x}'$ as

$$d(\mathbf{x}, \mathbf{x}') = \frac{1}{p} \sum_{i=1}^{p} d^*, \tag{5.14}$$

where d^* denotes the appropriate individual distance metric from equations (5.11)–(5.13).

As an example, let x and x' be two vectors of three features; specifically, $x =${w/ buy-it-now, Dell, 1 GB memory} and $x' =${w/o buy-it-now, IBM, 1 GB memory} in which the first, second, and third features are binary, categorical, and interval-scaled, respectively. Using equation (5.14), we have $d(x, x') = 1/3(d_1 + d_2 + d_3)$, where $d_1 = 1$ based on equation (5.11), $d_2 = 1$ based on equation (5.12) and $d_3 = 0$ based on equation (5.13). So the overall distance between x and x' is 2/3.

Note that the definition of d in (5.14) is flexible in the sense that one can use only subsets of the available information. For instance, d^{Static} would refer to the distance metric using only static information, while $d^{\mathrm{Time\text{-}Varying}}$ would refer to the metric with only time-varying information. One problem with distance metrics of this type is that they may overweigh different sources of information, depending on how elaborately each source is recorded. For instance, a data set with 100 different static features and only 10 time-varying features puts 10 times more weight on the information from static features. To overcome this potential bias, we follow the ideas of Becker et al. (1988) and first scale each individual distance metric by its mean root square (MRS). MRS is a statistical measure of the magnitude of a vector. For a vector $x = \{x_1, \ldots, x_p\}$, MRS is defined as $\sqrt{\frac{1}{p}\sum_{i=1}^{p} x_i^2}$ (Levinson, 1946). We apply the same scaling to each individual distance metric and obtain

$$d_s^{\text{Static}} = d^{\text{Static}}/\text{MRS}(d^{\text{Static}}), \tag{5.15}$$

$$d_s^{\text{Time-Varying}} = d^{\text{Time-Varying}}/\text{MRS}(d^{\text{Time-Varying}}), \tag{5.16}$$

$$d_s^{\text{Static\&Time-Varying}} = d_s^{\text{Static}} + d_s^{\text{Time-Varying}}. \tag{5.17}$$

Note that the combined metric $d_s^{\text{Static\&Time-Varying}}$ now puts equal weight on both static and time-varying information.

Distance Between Functional Data (Price Dynamics): We model an auction's price path and dynamics using the Beta cumulative distribution function (see also Chapter 4). In this fashion, we can measure the distance between two functional observations using the KL distance. Let (α, β) and (α', β') denote the Beta parameters for two different auction price paths, then their distance (when x is the focal auction) is defined as

$$d^{\text{F}} = \left| D_{\text{KL}}(x, x') \right|, \tag{5.18}$$

where $D_{\text{KL}}(x, x')$ is given by

$$D_{\text{KL}}(X, Y) = \ln \frac{B(\alpha', \beta')}{B(\alpha, \beta)} - (\alpha' - \alpha)\psi(\alpha) - (\beta' - \beta)\psi(\beta) + (\alpha' - \alpha + \beta' - \beta)\psi(\alpha + \beta), \tag{5.19}$$

where B and ψ denote the Beta and digamma function, respectively.

Note that d^{F} ranges within $[0, +\infty)$ as the KL distance assumes values on the real line. To make d^{F} comparable with the nonfunctional distance measures, we again scale it using the MRS transformation. Thus, we obtain

$$d_s^{\text{Dynamics}} = d^{\text{F}}/\text{MRS}(d^{\text{F}}). \tag{5.20}$$

The Optimal Distance Metric: To determine which combination of individual distance metrics leads to the best forecasting model, we investigate a series of different distance metrics. In particular, we investigate performance using five different metrics: d_s^{Static}, $d_s^{\text{Time-Varying}}$, d_s^{Dynamics}, $d_s^{\text{Static\&Time-Varying}}$, or $d_s^{\text{All}} = d_s^{\text{Static\&Time-Varying}} + d_s^{\text{Dynamics}}$. We first determine the optimal metric based on a validation set and then investigate the predictive accuracy of the resulting fKNN forecaster on a test set.

Choice of K The second important component of fKNN is the choice of K, the number of neighbors from which the forecasting is calculated. Too small a value will filter out relevant neighbors; too big a value will introduce noise and weaken the prediction.

Stone (1977) finds that the optimal value of K is data dependent, and it usually grows with the sample size. In addition, K may also vary as different distance metrics are used. Therefore, we select the optimal value of K separately for each distance metric and each data set. To do so, we again select the best value of K based on a validation set; we then apply the resulting model to the test set.

Forecasting Scheme Our complete forecasting process includes determining the optimal distance metric and the optimal value of K. We determine both based on a validation set. Then, using the optimal metric and K, we estimate the fKNN model based on the records in the training set. We investigate the performance of that model by predicting a new focal auction using auctions from a test set.

Comparison with Alternate Methods We benchmark the fKNN forecaster against two other very popular prediction methods: parametric regression models and non-parametric regression trees.

In a linear regression model, the closing price is modeled as a linear function of the observed predictor information. This information can include some or all the three types of data from Table 5.4. Note that in such models, all auctions from the training set are weighed equally when estimating the model coefficients.

Regression tree forecasting takes a hierarchical approach. It recursively partitions the data into smaller subgroups; the focal auction is then forecasted based on the average of the most relevant subgroup. While the regression tree, like KNN, uses neighboring information from similar auctions, it weighs each auction equally, which is one major aspect in which it differs from KNN.

We discuss differences in prediction performance next.

Case Study We discuss the predictive performance of the functional KNN forecaster when applied to two data sets of eBay auctions and compare it with competing approaches. We also investigate the optimal distance metric and the optimal value of K. The two data sets, Palm PDAs and laptops, are different in their level of heterogeneity. While the Palm PDA data set is very homogeneous, the laptop data are very heterogeneous. We also investigate different forecast horizons.

We partition each of the two data sets into a training set (50% of the auctions), a validation set (25%), and a test set (25%). Data partitioning is done chronologically, such that auctions in the training set transact prior to those in the validation set and auctions in the test set transact after those in the validation set. Therefore, our experiments mimic the prediction task that real bidders face.

For the competing models (regression and regression trees), we train the models on the combined training and validation set, and then test their predictive performance on the test set.

We evaluate all models using the MAPE measure, given by:

$$\text{MAPE} := \frac{1}{N} \sum_{i=1}^{N} \left| \frac{y_i - \widehat{y}_i}{y_i} \right|, \tag{5.21}$$

where y_i and $\widehat{y}_i$ denote the true and the estimated final price in auction i, respectively.

Selecting the Optimal K and the Optimal Distance Metric: We select the optimal value of K in the following way. Recall that we have five candidate metrics, $D \in \{d_s^{\text{Static}}, d_s^{\text{Time-Varying}}, d_s^{\text{Dynamics}}, d_s^{\text{Static\&Time-Varying}}, d_s^{\text{All}}\}$. For each metric, we select a value of K from the set $K \in \{1, 2, \ldots, 100\}$. For each combination of $(D \times K)$, we estimate the corresponding fKNN model on the training set, and then measure its predictive accuracy (in terms of MAPE) on the test set. Figure 5.6 shows the results. The left panel shows the results for the laptop auctions; the right panel shows the corresponding Palm PDA results. The top panel shows an overview and the bottom panel zooms in on the most relevant part.

From the left panel in Figure 5.6 (laptop auctions), we can see that d_s^{Dynamics} results in the worst model performance, regardless of the value of K. In other words, using only the dynamic information of the price path is not sufficient for achieving good prediction accuracy. We also see that, of the remaining four distance metrics, d_s^{All} yields the uniformly lowest prediction error. This suggests that for laptop auctions, due to their diversity in makes and models, every single piece of auction information is necessary to achieve good prediction accuracy. Moreover, we note that for d_s^{All}, the lowest prediction error is achieved at $K = 41$. We conclude that $D = d_s^{\text{All}}$ together with $K = 41$ results in the best predictions. The story is somewhat different for the Palm PDA data (right panel in Figure 5.6). For those data, $D = d_s^{\text{Time-Varying}}$ results in the uniformly lowest error (across all distances). Moreover, choosing $K = 2$ optimizes that distance.

It is interesting that the two different data sets result in very different choices for K and D. While for the laptop data we need all auction information (using the distance metric d_s^{All}) and a very large neighborhood (via $K = 41$), the Palm PDA auctions require only the time-varying information of the auction process (using $D = d_s^{\text{Time-Varying}}$) and a very small neighborhood (via $K = 2$). One possible explanation lies in the difference in heterogeneity between the two data sets. In the homogeneous data (Palm PDA), all products are the same and differences in auction outcome will be mostly due to differences in the current price and the level of competition for that product. The competition level is reflected in the number of bids and bidders, which, together with the price level, are captured in $d_s^{\text{Time-Varying}}$. Moreover, since products are very homogeneous, we need only a very small neighborhood, and thus $K = 2$. This is different for the laptop auctions, where products are very heterogeneous, and therefore the forecaster needs all available information (in d_s^{All}) to distinguish between samples. Since the products are very different, the method also requires a larger neighborhood that leads to a larger value of K. This suggests that, as expected, forecasting more heterogeneous auctions is a more difficult task.

Robustness of Optimal D and K to the Time Horizon: In the previous section, we investigated the interplay of K and D for a fixed forecast horizon of 1 min. That is, we assumed that we observe the auction until 1 min before its close. We now investigate the robustness of this choice for different forecast horizons. Specifically, we investigate the robustness of K and D for different forecast horizons (δ_T) in the set $\delta_T \in \{$ 2 h, 1 h, 30 min, 15 min, 5 min, 1 min$\}$.

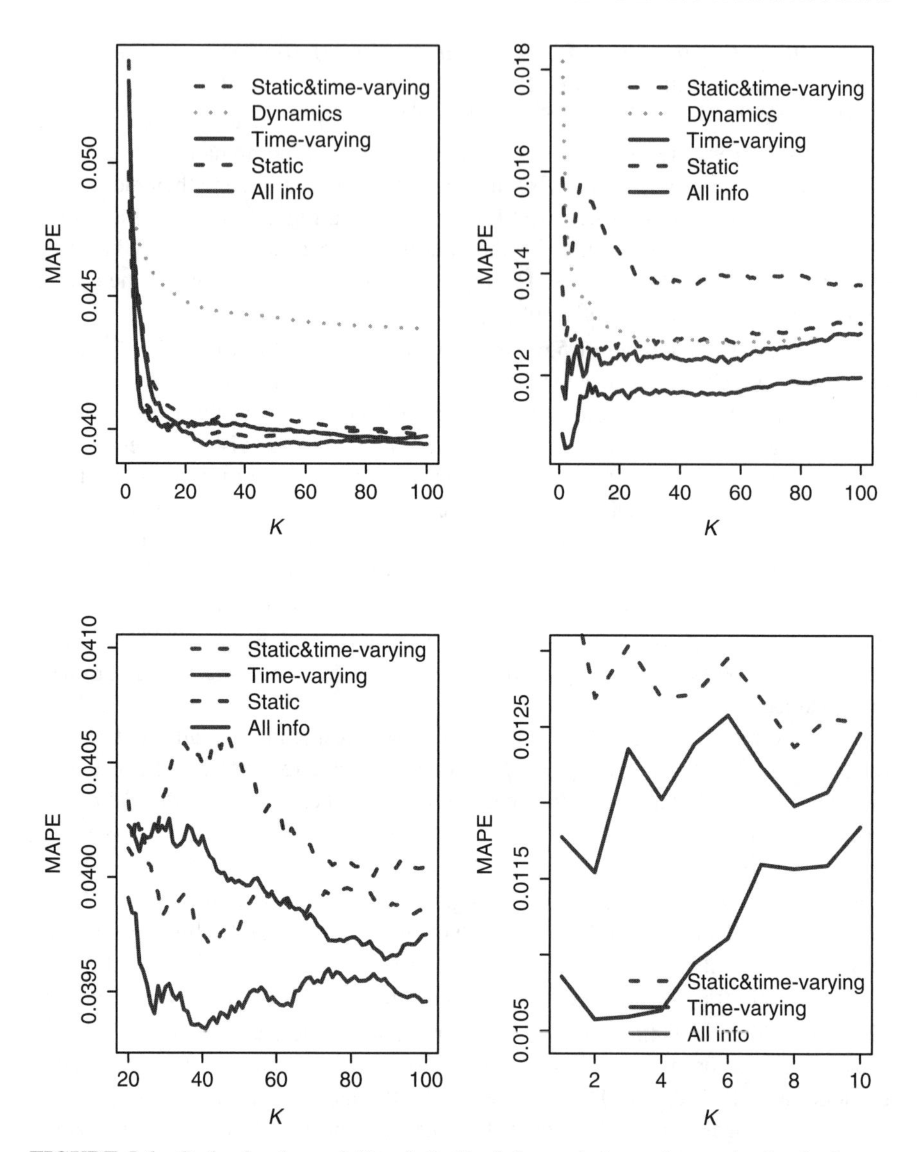

FIGURE 5.6 Optimal values of K and D. The left panel shows the results for the laptop auctions; the right panel shows the corresponding Palm PDA results. The top panel gives an overview and the bottom panel zooms in on the most relevant part.

Figure 5.7 investigates the robustness of K to the choice of δ_T. For a given value of K ($K \in \{20, 40, 60, 80, 100\}$ for the laptop data and $K \in \{2, 5, 10, 50, 100\}$ for the Palm PDA data), we investigate the prediction accuracy for different values of δ_T. We hold D fixed at $D = d_{\mathrm{s}}^{\mathrm{All}}$ for the laptop data and $D = d_{\mathrm{s}}^{\mathrm{Time-Varying}}$ for the Palm data. Figure 5.7 shows the *relative* prediction error $\mathrm{Rel.MAPE}_K := \mathrm{MAPE}_K/\mathrm{MAPE}_{K^*}$

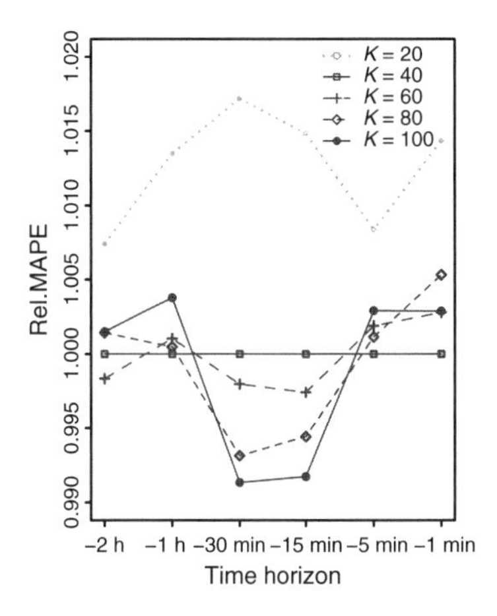

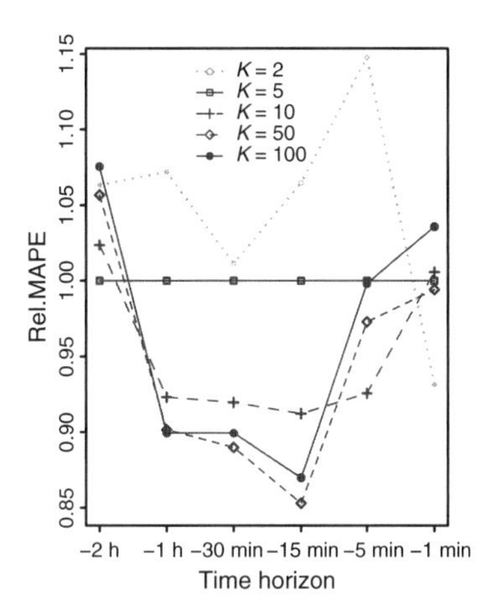

FIGURE 5.7 Relative prediction accuracy for different values of K at different forecast horizons δ_T. The top panel corresponds to the laptop data, and the bottom panel corresponds to the Palm PDA data.

relative to a benchmark value ($K^* = 40$ for the laptop data, and $K^* = 5$ for the Palm PDA data).

We see that for the laptop data (top), lower values of K ($K = 20$) lead to poor performance. We also see that while $K = 40$ generally leads to good forecasting accuracy, it is outperformed by higher K values when forecasting forecast horizons of 30 or 15 min. This suggests that the value of K is not very robust to the forecast horizon. It is even less robust for the Palm PDA data (bottom), where $K = 5$ leads to good forecasting performance only for very long forecast horizons ($\delta_T = 2\,\text{h}$); in contrast, choosing $K = 2$ leads to the best performance for very short horizons ($\delta_T = 1$ min). This suggests that the choice of K should be a function of δ_T. Table 5.5 lists the optimal value of K for each combination of δ_T and D.

We now investigate the impact of the forecast horizon δ_T on the choice of the distance metric D. Figure 5.8 shows the prediction accuracy as a function of the forecast horizon δ_T for different choices of D. Note that for each combination of D and δ_T, we use the optimal values of K from Table 5.5.

The top panel in Figure 5.8 corresponds to the laptop data; the bottom panel is for the Palm PDA data. Each line corresponds to a distance metric $D \in \{d_s^{\text{Static}}, d_s^{\text{Time-Varying}}, d_s^{\text{Dynamics}}, d_s^{\text{Static\&Time-Varying}}, d_s^{\text{All}}\}$. We can see that, for each data set, a single distance metric yields the consistently best result across all values of the forecast horizon. That is, d_s^{All} results in the best prediction accuracy for the laptop data, regardless of the value of δ_T; similarly, $d_s^{\text{Time-Varying}}$ yields the best results for all values of δ_T in the Palm PDA data. This suggests that the choice of the distance metric is very robust to the forecasting horizon, at least for a given data set. We also note that while $d_s^{\text{Time-Varying}}$ significantly outperforms all other distance metrics for the Palm PDA data, for the laptop data most choices of D (except for $d_{s^{\text{Dynamics}}}$) yield very comparable results, at least for short forecast horizons ($\delta_T \leq 30$ min).

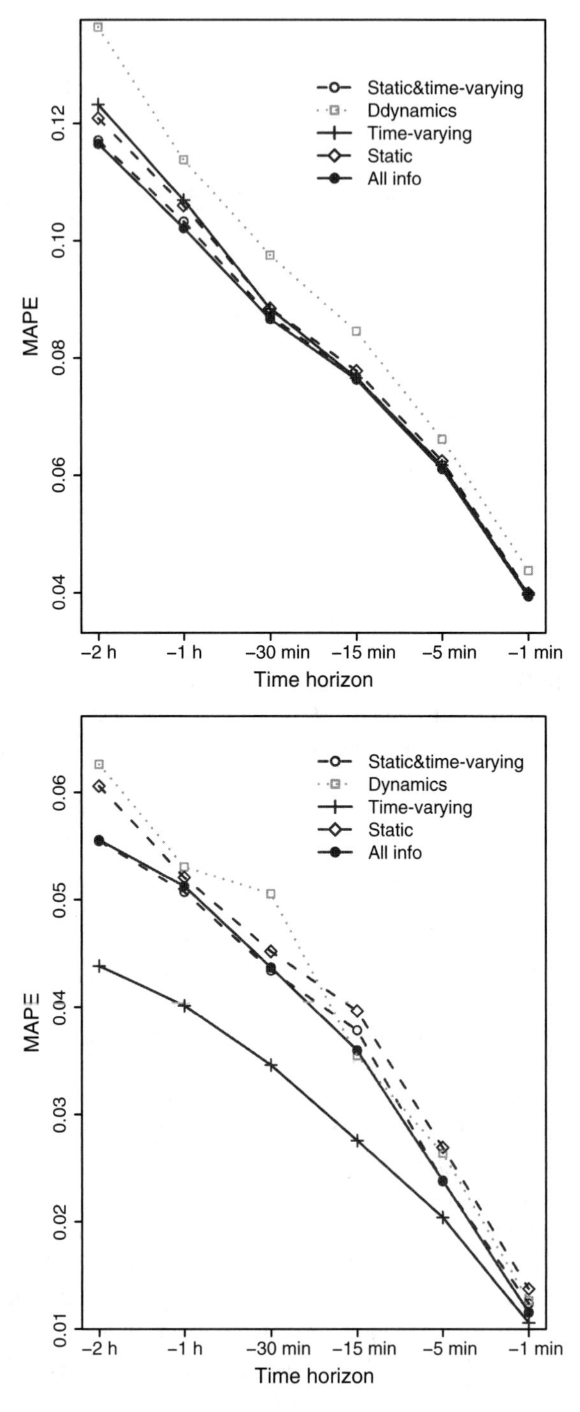

FIGURE 5.8 Comparison of different distance metrics. The top panel is for laptop auctions and the bottom panel is for palm auctions.

TABLE 5.5 Optimal Choice of K for Different Distance Metrics D and Different Forecast Horizons δ_T.

	Laptop Data					
Time Horizon	2 h	1 h	30 min	15 min	5 min	1 min
Static	95	94	99	99	81	14
Time-varying	31	27	97	100	91	89
Dynamics	100	100	100	100	100	100
Static&time-varying	40	79	100	96	47	44
All	33	77	98	100	44	41
	Palm PDA Data					
Time horizon	2 h	1 h	30 min	15 min	5 min	1 min
Static	52	69	63	63	61	37
Time-varying	3	10	4	1	1	2
Dynamics	94	95	95	29	29	68
Static&time-varying	7	18	32	52	12	8
All	6	40	30	61	11	2

The top panel corresponds to the laptop data; the bottom panel is for the Palm PDA data.

Comparison with Alternate Prediction Methods: We evaluate the performance of functional KNN by comparing its predictive accuracy to more classical approaches—linear regression models and regression trees.[1].

We study the performance of all methods on the test set. Recall that we partitioned our data into a training set (50%), validation set (25%), and test set (25%). While we estimated the fKNN forecaster on the training set and optimized its parameters K and D on the validation set, we now compare its performance (using the optimal parameter values) on the test set. That is, for each forecast horizon δ_T, we use the optimal combination of K and D from the previous section. To make fair comparisons, we apply regression and regression trees using the same information as for functional KNN.

Figure 5.9 shows the results. We display the *relative* prediction error between fKNN and the regression model (dotted line) and between fKNN and the tree model (dashed line). We can see that fKNN generally outperforms its two competitors. In particular, for the laptop data (top), fKNN outperforms the tree model by as much as 40%. While the gap between the regression model and fKNN is smaller, fKNN leads to improvements that range between 5% and 10%. The picture is similar for the Palm PDA data (bottom). While for this data set fKNN also leads to general improvements, it is interesting to see that only for the longest forecast horizon ($\delta_T = 2$ h) both alternate approaches are competitive.

It is revealing to compare performance on each of the two data sets. While for the laptop data, both fKNN and regression significantly outperform the tree, the gap is

[1] We used the software defaults for pruning in the *R* package *rpart*.

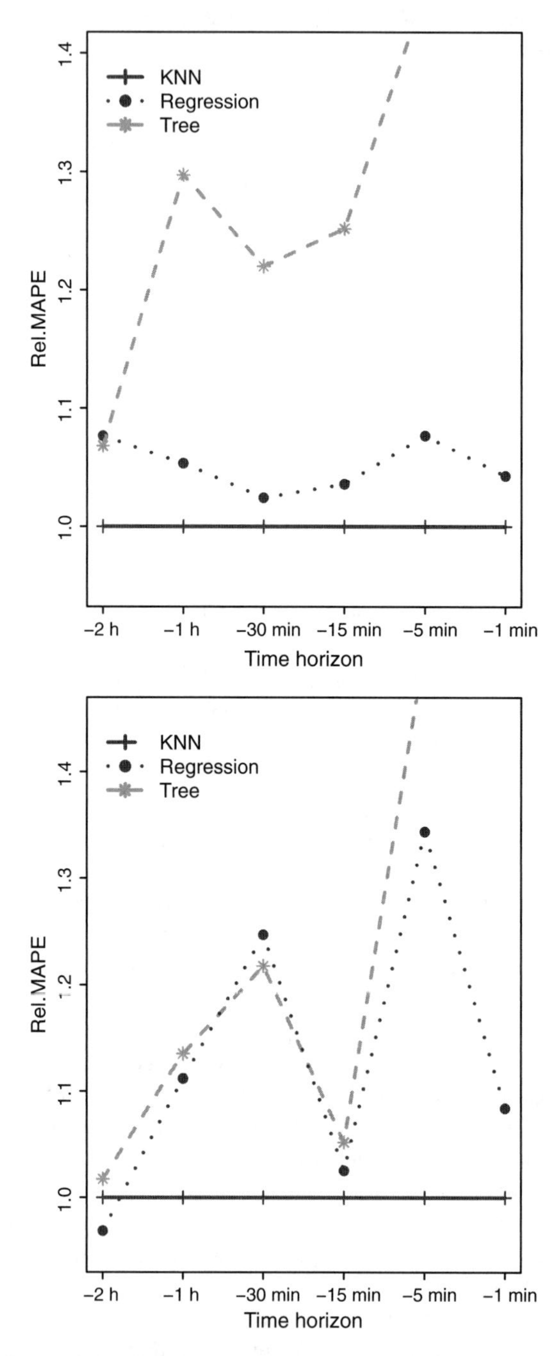

FIGURE 5.9 Comparison of different forecasting methods. The top panel corresponds to laptop auctions; the bottom panel is for the Palm PDA auctions.

not as large in the Palm PDA data; in fact, for the Palm PDA data, the regression tree and regression model are comparable for almost all forecast horizons. The poor performance of CART on the laptop data illustrates the general problem of the method with prediction: while it often fits the training set well, it has a tendency to overfit and thus perform poorly on the test set, especially in situations like the laptop data where the underlying population is very heterogeneous. On the other hand, functional KNN can handle heterogeneous populations well by selecting only those neighbors that are most relevant for the focal auction; in particular, compared to regression, it performs especially well for forecasting longer forecast horizons (1 h, 2 h), which is very relevant in practical situations.

Functional KNN also leads to improvements for less heterogeneous data sets such as the Palm PDA data. While the bottom panel in Figure 5.9 suggests that with a decreasing forecast horizon the improvement of fKNN over both competitors generally increases, there is an unexpected spike at $\delta_T \leq 15$ min. At that spike, both the regression and the tree model perform almost as well as fKNN. A closer investigation of this phenomenon reveals that for this time period, the optimal value of K (based on the validation set) equals 1 (see top panel in Figure 5.10); however, this value leads to very poor performance on the test set (bottom panel in Figure 5.10). This suggests that finding the right value of K is especially difficult for homogeneous data sets (such as the Palm PDA data). While the data homogeneity suggests very small values of K, slight perturbation of the homogeneity can lead to weaker results. This was already implied by the lack of robustness seen earlier.

Discussion We have presented a novel functional KNN forecaster for forecasting the final price of an ongoing online auction. Assuming that more similar auctions contain more relevant information for incorporation into forecasting models, Zhang et al. (2010) proposed a novel dissimilarity measure that takes into account both static and time-varying features, as well as the auction's price dynamics information. The latter is obtained via a functional representation of the auction's price path. We select both the optimal distance metric and the optimal number of neighbors based on a validation set. We find that weighting information unequally yields better forecasts compared to classical methods such as regression models or trees, and this result holds in auctions of varying levels of heterogeneity.

Although we observe improvement of KNN forecaster over regression and regression trees for auctions of varying levels of heterogeneity, our results show that the improvement is bigger for heterogeneous data set. This says that selecting the most useful information and making use of only most relevant neighbors is especially crucial for prediction accuracy in situations where objects are heterogeneous and information is noisy, a fact that is generally true for forecasting not only online auctions but also in many other situations (e.g., weather forecasting). Another finding worth noting is the robustness of the optimal distance to forecast horizons. The fact that the same distance metric is optimal regardless of the forecast horizon implies that the most important information for making price prediction is time invariant. This fact simplifies the decision-making process. To generate forecasts, we need to find only the optimal distance once that can be kept as the forecasting process goes on.

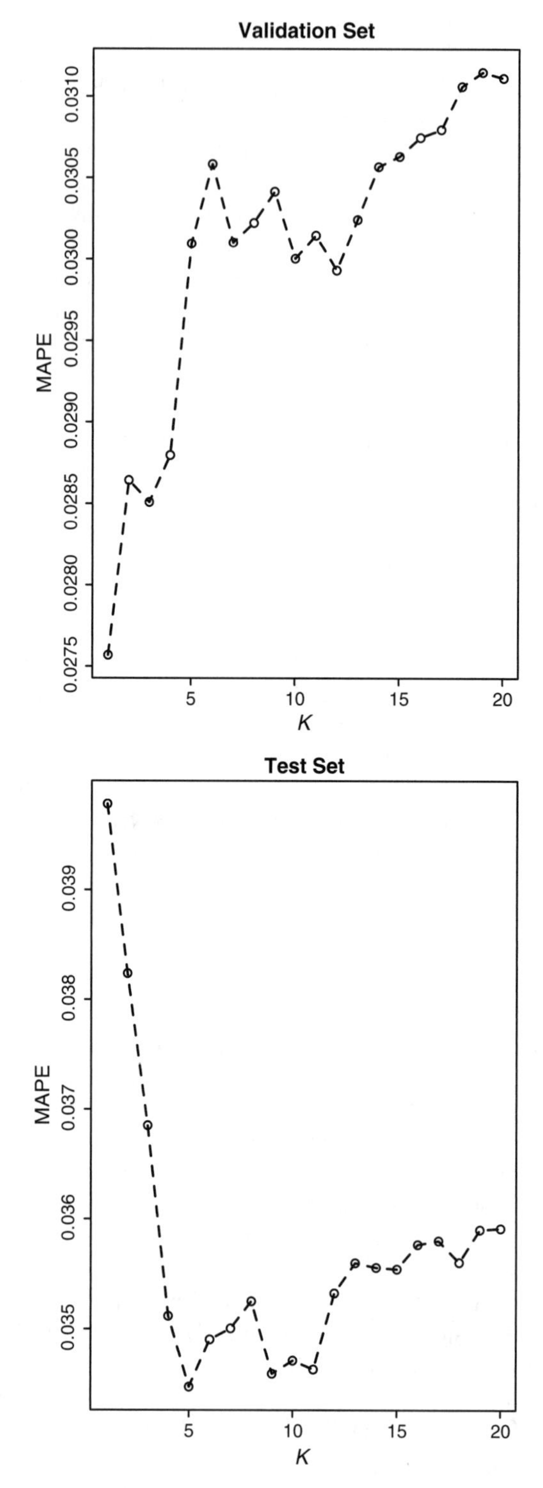

FIGURE 5.10 Optimal values of K for the Palm PDA data at $\delta_T = 15$ min. The top panel corresponds to the validation set; the bottom panel corresponds to the test set.

There are several ways to extend this forecasting approach. While we scale distance metrics for different information sources to achieve equal weighing across all metrics, one could alternatively assign individual weights to individual metrics and then optimize the weights. There are also other ways in which the distance metrics for different data types might be changed. Take categorical data, for example, the distance between objects in different categories need not be 1. Rather, we can set the distance to 0.5 for similar categories (e.g., U.S. brands), while assigning a bigger distance, 1 for instance, to more different categories. Another way to complement this study is by investigating alternates to classical regression and trees; for example, via weighted regression or weighted tree models that should lead to forecasting advantages especially for heterogeneous data.

5.2 FORECASTING COMPETING AUCTIONS

We now focus on a direct incorporation of the effect of competition into a forecasting model. We have argued at several places throughout this book that auctions form an interlinked network and that what takes place in one auction potentially influences the outcome of another auction. In that sense, accounting for this network effect should, at least in principle, result in more accurate forecasts (Jank and Zhang, 2009a). However, it is conceptually much more complicated to measure the effect of competition. This leads us to the third forecasting model of this chapter that builds upon some of the features discussed in earlier models. In particular, in addition to price dynamics of the auction itself, this model explicitly accounts for the effect of competing auctions and in particular their price dynamics.

5.2.1 Capturing Competition

To capture the effect of what takes place in other simultaneous auctions, we must first define meaningful measures for competition. There are many different ways of defining competition measures and we discuss several alternatives below. All measures are driven by the same general principle, illustrated in Figure 5.11, where we define a *focal auction* (indicated by the solid line) as the auction for which a bidder wants to decide whether or not to bid on and *concurrent auctions* – the auctions that take place simultaneously at the time T of decision making (indicated by the dotted lines).

One meaningful measure of competition is the price level in the concurrent auctions. In our example, there are four different price levels at time T, varying from high (p_1) to low (p_4). The price level in the focal auction at that time is p3. Thus, a possible measure for the price competition is given by the *average price* in concurrent auctions (which we denote by $c.avg.price$), that is, by the average of p_1, p_2, and p_4. In a similar fashion, the *average price velocity* ($c.avg.vel$) in concurrent auctions is defined as the average of the corresponding price velocities.

We now investigate several different competition features and their impact on the price of the focal auction. Table 5.6 categorizes these features by the information that

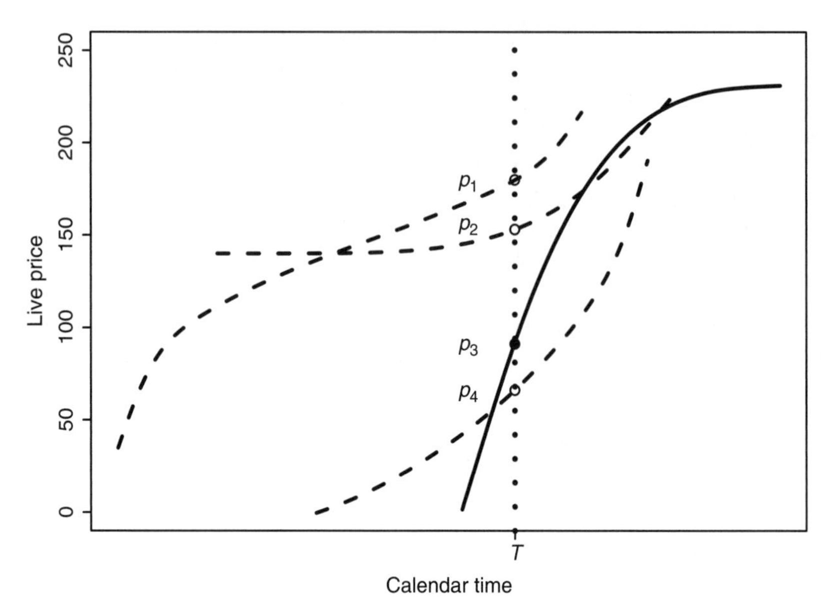

FIGURE 5.11 Illustrating competition: The solid black line denotes the focal auction; the dashed lines denote competing auctions; *T* denotes the time of decision making.

they carry: *Static competition features* are known at the outset of the auction and do not change during the auction; examples include the opening price of concurrent auctions (a high opening price in other auctions could discourage bidders and make them participate in the focal auction) or the duration of concurrent auctions (if competing auctions have a shorter duration, then bidders with an immediate desire may be attracted to those auctions); *evolving competition features* change during the auction, such as the current price of concurrent auctions (if the price is low in other auctions, bidders may leave the focal auction) or the number of bidders in concurrent auctions (bidders may feel that their chances of winning are higher in auctions with lower competition); and *price dynamic competition features* capture the effect of changing dynamics in competing auctions (if the price dynamics increase in competing auctions, for example, due to increased bidding activity in those auctions, then the price velocity in the focal auction is likely to slow down).

In Figure 5.12, we explore the relationship between some of the competition features from Table 5.6 and the future price in the focal auction. We see that some features have a strong relationship with price (e.g., the average price and its velocity in competing auctions), while others do not (e.g., the average opening bids or the shipping fee in simultaneous auctions). Pairwise correlation analysis (not shown here) also shows that, unsurprisingly, many of the features in Table 5.6 are collinear. Thus, a good modeling strategy is to start with a suitable variable selection procedure. We use the initial observations from Figure 5.12 for guidance when selecting the most relevant set of competition features in the next section.

TABLE 5.6 Candidate Competition Features

Name	Description
Static features	
c.openbid.avg	Average opening price of concurrent auctions
c.dura.avg	Average duration of concurrent auctions
c.ship.avg	Average shipping fee of concurrent auctions
c.feedback.avg	Average sellers' feedback of concurrent auctions
c.power.avg	Average number of power seller in concurrent auctions
c.store.avg	Average number of eBay stores in concurrent auctions
c.pic.avg	Average number of pictures in concurrent auctions
Evolving features	
c.price.avg	Current average price in concurrent auctions
c.price.vol	Current price volatility (StDev) in concurrent auctions
c.price.disc	Price discount (difference) between focal auction and highest concurrent price
c.t.left.avg	Average time left in concurrent auctions
c.t.left.vol	Volatility (StDev) of time left in concurrent auctions
c.nbids.avg	Average number of bids in concurrent auctions
c.nbids.vol	Volatility (StDev) of number of bids in concurrent auctions
c.nbidders.avg	Average number of bidders common to focal and concurrent auctions
c.nbidders.vol	Volatility (StDev) of number of bidders common to focal and concurrent auctions
Price dynamic features	
c.vel.avg	Average price velocity in concurrent auctions
c.vel.vol	Volatility (StDev) of price-velocity in concurrent auctions
c.acc.avg	Average price acceleration in concurrent auctions
c.acc.vol	Volatility (StDev) of price-acceleration in concurrent auctions

5.2.2 Variable Selection

Many different pieces of information can affect price in online auctions. We differentiate between two main components, that is, information from *within* the focal auction versus information from concurrent, *competing* auctions that take place simultaneously. Within each component, information can be further segmented into static, evolving, and price dynamic, similar to Table 5.6. Table 5.7 lists all the different pieces of information that are candidates for our forecasting model, revealing over 30 such variables. Thus, an important first step is the selection of a parsimonious subset of relevant predictors. Variable selection has been extensively researched in the statistics literature (Berk, 1978), and it is receiving increasing attention today with the availability of more and more data sets featuring larger and larger number of variables (George, 2000). A complicating factor in our situation is the time-varying nature of our model. Our goal is to find a model that predicts well at time $T+1$, *universally* across all time periods $T = 1, 2, 3, \ldots, N_T$. Classical variable selection procedures focus on cross-sectional data that corresponds to a *single* time period only. Since our

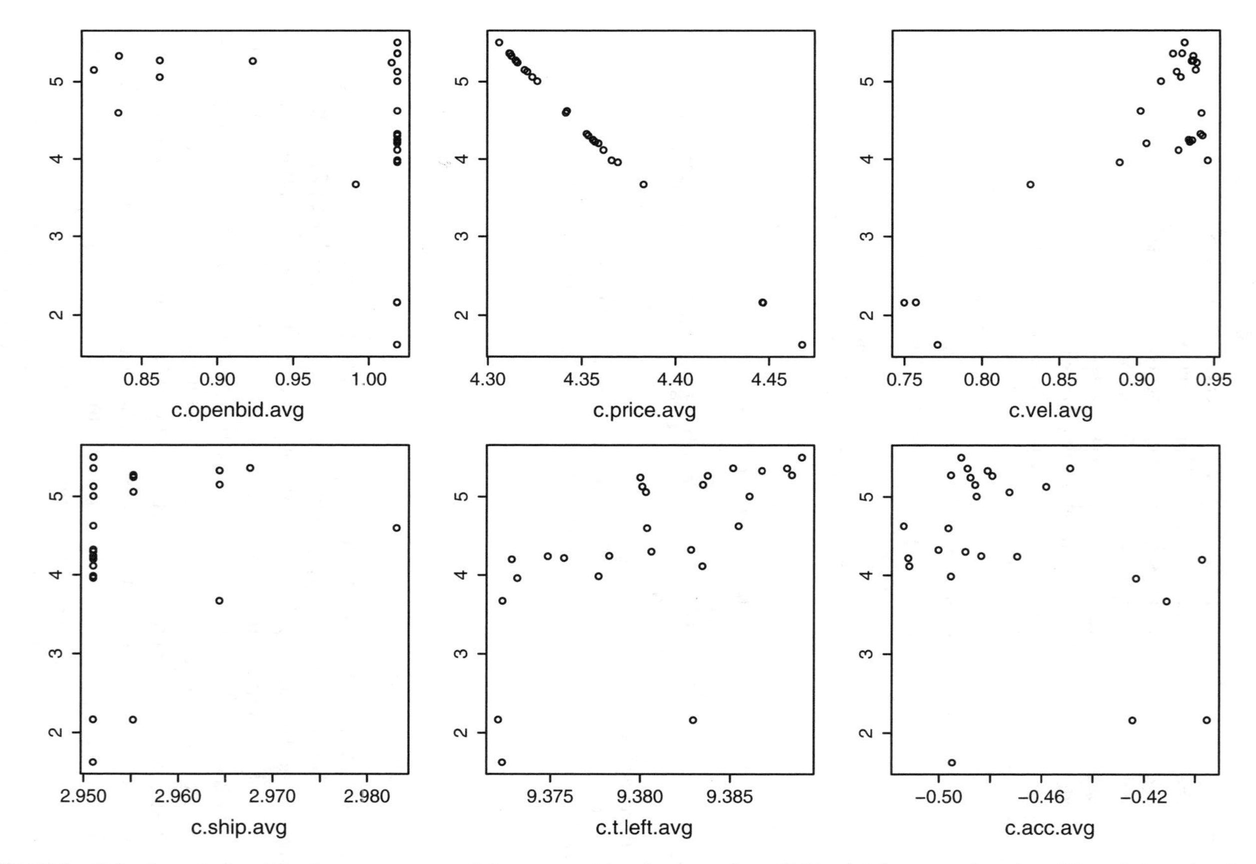

FIGURE 5.12 Pairwise relationships between some of the competition features from Table 5.6 (measured at time T) and the price (measured at $T + 1$) in the focal auction.

TABLE 5.7 All Candidate Information for Our Forecasting Model

Information from within the focal auction	
Static information	Opening bid, auction duration, shipping fee, seller's feedback, power seller, eBay store, picture
Evolving information	Current price, time left, current number of bids, current number of bidders
Price dynamic information	Price velocity, price acceleration
Information from competing auctions	
Static information	c.openbid.avg, c.dura.avg, c.ship.avg, c.feedback.avg, c.power.avg, c.store.avg, c.pic.avg
Evolving information	c.price.avg, c.price.vol, c.price.disc, c.t.left.avg, c.t.left.vol,c.nbids.avg, c.nbids.vol, c.nbidders.avg, c.nbidders.vol
Price dynamic information	c.vel.avg, c.vel.vol, c.acc.avg, c.acc.vol

data vary over time, it is quite plausible that there exists one model that best predicts at time $T + 1$, while another best predicts at a different time $T' + 1$. Our goal is to find a model that is not geared to a single time period, but rather applies universally to the auction market over a longer time window. To that end, we choose a model that has good *average* performance, averaged over all time periods T of interest.

Our model has the general form

$$\mathbf{y}_{T+1} = \boldsymbol{\beta}_T' \mathbf{x}_T, \tag{5.22}$$

where $\mathbf{y}_{T+1}$ denotes the auction price at $T + 1$, $\mathbf{x}_T = (x_{T1}, \ldots, x_{Tp})'$ is a vector of predictors, and $\boldsymbol{\beta}_T = (\beta_{T1}, \ldots, \beta_{Tp})'$ is a vector of coefficients to be estimated from the data. The goal is to select only those predictors that are important for predicting the price $\mathbf{y}_{T+1}$, across all time periods $T = 1, 2, 3, \ldots, N_T$.

We accomplish this in several steps. In the first step, we run simple regressions (i.e., $p = 1$) between each individual predictor from Table 5.7 and the response $\mathbf{y}_{T+1}$ at *each* time period $T(T = 1, 2, 3, \ldots, N_T)$. We then calculate the percentage of time points a predictor is significant (at the 5% significance level). That is, for each predictor $x_k = (x_{1k}, \ldots, x_{N_T k})'$, $k = 1, \ldots, p$, we compute the average[2]

$$p.sig_k := \frac{1}{N_T} \sum_T \mathbf{1}_{\{x_{Tk} \text{ significant at 5\% level}\}}. \tag{5.23}$$

Table 5.8 shows the results for a fine grid of hourly forecasts (i.e. $(T + 1) - T = 1$ h) that results in $N_T = 1754$ different time periods. We can see that the predictors that *individually* have a strong effect on $\mathbf{y}_{T+1}$ (consistently across all time periods T)

[2] While we use an unweighted average, a possible alternative would be to weight each time point according to its distance from the close of the auction.

TABLE 5.8 Percentage of Significant Time Points

Focal Auction	*p.sig*	Competing Auctions	*p.sig*
openbid	0.199	c.openbid.avg	0.193
duration	0.032	c.dura.avg	0.032
shipping	0.039	c.ship.avg	0.046
sellerfeed	0.055	c.feedback.avg	0.044
powerseller	0.061	c.power.avg	0.076
store	0.092	c.store.avg	0.104
picture	0.028	c.pic.avg	0.028
currenprice	1.00	c.price.avg	1.00
		c.price.vol	0.886
		c.price.disc	1.00
timeleft	0.775	c.t.left.avg	0.771
		c.t.left.vol	0.758
numbids	0.780	c.nbids.avg	0.777
		c.nbids.vol	0.509
numbidders	0.197	c.nbidders.avg	0.188
		c.nbidders.vol	0.086
price-velocity	0.762	c.vel.avg	0.762
		c.vel.vol	0.624
price-acceleration	0.308	c.acc.avg	0.309
		c.acc.vol	0.306

The two leftmost columns refer to predictors from within the focal auction; the two rightmost columns refer to predictors from competing auctions.

are the current price, price velocity and price-acceleration, time left and the number of bids (from within the focal auctions), and c.price.avg, c.price.vol, c.price.disc, c.t.left.avg, c.t.left.vol, c.nbids.avg, c.nbids.vol, c.vel.avg, c.vel.vol, c.acc.avg, and c.acc.vol (from competing auctions). It is interesting that most of these variables relate to price (or price movement) from the focal auction relative to competing auctions. This suggests that information about price and its dynamics effectively captures much of the relevant auction information such as information about the product, the auction format, the seller, and the competition between bidders. However, also note that the results thus far are based only on simple regressions ($p = 1$) and thus may not fully reflect the *joint* effect of a predictor in the presence of other predictor variables. To that end, we investigate pairwise correlations (again, averaged across all time periods, $T = 1, \ldots, N_T$; correlation table not reported here) and find high collinearity between 10 specific pairs: the current price with each of {c.price.avg, c.price.vol, c.price.disc, time left, c.t.left.avg, number of bids, c.nbids.avg} and the three pairs {price velocity, c.vel.avg}, {price velocity,c.vel.vol}, and {price acceleration, c.acc.avg}. This high collinearity is not surprising since many of these predictors carry similar information, only coded in a slightly different way. We therefore eliminate the nonprice predictors that are involved in high pairwise correlations (above 0.7) and retain only the corresponding price-related predictor.

TABLE 5.9 Average BIC Computed Across All Time Periods T

Model	avg.BIC
Our model from eq. (5.24)	−381.59
Full model (all 33 predictors from Table 5.7)	−147.08
All 13 predictors from the focal auction (Table 5.7)	−319.82
Only 2 focal auction dynamics	37.77
Only 4 focal auction evolving predictors	−83.87
All 20 predictors from competing auctions (Table 5.7)	−313.52
Only 4 competing auction dynamics	37.58
Only 9 competing auction evolving predictors	−84.29

The first row shows the value of *avg.BIC* for our model in (5.24); the remaining rows show the corresponding values of several competing models.

Next, we derive our final model using the Bayesian information criterion (BIC). In a similar fashion to (5.23), one can compute the *average* BIC across all time periods.[3] That is, let avg.BIC $:= 1/N_T \sum_T \text{BIC}(T)$, where BIC($T$), $T = 1, 2, 3, \ldots, N_T$, denotes the Bayesian information criterion (George, 2000) of a model computed at time period T. By comparing all possible subsets of noncollinear predictors, we arrive at our final forecasting model as

$$\begin{aligned} \mathbf{y}_{T+1} &= \alpha_T + \beta_{1T}\text{current price}_T + \beta_{2T}\text{velocity}_T + \beta_{3T}\text{acceleration}_T \\ &\quad -\beta_{4T}\text{c.acc.vol}_T. \end{aligned} \tag{5.24}$$

Table 5.9 shows the avg.BIC of our final forecasting model (5.24) compared to several competitor models. We can see that our model results in the lowest avg.BIC. It is also interesting to see that models with information *only* from competing auctions perform almost as well as models with the corresponding information only from within the focal auction. This is yet another piece of evidence for the tight connectivity of the auction marketplace.

A few comments about the final model in (5.24) are in order. It is interesting to see that the model relies only on price-related information. In particular, many variables that have been found significant in previous studies have been dropped from our model. For instance, Lucking-Reiley et al. (2007) found, among other things, a significant effect of the seller's rating on the final price. One key difference between previous studies and our study is that while they take a static look at online auctions, our model captures the dynamic nature of the auction process. In other words, previous studies typically look only at the static, pre-auction information that is available before the start of the auction (such as the auction length, the opening bid, or a seller's rating). In such a static view, the effect of the seller's rating is highly significant (since

[3] Jank and Zhang (2009b) also investigate alternate ways for model selection across an entire range of time periods. They consider the "distribution" of model selection criteria such as AIC or BIC and different ways of summarizing important features of that distribution.

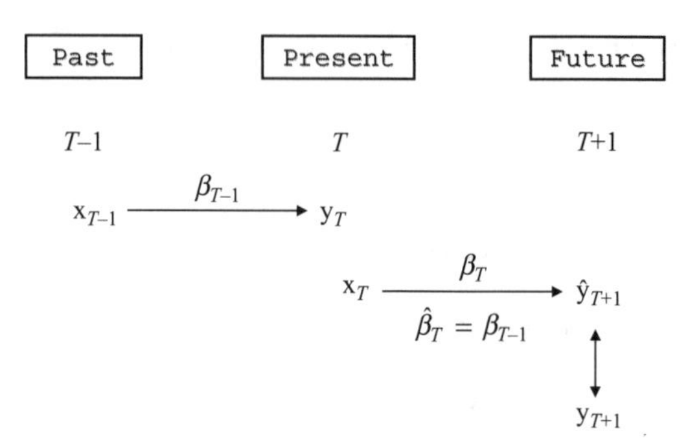

FIGURE 5.13 Illustration of the updating scheme of the forecasting model.

the seller's reputation and trustworthiness will impact the final price). However, our model is dynamic in the sense that all previous price considerations and bidding decisions have already been factored into the current price and its current dynamics (current price$_T$, velocity$_T$, acceleration$_T$). In this sense, price dynamics reflect the expectations of all bidders about the product, the seller, and the bidding competition up to this time point. It is therefore not too surprising that all static variables drop out from our final model. The effect that captures concurrency is more intriguing. Note that the information from concurrent auctions is captured in a single variable, which is the volatility of dynamics from competing auctions (c.acc.vol$_T$). As there has not been much prior research on the effect of concurrent auctions, it is hard to formulate an expectation about c.acc.vol$_T$. However, the negative sign indicates that higher variance in the price movements of competing auctions will result in smaller price advances of the focal auction. In other words, more price activity in different parts of the market will lead to price stallings of the focal auction.

5.2.3 Model Updating

The goal of our model is to predict price at a future time $T + 1$ using information only from the present (i.e., time T) and the past ($T - 1$, $T - 2$, etc.). We accomplish this by estimating the functional relationship between $T - 1$ and T and then applying this relationship to predict $T + 1$ from T. Figure 5.13 illustrates this updating scheme.

At time T (present), we wish to make a prediction about the future price at time $T + 1$. Per our model, $\mathbf{y}_{T+1}$ is given by $\boldsymbol{\beta}'_T\mathbf{x}_T$, where $\mathbf{x}_T$ contains information observed in the present (or past). Note that we cannot estimate $\boldsymbol{\beta}_T$ directly since the response ($\mathbf{y}_{T+1}$) is yet unobserved. We therefore estimate the relationship from the past: We estimate $\boldsymbol{\beta}_{T-1}$ for the price at T ($\mathbf{y}_T$) and then estimate $\boldsymbol{\beta}_T$ via $\hat{\boldsymbol{\beta}}_T := \boldsymbol{\beta}_{T-1}$. In that sense, we "roll" the relationship from the past one time period forward. We also investigated alternate updating approaches (such as estimating $\boldsymbol{\beta}_T$ via a moving average of prior relationships, $\hat{\boldsymbol{\beta}}_T := \text{MA}\{\boldsymbol{\beta}_{T-1}, \boldsymbol{\beta}_{T-2}, \boldsymbol{\beta}_{T-3}, \ldots\}$) but did not find significant improvements in model performance.

5.2.4 Model Estimation and Prediction Performance

Our data contain eBay auctions that transacted between March 14 and May 25. For estimation and prediction, we divide our data into a training set (80% of the data) and a validation set (remaining 20% of the data). Since our data vary over time (and since we are primarily interested in making accurate predictions of the future), our training set consists of all auctions that complete during the first 80% of our data's time span (i.e., between March 14 and May 10); the validation set contains all remaining auctions (i.e., between May 11 and May 25). In that sense, we first estimate our model on the training set; results of model estimation and model fit are discussed below. We then apply the estimated model to the validation set to gauge its predictive capabilities; this is discussed Section 5.2.4.3.

Model Estimation Figure 5.14 shows the estimated coefficients for the parameters of our forecasting model (5.24). Recall that we estimate the model at every time point T, $T = 1, 2, 3, \ldots, N_T$ in the training set. In our application, we consider time intervals of 1 h over the time period between March 14 and May 10, and hence the coefficients also vary over that time period. Figure 5.14 shows the resulting trend of the coefficients together with 95% confidence bounds.

We can see that information from within the focal auction (current price, price velocity, and price acceleration) has a positive relationship with the future price $\mathbf{y}_{T+1}$; in contrast, information from competing auctions (c.acc.vol) has a negative relationship. In other words, both the current level of price and its dynamics are positive indicators of future price. In contrast, the volatility of price acceleration in competing auctions is a negative indicator. Price acceleration in competing auctions will be high if many bidders bid in auctions *different* from the focal auction. A high volatility in price acceleration may suggest high uncertainty in the marketplace, with some auctions experiencing large price jumps and others experiencing no price movements at all. This high uncertainty results in depressed prices of the focal auction.

Model Fit and Varying Time Intervals Figure 5.14 shows the estimated model coefficients for a 1 h forecast horizon ($(T + 1) - T = 1$ h). Alternatively, one could consider models with longer forecast horizons. Intuitively, since forecasting further into the future is harder, such models should not perform as well. Figure 5.15 (top panel) shows model fit for the forecast horizons $(T + 1) - T = 1, 2, 3, \ldots, 14$ h. We measure strength of fit by the average R^2 value, avg.$R^2 := 1/N_T \sum_T R^2(T)$, where $R^2(T)$, $T = 1, 2, 3, \ldots, N_T$, denotes the R^2 of a model computed at time period T. We can see that, as expected, model fit and strength of fit decrease as the forecast horizon get longer. Note though that even for the longest forecast horizon (14 h), the value of avg.R^2 is still larger than 99%.

Prediction Performance As pointed out above, we estimate the model on the training set, and then we gauge its predictive performance on the validation set. We measure predictive performance of a model in terms of its MAPE. For each time

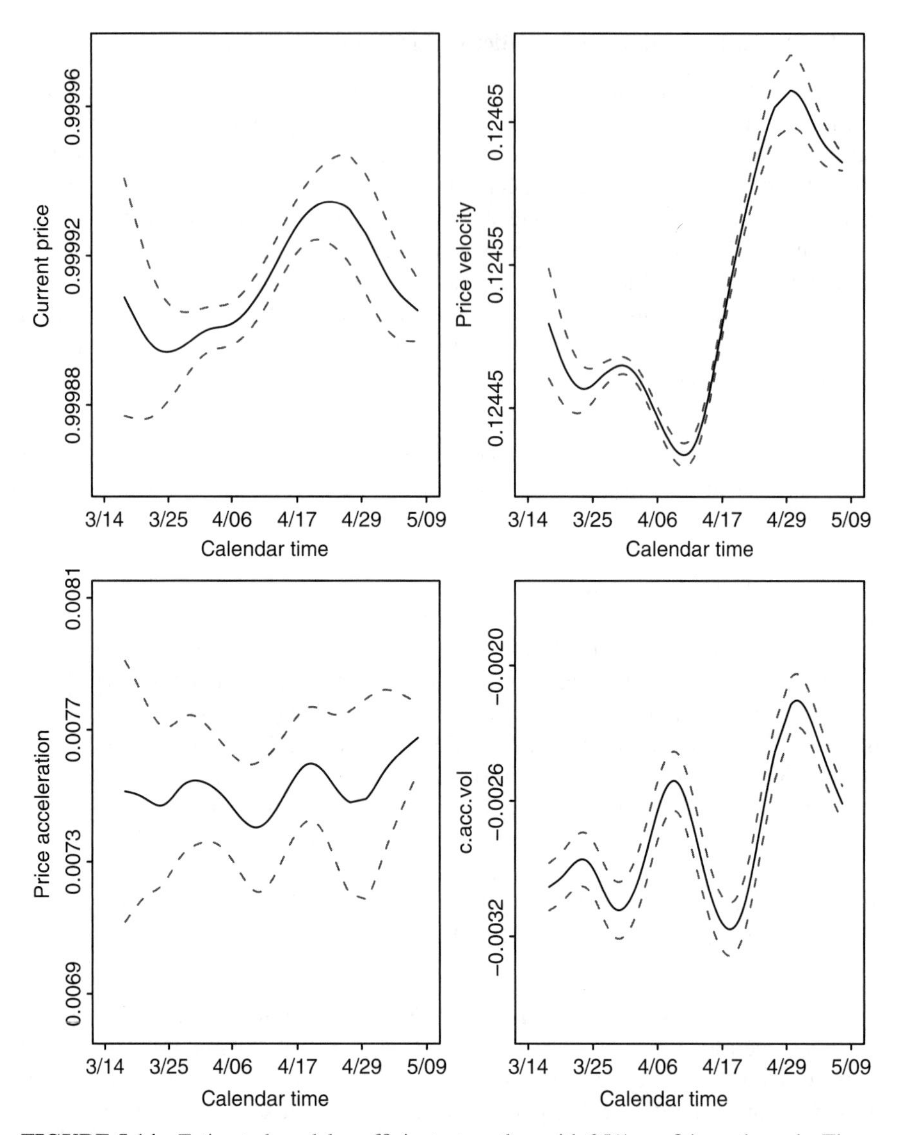

FIGURE 5.14 Estimated model coefficients, together with 95% confidence bounds. The x-axis denotes calendar time; the y-axis denotes the magnitude of the coefficient. The panels show (from left to right) current price, price velocity, price acceleration, and the acceleration volatility of competing auctions (c.acc.vol).

period T, $T = 1, 2, 3, \ldots, N_T$, in the validation set, we compute

$$\text{MAPE}(T) = \frac{1}{m_{T+1}} \sum_{i=1} \frac{|y_{T+1,i} - \hat{y}_{T+1,i}|}{|y_{T+1,i}|}, \tag{5.25}$$

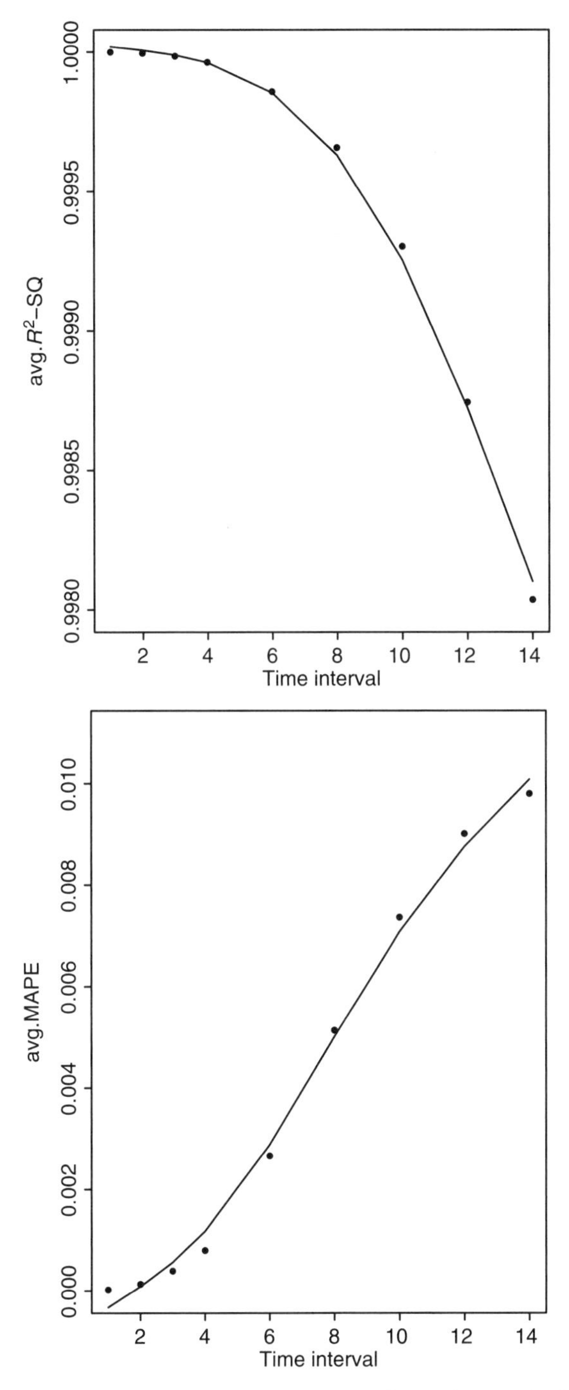

FIGURE 5.15 Model fit and prediction accuracy for different forecast horizons. The x-axis represents the forecast horizon (in hours; ranging from 1–14 h); the y-axis represents the value of avg.R^2 (top panel) and the value of avg.MAPE (bottom panel).

where $y_{T+1,i}$ and $\hat{y}_{T+1,i}$ denote the true and predicted values of auction #i at time $T+1$, respectively, and m_{T+1} denotes the number of auctions available at time $T+1$. We compute the *average MAPE* across all time periods as avg.MAPE $:= 1/N_T \sum_T$ MAPE(T). In a similar fashion to Section 4.2, we investigate avg.MAPE for different forecast horizons, $(T+1)-T = 1, 2, 3, \ldots, 14$ h. The bottom panel in Figure 5.15 shows the results.

Unsurprisingly, we see that as we predict further into the future (i.e., as forecast horizon gets larger), the predictive performance decreases (i.e., avg.MAPE increases). It is interesting to see that for predictions up to 4 h into the future, the prediction error is less than 0.1%. For forecast horizons larger than 4 h, the prediction error increases at a faster rate. However, even for predictions as far as 14 h into the future, the error is still less than 1%. This predictive accuracy is quite remarkable as we will see in the next section, where we benchmark our approach against several competitor approaches. We also note that while we cannot claim generalizability to all eBay auctions, there has been prior evidence that real-time forecasting models can provide superior predictive accuracy, especially for books and electronics (Wang et al., 2008a; Jank et al., 2006; Jank and Shmueli, 2010) as well as for other auction websites such as SaffronArt.com (Dass et al., 2009).

Comparison with Alternate Models We benchmark our model against five alternate models, the generalized additive model (GAM), classification and regression trees (CART), neural networks, and two simpler linear models: a purely static and an evolving linear model.

GAMs relax the restrictive linear model assumption between the response and predictors by a more flexible nonparametric form (Hastie and Tibshirani, 1990). Regression trees (Breiman et al., 1984) provide a data-driven way to partition the variable space and are thus often viewed as alternatives to formal variable selection. Neural networks also provide a technique that can approximate nonlinear functional relationships. In addition, we consider two linear models that use a subset of the variables from Table 5.7: one that uses only static information from the focal auction and the other that uses static and evolving information from the focal auction; we refer to these two models as "STATIC" and "EVOLVING," respectively. The static model corresponds to the information of many prior eBay studies (e.g., Lucking-Reiley et al., 2007) in that it considers only pre-auction information. The evolving model accounts for changes due to the process of bidding, but it does not account for price dynamics or competition.

Figure 5.16 shows avg.MAPE (similar to Figure 5.15) for forecast horizons $(T+1)-T = 1, 2, 3, \ldots, 14$ h, for all six different models. We refer to our model (5.24) as "DYN&COMP," since it includes dynamics and competition features. We can see that STATIC and the tree have the worst prediction performance, with an error universally larger than 10%. While our model performs the best, GAM, EVOLVING, and neural nets are competitive, at least for shorter forecast horizons. In other words, for predicting less than 4 h into the future, both GAM and EVOLVING pose alternatives with prediction errors not too much larger compared to DYN&COMP. However, their predictive performance breaks down for longer forecast horizons. In fact, the

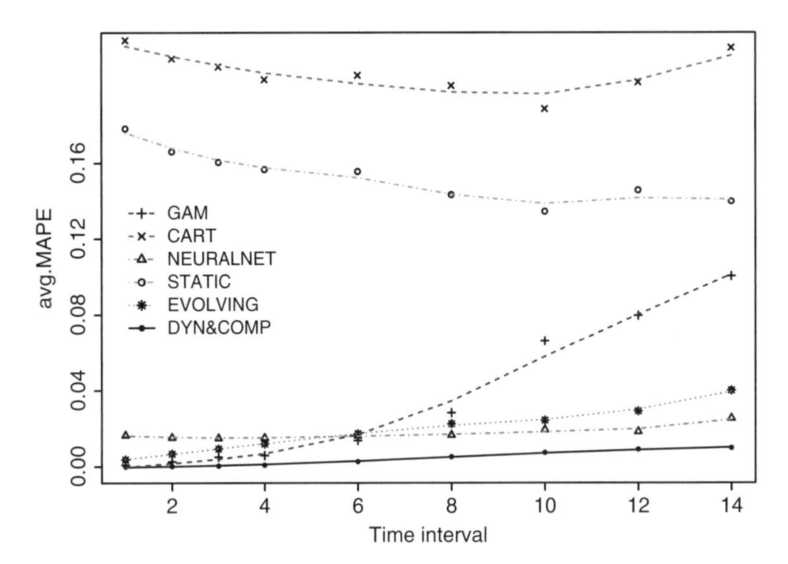

FIGURE 5.16 Prediction accuracy for competing models. The x-axis represents the forecast horizon (in hours); the y-axis represents the value of avg.MAPE.

error of GAM is as large as 10% for predicting 14 h into the future, which is 10-times more than the corresponding prediction error of DYN&COMP. While the performance of EVOLVING is somewhat better, its prediction error is four times more than DYN&COMP for a 14 h forecast horizon (similar for the neural network). In Section 5.3, we use the excellent forecasting performance of the DYN&COMP model and build an automated bidding strategy around it.

5.3 AUTOMATED BIDDING DECISIONS

We now discuss a practical application of auction forecasting models: making automated bidding decisions. In particular, we build an automated and data-driven decision rule around the forecasting model developed in the previous section. We are quick to point out though that this decision rule applies to *any* forecasting model; however, the better the model, the better the ultimate bidding decisions.

5.3.1 Motivation

Buyers in online auctions face many different decisions. They have to decide whether to bid early or late, whether to place a single bid or multiple updates, and whether to bid high or low. Bidding is further complicated by the existence of many auctions that offer the same or similar item simultaneously. In this case, one's bidding strategy has to be expanded to include decisions on which auction to bid on, when to bid on that auction, and how much.

There exist two very well documented bidding strategies widely employed in practice: *early bidding* and *last-minute bidding* . By signaling their commitment early, early bidders discourage competitors from entering the same auction (Bapna et al., 2004). In contrast, last-minute bidders wait until the very last moment as the chances of being outbid decrease with the time left in the auction (Roth and Ockenfels, 2002; Shmueli et al., 2007). However, both bidding strategies suffer from limitations since neither takes into account the effect of competition (Haruvy et al., 2008). In other words, neither strategy considers the information from simultaneous auctions offering the same (or similar) products. While Zeithammer (2006) recently suggested that bidders should shade their bids in the presence of sequential auctions with identical goods, the precise amount of the optimal shading on an auction-by-auction basis is not quite clear.

We develop a decision rule that provides answers to two basic bidding questions: which auction to bid on and how much to bid. The idea is based on maximizing consumer surplus, which refers to the difference between the bidders' *willingness to pay* and the price *actually paid*. We formulate an automated algorithm for selecting the best auction to bid on and for determining the best bid amount. The best auction provides bidders with the highest surplus, and the best bid amount equals the predicted closing price. This strategy automates the decision-making process, and, in contrast to early or late-minute manual bidding strategies, it frees the bidder of time constraints since bidding can occur automatically.

5.3.2 Decision Framework

Our decision framework is built upon the principles of maximizing a bidder's surplus (e.g., Bapna et al., 2004, 2008a). Surplus is the difference between the actual price paid and the bidder's willingness-to-pay amount (WTP) for an item. Therefore, the lower the price, the higher the winner's surplus.

For each individual auction, our forecasting model in equation (5.24) provides bidders with that auction's forecasted price; combining this with a bidder's WTP leads to the forecasted surplus. For a set of competing auctions, a plausible decision rule is to bid on that auction with the highest estimated surplus. Moreover, to avoid negative surplus, bidders should only bid on an auction if the predicted price is lower than their WTP.

Note that our forecasting model depends on the length of the forecast horizon $(T+1)-T$. Our model can predict only the final price of an auction that ends at or before time $(T+1)$. Therefore, longer forecast horizons result in a larger number of candidate auctions, that is, in a larger supply of *potential* auctions to bid on. On the other hand, we have also seen earlier that a longer forecast horizon leads to increased prediction error. Therefore, our decision rule faces a trade-off between supply of candidate auctions and prediction accuracy for each individual auction.

Our decision rule picks the auction with the highest estimated surplus, as long as the surplus is positive. After choosing an auction, the next two questions are with respect to the *time* and the *amount* of the bid. Since our forecasting model is based on a fixed forecast horizon, nothing is gained by waiting. So we suggest placing the bid as soon as an auction is chosen. Moreover, since our model predicts an auction's closing price at $\hat{y}_{T+1}$, we would expect to lose for bids lower than $\hat{y}_{T+1}$. Similarly, bids higher than $\hat{y}_{T+1}$ are expected to overpay. Therefore, we suggest bidding *exactly* the expected (or predicted) closing price $\hat{y}_{T+1}$. In summary, our decision rule chooses the auction with the highest predicted surplus, it bids the predicted price, and it places the bid immediately.

There are similarities between our approach and the bidding rule proposed by Peters and Severinov (2006). In Peters and Severinov's rule, a bidder places a bid on the auction with the lowest standing bid (or current price). In the case of two auctions with the same lowest standing bid, the bid is placed on the auction where the standing bid has changed after a change in the identity of the winning bidder, and the authors show that following this rule constitutes a perfect Bayesian equilibrium. Note that, similar to our forecasting model in (5.24), Peters and Severinov's rule uses only information about the price and its change (i.e., its dynamics). However, there are also important differences. First, in contrast to our model, Peters and Severinov's rule does not allow to derive explicit forecasts of an auction's final price. As a result, their rule can only recommend to "raise the bid as slowly as possible," while our approach also equips the bidder with a concrete and actionable recommended bid amount. Such a *focal point* may be important to bidders unsure of their true willingness to pay, or to those who often find themselves engaged in strong competition with other bidders, leading to bidding frenzy and other types of irrational behavior (Ariely and Simonson, 2003). Moreover, our approach promotes one-shot bidding that is less time-consuming than the incremental bidding of Peters and Severinov, and it may also share some of the psychological advantages of early bidding in which high, early bids often deter other bidders from competing. In addition, the ability to conduct one-shot bidding with a clear, objective understanding of the auction's predicted outcome may allow bidders to better distance themselves from the auction process and avoid emotional attachments that often result in irrational bidding decisions (Ariely and Simonson, 2003).

5.3.3 Assumptions and Limitations

Our decision framework makes several important assumptions. In particular, we assume that our bidding agent is available only to a few select bidders at a time (and not to the entire marketplace). This can be accomplished via a subscription service or, even more restrictive, via a consulting model in which bidders pay for services rendered on a case-by-case basis. The restrictive approach gives rise to a business model similar to the proprietary trading models employed by many investment firms. While it is unlikely that the entire eBay marketplace would immediately adopt a bidding agent that was freely available, there are important consequences of such a theoretical scenario. If all bidders had access to the same bidding agent (and would exclusively use that agent for all their bidding decisions), then this could impact the overall market allocative efficiency. In fact, if multiple high-value bidders congregate on the auction with the lowest forecasted price, this could leave other auctions with correspondingly less attention and the overall matching of buyers to sellers would not be done in the best interest of the market. While this is a valid concern, the goal of this chapter is simply a first attempt to help (one or more) select bidders choose the right auction and the right amount to bid. One could argue that once such an agent becomes successful with a group of bidders, it would give rise to other competing agents,[4] and the market would attain equilibrium under allocative efficiency. Moreover, it is unlikely that every single bidder on eBay would rely on such agents, and even if they did, some bidders might use multiple agents.

In the same vein, our bidding agent also does not incorporate reactions of other bidders as a result of the agent's bid. We do not believe that this is a severe limitation since, as pointed out above, we see our agent as being available to only a select number of bidders and, as a result, most of the market would not even be aware of its existence. In addition, modern technology allows agent-placed bids to be undistinguishable from manual bids, which makes it impossible for bidders to infer about the origin of a bid.

5.3.4 Experimental Setup

We conduct a simulation study to compare our automated bidding strategy to two alternate (and popular) bidding approaches: *early bidding* (Bapna et al., 2004) and *last-minute bidding* (Roth and Ockenfels, 2002). Early bidding is often viewed as a bidder's strong commitment and intends to deter others from entering the auction. Last-minute bidding is popular because it does not allow much time for other bidders to react. In our simulation, we assume that a bidder's WTP is drawn from a uniform distribution (Adamowicz et al., 1993) distributed symmetrically around the market value ($230 at the time of data collection); that is, we assume WTP $\sim$ Uniform($220, $240). Our experiment then proceeds as follows. We randomly draw a WTP from that distribution. We also draw an auction from the validation set (i.e., we compare the bidding strategies on the same real-world data that we compare the forecasting models). The bidder then makes a bidding decision (whether or not to bid and how much to bid) with each of the three bidding strategies (described in detail below). We repeat this experiment for all auctions in the validation set and for 20 different random draws from a bidder's WTP distribution.

Early Bidding Strategy We assume that early bidders bid at the end of the first auction day (Bapna et al., 2004). In fact, we find that slightly earlier or later bid times barely affect the outcome of the experiment. The process of early bidding is illustrated in the top panel of Figure 5.17. Bidders compare their WTP with the auction's current price at the end of the first day (p_{early}); if the WTP is higher, then they place a bid; otherwise they do not. If they do place a bid, then the bid amount equals the WTP. Note though that due to eBay's proxy bidding system that incrementally bids up to the WTP on behalf of the bidder, the final price may be lower than the WTP. As a consequence, the bidder pays only the amount of the second highest bid (plus a

[4] There has been a similar evolution with *sniping* agents of which there exist likely over 100 of them.

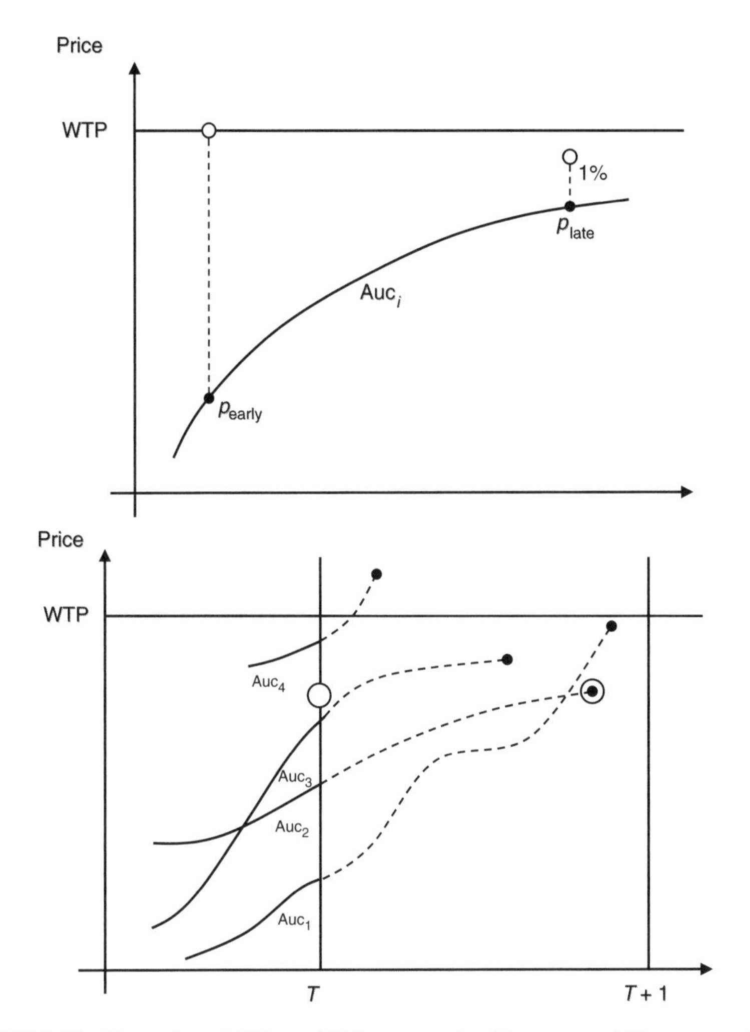

FIGURE 5.17 Illustration of different bidding strategies. The top panel illustrates early and last-minute bidding; the bottom panel illustrates our automated bidding strategy.

prespecified bid increment if the winning bid was submitted later than the second highest bid; this increment ranges between \$2.5 and \$5 in our case).[5]

Last-Minute Bidding Strategy We assume that last-minute bidders place their bid 1 min before the auction closes (Roth and Ockenfels, 2002). Last-minute bidding carries the danger that the bid does not go through due to network congestion (Shmueli et al., 2007), but we will not explicitly consider this disadvantage in our simulations.

[5] Jank and Zhang (2009a) also investigated alternate bidding heuristics; however, none of these heuristics beat last-minute bidding or our automated bidding strategy.

The process of last-minute bidding is again illustrated in the top panel of Figure 5.17. A bidder compares his WTP with the auction's current price 1 min before closing (p_{late}). Similar to early bidding, if his WTP is higher, then he places a bid, otherwise he does not place a bid and moves on to another auction. If he does place a bid, then the bid amount is only incrementally higher than the current price, since the chances of being outbid within the last 60 s are small. In our simulations, we bid an increment of 2% over the current price p_{late}.[6]

Our Automated Data-Driven Bidding Strategy In contrast to early and last-minute bidding, where a bidding decision is reached for each auction individually, our strategy requires a bidding decision for each forecast horizon. Consider the bottom panel of Figure 5.17. At time T of decision making, there are four competing auctions, denoted by Auc_1–Auc_4, which all close before time $T + 1$. The solid lines correspond to the observed part of the auction history; the dotted lines denote the future (and yet unobserved) price path. Since all auctions close before $T + 1$, our model yields predictions of their final prices (denoted by the solid black circles). Note that Auc_4 has the highest predicted price; moreover, its predicted price is higher than the bidder's WTP; hence the bidder will not consider this auction. Auc_3 has the smallest predicted price; since the predicted price is also smaller than the bidder's WTP, he places a bid on this auction. He bids the predicted price[7] and he bids immediately, that is, at time T. If he wins, then the bidder's surplus will be the difference between his WTP and the *actual* closing price.

5.3.5 Simulation Results

Similar to Bapna et al. (2004), we compare all bidding strategies on two dimensions: the probability of winning the auction and the average surplus accrued. We compute the probability of winning (p.win) as the number of auctions won divided by the total number of auctions that the bidder placed a bid in. We compute the average surplus (avg.sur) as the corresponding difference between the WTP and the actual price paid for an auction. The results are shown in Table 5.10.

We see that last-minute bidders have the highest probability of winning (95%, compared to 61% for our automated bidding strategy). This is not surprising since last-minute bidding is geared toward outbidding the competition in the very last moment. However, we also see that last-minute bidding accrues a significantly lower surplus compared to automated bidding (\$18 versus \$32). Another way of comparing the two bidding strategies is via their expected surplus, that is, the product ($p.win \times avg.sur$). We find that last-minute bidding yields an expected surplus of \$17.11 while that of our automated bidding strategy is higher: \$19.72. Moreover, while early bidders have a probability of 53% of winning the auction, their expected surplus is significantly

[6] Jank and Zhang (2009a) also study the robustness to different increments and find that bid increments of 1%, 2%, or \$2 yield comparable results.

[7] We also investigate bidding the WTP rather than the predicted price since in the proxy bidding system, the winner pays only the second highest price.

TABLE 5.10 Comparison of Different Bidding Strategies

	p.win	avg.sur	exp.sur
Last-moment bidding	95% (.5%)	$17.97 ($0.35)	$17.11
Early bidding	53% (2%)	$18.85 ($0.57)	$9.99
Automated bidding (predicted price)	61% (1%)	$32.33 ($1.95)	$19.72
Automated bidding (WTP)	78% (4%)	$25.05 ($1.85)	$19.60

The first column denotes the probability of winning; the second denotes the average accrued surplus, and the last column denotes the expected surplus, that is, exp.sur = (p.win × avg.sur). Standard errors are reported in parentheses.

lower: $9.99 (=53% × $18.85). The last row of Table 5.10 shows our automated bidding strategy where bidders bid their WTP rather than the predicted price. Bidding the WTP increases the probability of winning; however, it also slightly reduces the average surplus. In the end, the expected surplus is almost identical to that of bidding the predicted price. All in all, we can conclude that bidding strategies that reach across competing auctions lead to measurable gains for bidders. This finding is supported by earlier research by Anwar et al. (2006), who report that bidders who actively bid across competing auctions pay around 91% of the price paid by bidders who do not cross-bid.

Effect of the Forecast Horizon The results from the previous section (Table 5.10) are based on a forecast horizon of 12 h, and we have seen that it yields an expected surplus of $19.72 for our automated bidding strategy. We have pointed out earlier that the length of the forecast horizon has an effect on the outcome of our automated bidding strategy. On the one hand, longer forecast horizons yield a larger number of candidate auctions and as such a larger probability of including an auction with a lower price (and hence a higher surplus). On the other hand, longer forecast horizons also lead to less accurate predictions. Less accurate predictions can lead to overpayment (if the predicted price, and hence our bid, is higher than the actual price) and overpayment leads to lower surplus. Less accurate predictions can also lead to a reduced probability of winning the auction (if the predicted price, and hence our bid, is lower than the actual price). Thus, a change in the forecast horizons affects both the probability of winning and the average accrued surplus, and it is not quite clear how it affects the overall *expected surplus*. To that end, we repeat the simulation study from Table 5.10 for forecast horizons of different lengths. Table 5.11 shows the results.

We see that longer forecast horizons result in a higher average surplus, which suggests that the effect of having a larger pool of candidate auctions outweighs the effect of overpayment. However, we also see that longer horizons result in a smaller probability of winning because the less accurate predictions more frequently yield bids below the auction's actual closing price and hence an unsuccessful auction. Interestingly, the *expected surplus* is maximized for a forecast horizon of 9 h. While our results do not prove optimality, they suggest an interesting global optimization problem for future research.

TABLE 5.11 Trade-Off Between the Length of the Forecast Horizon and Expected Surplus

Forecast Horizon	p.win	avg.sur	exp.sur
14 h	59.01%	$35.29	$20.82
12 h	61.44%	$32.33	$19.80
9 h	67.82%	$30.99	$21.02
6 h	69.43%	$29.60	$20.55
3 h	75.77%	$27.21	$20.62

5.3.6 Discussion

It is important to understand that our automated bidding approach relies on a number of key ingredients. We assume that appropriate bidding records are available and we focus only on deriving a model from these bidding records and subsequently designing a bidding strategy around that model. Before deploying our automated bidding approach, one also needs to put in place methods for *searching* and *selecting* the right bidding records. Finding suitable bidding records can be accomplished in several ways, for example, using automated agents such as web crawlers, obtaining data via Web services, or directly purchasing bidding data (see Chapter 2).

Given a pool of bidding records, the next challenge is to select the right subset of *most relevant* bidding records. One could find this set by listing a desired set of product features (e.g., "iPod Nano, 8 GB, yellow") and then selecting only those bidding records that are most similar to the feature list. Deriving a suitable similarity metric can be done using, for example, the spatial feature model presented in Section 4.3. Isolating product features from bidding records is made possible via eBay's effort of standardizing certain product descriptions (e.g., product descriptions for MP3 players require fields such as brand, product line, storage capacity, color, or condition); additional information is often contained in the unstructured descriptive text that may take more effort to mine.

Related to the issue of search and selection is the issue of incorporating individual user preferences or risk tolerance into the bidding process. While some bidders may consider all relevant auctions as potential candidates, others may be more selective and wish to filter auctions based on certain constraints (e.g., filter out auctions with seller ratings lower than a certain threshold or for red iPods). In fact, when applying our method to high-reputation sellers only, the expected surplus increases to $20.05. Alternatively, rather than filtering auctions, users may incorporate their risk tolerance in a varying willingness-to-pay function. For instance, a bidder may be willing to pay as much as $240 from a seller with a high reputation, but no more than $220 from a seller with a low reputation. In our simulations, the expected surplus drops from $18.03 to $7.67 to $2.30 when reducing the WTP function for low-reputation sellers from Uniform($220, $240) to Uniform($200, $220) to Uniform($180, $200), respectively. One reason is that the decreased risk tolerance (i.e., smaller WTP function) results in a smaller number of candidate auctions and hence a reduced choice for the bidder.

Another is that, given a constant "market value" ($230 in our case), a lower WTP must, at least on average, reduce a bidder's surplus.

We also point out that in practice one would apply this method repeatedly until a consumer's demand is satisfied. We assume here that a consumer has demand for only a single unit and that they do not have any time constraints. Then, our automated system would place a bid while continuously monitoring the remaining market—which could be done at no extra cost for the bidder using automated agents. Once the outcome of the first bid is known, the system would then decide whether and (if the previous bid was unsuccessful) where to place the next bid, and so on. While a bidder could also decide to place more than one bid simultaneously, this runs the risk of winning multiple auctions, which is undesirable in the case of single unit demand.

It is also important to note that the proposed system is modular in the sense that individual modules can be exchanged. For instance, one can replace the dynamic forecasting model by alternatives (such as GAM or regression trees); similarly, one can replace the bidding decisions by an alternate set of rules. All in all, in order for the approach to be deployed, one will ultimately have to rely on agent-based technologies, similar to those currently in place for bid sniping (e.g., Cniper.com).

We point out that our simulations are based on a static evaluation of the bidding data; that is, the bid history does not change in response to a new bidder's actions. The problem with such a static evaluation is that, in reality, when a new bidder submits a bid, all other participants have an opportunity to respond to this bid, which may result in a different outcome. While such a scenario is, unfortunately, not reproducible using historical bidding data, it would take an experimental setting (either a costly field experiment or an oftentimes not all realistic classroom experiment) to investigate the true life impact of our bidding strategy.

While we consider only single unit auctions in this research, one could expand the scope of our bidding strategy to multi-unit auctions. Let us assume that a seller sells n items (of identical product specification and quality) in the same auction; then the bidders with the top n bids each win one item. To apply our bidding strategy to this scenario, bidders need to know the lowest transacting bid; that is, they need to predict the price at which the nth item sells. Given a set of relevant bidding records, one solution is to apply our model to the price of the nth item; that is, we would train our model to predict the lowest transacting bid.

A related issue is multi-unit demand. If a bidder requires more than one unit, then the current bidding strategy could still be employed if the bidder has no time constraints and decides to bid sequentially, and if the bidder's WTP is the same for all units (assuming that there is a practically unlimited supply, which is realistic for many items in large auction markets). However, if the bidder needs to purchase n units within a short period of time and places m bids simultaneously, then each bid should be discounted relative to the size of m; on the other hand, bids may be inflated with decreasing time periods to assure that all units are available on time. This calculation may change further for varying WTP distributions.

There are additional ways in which this research can be expanded. We have already pointed to the problem of selecting the optimal forecast horizon earlier. Another way to expand this research is via allowing closing *and* continuing auctions. Recall that our current approach considers only auctions that close within the given forecast horizon. The reason is that our forecasting model is geared to a fixed forecast horizon $(T + 1) - T$, so we can only predict the final price of auctions that end within that interval. Of course, one can roll the model forward one additional time period to make predictions at $T + 2$, based on the predicted values at $T + 1$; however, predicting two steps ahead (i.e., $T \rightarrow T + 2$) is more uncertain than one-step-ahead prediction (i.e., $T \rightarrow T + 1$). It is not quite clear how to discount the additional prediction uncertainty in our decision framework. Another way to expand this research is to allow for variable and adaptive WTP distributions. In our simulations, we assume that both early and last-minute bidders have the same WTP distribution. It may be possible that bidders with different strategies also have different product valuations. Moreover, we assume that the WTP distribution remains constant over the forecast horizon. While this may be realistic for short horizons (up to several hours), a bidder who wants an item immediately may have a different valuation compared to a bidder who is willing to wait several weeks.

Finally, another important issue is the potential effect of a forecasting model on the market as a whole. If every bidder had access to the same model and bid on the same auction (with the lowest forecasted price), then forecasts, and as a consequence bidding decisions, would become unstable. This is very similar to the stock market where investment houses deploy complex math models to guide investment decisions. In such a scenario, there is a risk that if all investors base their decisions on the same model, the model—and not the investments' performance—could eventually drive the market. In this research, we are much less ambitious. While, at least in theory, one single model could eventually drive all bidding decisions on eBay, it is unlikely that it ever will. Rather, we view our automated bidding strategy, if ever deployed, as a decision tool that would be made available only to a few select bidders and thus not destabilize the market.

BIBLIOGRAPHY

Abraham, C., Cornillion, P. A., Matzner-Lober, E., and Molinari, N. (2003). Unsupervised curve-clustering using B-spline. *Scandinavian Journal of Statistics*, 30(3):581–595.

Aczel, J. (1966). *Lectures on Functional Equations and Their Applications*. Academic Press, New York.

Adamowicz, W., Bhardwaj, V., and Macnab, B. (1993). Experiments on the difference between willingness to pay and willingness to accept. *Land Economics*, 69(4):416–427.

Afshartous, D. and de Leeuw, J. (2005). Prediction in multilevel models. *Journal of Educational and Behavioral Statistics*, 30(2):109–139.

Agarwal, D. (2008). Statistical challenges in Internet advertising. In Jank, W. and Shmueli, G. (eds.), *Statistical Methods in eCommerce Research*, Wiley, New York.

Allen, G. N., Burke, D. L., and David, G. B. (2006). Academic data collection in electronic environments: defining acceptable use of Internet resources. *MIS Quarterly*, 30(3):599–610.

Allen, M. T. and Swisher, J. (2000). An analysis of the price formation process at a HUD auction. *Journal of Real Estate Research*, 20:279–298.

Almgren, R. (2002). Financial derivatives and partial differential equations. *American Mathematical Monthly*, 109(1):1–12.

Anwar, S., McMillan, R., and Zheng, M. (2006). Bidding behavior at competing auctions: evidence from eBay. *European Economic Review*, 50(2):307–322.

Ariely, D., Ockenfels, A., and Roth, A. E. (2005). An experimental analysis of ending rules in Internet auctions. *RAND Journal of Economics*, 36(4):890–907.

Ariely, D. and Simonson, I. (2003). Buying, bidding, playing, or competing? Value assessment and decision dynamics in online auctions. *Journal of Consumer Psychology*, 13(1–2):113–123.

Aris, A., Shneiderman, B., Plaisant, C., Shmueli, G., and Jank, W. (2005). Representing unevenly-spaced time series data for visualization and interactive exploration. In *International Conference on Human Computer Interaction (INTERACT 2005)*.

Aurell, E. and Hemmingsson, J. (1997). Bid frequency analysis in liquid markets. TRITA-PDC Report, pp. 1401–2731.

Avery, C. N., Jolls, C., Posner, R. A., and Roth, A. E. (2001). The market for federal judicial law clerks. *University of Chicago Law Review*, 68:793–902.

Bajari, P. and Hortacsu, A. (2003). The winner's curse, reserve prices and endogenous entry: empirical insights from eBay auctions. *RAND Journal of Economics*, 3(2):329–355.

Banks, D. and Said, Y. (2006). Data mining in electronic commerce. *Statistical Science*, 21(2):234–246.

Bapna, R., Goes, P., Gopal, R., and Marsden, J. (2006). Moving from data-constrained to data-enabled research: experiences and challenges in collecting, validating, and analyzing large-scale e-commerce data. *Statistical Science*, 21:116–130.

Bapna, R., Goes, P., and Gupta, A. (2003). Analysis and design of business-to-consumer online auctions. *Management Science*, 49:85–101.

Bapna, R., Goes, P., Gupta, A., and Jin, Y. (2004). User heterogeneity and its impact on electronic auction market design: an empirical exploration. *MIS Quarterly*, 28(1):21–43.

Bapna, R., Goes, P., Gupta, A., and Karuga, G. (2002a). Optimal design of the online auction channel: analytical, empirical and computational insights. *Decision Sciences*, 33(4):557–577.

Bapna, R., Goes, P., Gupta, A., and Karuga, G. (2002b). Predictive calibration of online multi-unit ascending auctions. In *Proceedings of WITS-2002,* Barcelona, Spain, December 2002.

Bapna, R., Jank, W., and Shmueli, G. (2008a). Consumer surplus in online auctions. *Information Systems Research*, 19:400–416.

Bapna, R., Jank, W., and Shmueli, G. (2008b). Price formation and its dynamics in online auctions. *Decision Support Systems*, 44:641–656.

Basseville, M. (1989). Distance measure for signal processing and pattern recognition. *Signal Processing*, 18(4):349–369.

Beam, C., Segev, A., and Shanthikumar, J. G. (1996). Electronic negotiation through Internet-based auctions. Technical Report, CITM Working Paper 96-WP-1019.

Beck, N., Glreditsch, K. S., and Beardsley, K. (2006). Space is more than geography: using spatial econometrics in the study of political economy. *International Studies Quarterly*, 50(1):27–44.

Becker, R., Chambers, J., and Wilks, A. (1988). *The New S Language*, Wadsworth & Brooks, Pacific Grove, CA.

Bederson, B., Shneiderman, B., and Wattenberg, M. (2002). Ordered and quantum treemaps: making effective use of 2D space to display hierarchies. *ACM Transactions on Graphics*, 21:833–854.

Berk, K. (1978). Comparing subset regression procedures. *Technometrics*, 20(1):1–6.

Bock, H. H. (1974). *Automatische Klassifikation.* Vandenhoeck & Rupprecht, Goettingen.

Borle, S., Boatwright, P., and Kadane, J. B. (2003). The common/private-value continuum: an empirical investigation using eBay online auctions. Technical Report, Rice University.

Borle, S., Boatwright, P., and Kadane, J. B. (2006). The timing of bid placement and extent of multiple bidding: an empirical investigation using eBay online auctions. *Statistical Science*, 21(2):194–205.

Breiman, L., Friedman, J., Olshen, R., and Stone, C. (1984). *Classification and Regression Trees*, Wadsworth & Brooks, Pacific Grove, CA.

Brown, J. and Morgan, J. (2006). Reputation in online auctions: the market for trust. *California Management Review*, 49(1):61.

Bruschi, D., Poletti, G., and Rosti, E. (2002). E-vote and PKI's: a need, a bliss or a curse? In Gritzalis, D. (ed.), *Secure Electronic Voting*, Kluwer Academic Publishers.

Buono, P., Plaisant, C., Simeone, A., Aris, A., Shneiderman, B., Shmueli, G., and Jank, W. (2007). Similarity-based forecasting with simultaneous previews: a river plot interface for time series forecasting. In *11th International Conference on Information Visualization*.

Bzik, T. J. (2005). Overcoming problems associated with the statistical reporting and analysis of ultratrace data. Available at `http://www.micromanagemagazine.com/archive/05/06/bzik.html` (January 8, 2007).

Caccetta, L., Chow, C., Dixon, T., and Stanton, J. (2005). In Modelling the structure of Australian wool auction prices. *MODSIM 2005 International Congress on Modelling and Simulation*, Modelling and Simulation Society of Australia and New Zealand, pp. 1737–1743.

Chan, K. Y. and Loh, W. Y. (2004). Lotus: an algorithm for building accurate and comprehensible logistic regression trees. *Journal of Computational and Graphical Statistics*, 13(4):826–852.

Coddington, E. A. and Levinson, L. (1955). *Theory of Ordinary Differential Equations*. McGraw-Hill, New York.

Cover, T. M. and Hart, P. (1967). Nearest neighbor pattern classification. *IEEE Transactions on Information Theory*, 13:21–27.

CRAN (2003). *The R Project for Statistical Computing*. Available at `http://www.r-project.org/index.html`.

Cressie, N. (1993). *Statistics for Spatial Data*. Wiley.

Croson, R. T. A. (1996). Partners and strangers revisited. *Economics Letters*, 53:25–32.

Cuesta-Albertos, J. A., Gordaliza, A., and Matran, C. (1997). Trimmed k-means: an attempt to robustify quantizers. *The Annals of Statistics*, 25:553–576.

Dass, M., Jank, W., and Shmueli, G. (2009). Dynamic price forecasting in simultaneous online art auctions. In Casillas, J. and Martnez-López, F. J. (eds.), *Marketing Intelligent Systems Using Soft Computing, Springer*.

Dass, M. and Reddy, S. K. (2008). An analysis of price dynamics, bidder networks and market structure in online auctions. In Jank, W. and Shmueli, G. (eds.), *Statistical Methods in eCommerce Research*. Wiley, New York.

Dellarocas, C. and Narayan, R. (2006). A statistical measure of a population's propensity to engage in post-purchase online word-of-mouth. *Statistical Science*, 21(2):277–285.

Deltas, G. (1999). Auction size and price dynamics in sequential auctions. Working Paper, University of Illinois.

Devroye, L. P. (1981). On the almost everywhere convergence of nonparametric regression function estimates. *The Annals of Statistics*, 9:1310–1319.

Donkers, B., Verhoef, P. C., and De Jong, M. (2003). Predicting customer lifetime value in multi-service industries. Technical Report, ERIM Report Series Reference No. ERS-2003-038-MKT. Available at SSRN: http://ssrn.com/abstract=411666.

Engle, R. F. and Russel, J. R. (1998). Autoregressive conditional duration: a new model for irregularly spaced transaction data. *Econometrica*, 66(5):1127–1162.

Etzion, H., Pinker, E., and Seidmann, A. (2003). Analyzing the simultaneous use of auctions and posted prices for on-line selling. Working Paper CIS-03-01, Simon School, University of Rochester.

Fader, P. and Hardie, B. (2006). How to project customer retention. Technical Report. Available at SSRN: http://ssrn.com/abstract=801145.

Fader, P., Hardie, B., and Lee, K. L. (2006). CLV: more than meets the eye. Technical Report. Available at SSRN: http://ssrn.com/abstract=913338.

Fienberg, S. E. (2006). Privacy and confidentiality in an e-commerce world: data mining, data warehousing, matching and disclosure limitation. *Statistical Science*, 21(2):143–154.

Fienberg, S. E. (2008). Is privacy protection for data in an eCommerce world an oxymoron? In Jank, W. and Shmueli, G. (eds.), *Statistical Methods in eCommerce Research*, Wiley, New York.

Forman, C. and Goldfarb, A. (2008). How has electronic commerce research advanced understanding of the offline world? In Jank, W. and Shmueli, G. (eds.), *Statistical Methods in eCommerce Research*, Wiley, New York.

Forsythe, R., Rietz, T. A., and Ross, T. W. (1999). Wishes, expectations, and actions: a survey on price formation in election stock markets. *Journal of Economic Behavior & Organization*, 39:83–110.

Fouque, J. P., Papanicolaou, G., and Sircar, K. R. (2000). *Derivatives in Financial Markets with Stochastic Volatility*. Cambridge University Press.

Foutz, N. and Jank, W. (2009). Pre-release demand forecasting for motion pictures using functional shape analysis of virtual stock markets. *Marketing Science*, in press. Published online in Articles in Advance, December 2, 2009 DOI: 10.1287/mksc.1090.0542.

Fox, J. (1997). *Applied Regression, Linear Models, and Related Methods*. Sage Publications.

French, J. L., Kammann, E. E., and Wand, M. P. (2001). Comment on Ke and Wang. *Journal of the American Statistical Association*, 96:1285–1288.

Gamma, J. (2004). Functional trees. *Machine Learning*, 55:219–250.

George, E. (2000). The variable selection problem. *Journal of the American Statistical Association*, 95(452):1304–1308.

Ghani, R. (2005). Price prediction and insurance for online auctions. In *11th ACM SIGKDD International Conference on Knowledge Discovery and Data Mining*, Chicago, IL.

Ghani, R. and Simmons, H. (2004). Predicting the end-price of online auctions. In *Proceedings of the International Workshop on Data Mining and Adaptive Modeling and Methods for Economics and Management*.

Ghose, A. (2008). The economic impact of user-generated and firm-published online content: directions for advancing the frontiers in electronic commerce research. In Jank, W. and Shmueli, G. (eds.), *Statistical Methods in eCommerce Research*, Wiley, New York.

Ghose, A., Smith, M. D., and Telang, R. (2006). Internet exchanges for used books: an empirical analysis of product cannibalization and welfare impact. *Information Systems Research*, 17(1):3–19.

Ghose, A. and Sundararajan, A. (2006). Evaluating pricing strategy using e-commerce data: evidence and estimation challenges. *Statistical Science*, 21(2):131–142.

Goldfarb, A. and Lu, Q. (2006). Household-specific regressions using clickstream data. *Statistical Science*, 21(2):247–255.

Goldstein, M. (1972). *K*-nearest neighbor classification. *IEEE Transactions on Information Theory*, 18(5):627–630.

Good, I. J. and Gaskins, R. J. (1980). Density estimation and bump hunting by the penalized maximum likelihood method exemplified by scattering and meteorite data. *Journal of the American Statistical Association*, 75(369):42–56.

Goyvaerts, J. and Levithan, S. (2009). *Regular Expressions Cookbook*. O'Reilly Media.

Guerre, E., Perrigne, I., and Vuong, Q. (2000). Optimal nonparametric estimation of first-price auctions. *Econometrica*, 68(3):525–574.

Gwebu, K., Wang, J., and Shanker, M. (2005). A simulation study of online auctions: analysis of bidders' and bid-takers' strategies. In *Proceedings of the Decision Science Institute*.

Harel, A., Kenett, R., and Ruggeri, F. (2008). Modeling web usability diagnostics based on usage statistics. In Jank, W. and Shmueli, G. (eds.), *Statistical Methods in eCommerce Research*, Wiley, New York.

Harter, H. L. (1961). Expected values of normal order statistics. *Biometrica*, 48(1–2):151–165.

Haruvy, E. and Popkowski Leszczyc, P. (2009). What does it take to make consumers search? Working Paper, Department of Marketing, Business Economics and Law, University of Alberta.

Haruvy, E. and Popkowski Leszczyc, P. (2010). The impact of online auction duration. *Decision Analysis*, 7(1):99–106.

Haruvy, E., Popkowski Leszczyc, P., Carare, O., Cox, J., Greenleaf, E., Jap, S., Jank, W., Park, Y., and Rothkopf, M. (2008). Competition between auctions. *Marketing Letters*, 19(3–4):431–448.

Hastie, T. J., Buja, A., and Tibshirani, R. J. (1995). Penalized discriminant analysis. *The Annals of Statistics*, 23:73–102.

Hastie, T. and Tibshirani, R. (1990). *Generalized Additive Models*. Chapman & Hall, London.

Hastie, T., Tibshirani, R., and Friedman, J. (2001). *The Elements of Statistical Learning*. Springer, New York.

Haubl, G. and Popkowski Leszczyc, P. T. L. (2003). Minimum prices and product valuations in auctions. *Marketing Science Institute Reports*, Issue 3, No. 03-117, pp. 115–141.

Henry, G. T. (1990). *Practical Sampling. Applied Social Research Methods*, Vol. 21, Sage Publications.

Hill, S., Provost, F., and Volinsky, C. (2006). Network-based marketing: identifying likely adopters via consumer networks. *Statistical Science*, 21(2):256–276.

Hlasny, V. (2006). Testing for the occurrence of shill bidding (in Internet auctions). *Graduate Journal of Social Science*, 3(1):61–81.

Holland, J. H. (1975). *Adaptation in Natural and Artificial Systems*. The University of Michigan Press, Ann Arbor, MI.

Huberman, B. A. and Adamic, L. A. (1999). Growth dynamics of the World Wide Web. *Nature*, 401:131–132.

Hyde, V., Jank, W., and Shmueli, G. (2006). Investigating concurrency in online auctions through visualization. *The American Statistician*, 60:241–250.

Hyde, V., Jank, W., and Shmueli, G. (2008). A family of growth models for representing the price process in online auctions. In Jank, W. and Shmueli, G. (eds.), *Statistical Methods in eCommerce Research*, Wiley, New York.

Jain, A., Murty, M., and Flynn, P. (1999). Data clustering: a review. *ACM Computing Surveys*, 31(3):264–323.

James, G. M. (2002). Generalized linear models with functional predictor variables. *Journal of the Royal Statistical Society, Series B*, 64:411–432.

James, G. M. and Hastie, T. (2001). Functional linear discriminant analysis for irregularly sampled curves. *Journal of the Royal Statistical Society, Series B*, 63:533–550.

James, G. M. and Sugar, C. A. (2003). Clustering sparsely sampled functional data. *Journal of the American Statistical Association*, 98(462):397–408.

James, G. M., Wang, J., and Zhu, J. (2009). Functional linear regression that's Interpretable. *The Annals of Statistics* 37:2083–2108.

Jank, W. and Kannan, P. K. (2008). Spatial models for online mortgage leads. In Jank, W and Shmueli, G. (eds.), *Statistical Methods in eCommerce Research*, Wiley, New York.

Jank, W. and Shmueli, G. (2005). Visualizing online auctions. *Journal of Computational and Graphical Statistics*, 14(2):299–319.

Jank, W. and Shmueli, G. (2006). Functional data analysis in electronic commerce research. *Statistical Science*, 21(2):155–166.

Jank, W. and Shmueli, G. (2007). Modelling concurrency of events in on-line auctions via spatiotemporal semiparametric models. *Journal of the Royal Statistical Society, Series C*, 56(1):1–27.

Jank, W. and Shmueli, G. (2008a). *Statistical Methods in eCommerce Research*, Wiley, New York.

Jank, W. and Shmueli, G. (2008b). Studying heterogeneity of price evolution in eBay auctions via functional clustering. In Adomavicius, G. and Gupta, A. (eds.), *Handbook of Information Systems Series: Business Computing*, Emerald, pp. 237–261.

Jank, W. and Shmueli, G. (2009). Studying heterogeneity of price evolution in eBay auctions via functional clustering. In *Handbook of Information Systems Series: Business Computing*, Emerald, pp. 237–261.

Jank, W. and Shmueli, G. (2010). Forecasting online auctions using dynamic models. In Soares, C. and Ghani, R. (eds.), *Data Mining for Business Applications*, IOS Press, in press.

Jank, W. Shmueli, G. Dass, M. Yahav, I., and Zhang, S. (2008a). Statistical challenges in eCommerce: modeling dynamic and networked data. *INFORMS Tutorials in Operations Research*, 2008 edition, pp. 31–54.

Jank, W., Shmueli, G., and Wang, S. (2006). Dynamic, real-time forecasting of online auction via functional models. In *Proceedings of the 12th ACM SIGKDD International Conference on Knowledge Discovery and Data Mining (KDD2006)*, Philadelphia, PA, August 20–23, 2006.

Jank, W., Shmueli, G., and Wang, S. (2008b). Modeling price dynamics in online auctions via regression trees. In Jank, W. and Shmueli, G. (eds.), *Statistical Methods in eCommerce Research*, Wiley, New York.

Jank, W., Shmueli, G., and Zhang, S. (2009). A flexible model for price dynamics in online auctions. Technical Report, RH Smith School of Business, University of Maryland.

Jank, W. and Yahav, I. (2010). E-loyalty networks in online auctions. *The Annals of Applied Statistics*, in press.

Jank, W. and Zhang, S. (2009a). An automated and data-driven bidding strategy for online auctions. Technical Report, RH Smith School of Business, University of Maryland. Available at SSRN: http://ssrn.com/abstract=1427212.

Jank, W. and Zhang, S. (2009b). Competition in online markets: model selection for improved forecasting. Technical Report, RH Smith School of Business, University of Maryland.

Jap, S. and Naik, P. (2008). BidAnalyzer: a method for estimation and selection of dynamic bidding models. *Marketing Science*, 27:949–960.

Karatzas, I. and Shreve, S. E. (1998). *Methods of Mathematical Finance*. Springer, New York.

Kauffman, R. J., March, S. T. and Wood, C. (2000). Mapping out design aspects for data-collecting agents. *International Journal of Intelligent Systems in Accounting, Finance, and Management*, 9(4):217–236.

Kauffman, R. and Wang, B. (2008). Developing rich insights on public internet firm entry and exit based on survival analysis and data visualization. In Jank, W. and Shmueli, G. (eds.), *Statistical Methods in eCommerce Research*, Wiley, New York.

Kauffman, R. J. and Wood, C. A. (2000). Running up the bid: modeling seller opportunism in internet auctions. In *Proceedings of the 2000 Americas Conference on Information Systems*.

Kauffman, R. J. and Wood, C. A. (2005). The effects of shilling on final bid prices in online auctions. *Electronic Commerce Research and Applications*, 4(2):21–34.

Kaufman, L. and Rousseeuw, P. J. (1987). Clustering by means of medoids. In Dodge, Y. (ed.), *Statistical Data Analysis Based on the* L_1*-Norm and Related Methods*. North Holland, Amsterdam, pp. 405–416.

Kim, H. and Loh, W. Y. (2001). Classification trees with unbiased multiway splits. *Journal of the American Statistical Association*, 96(454):589–604.

Klemperer, P. (1999). Auction theory: a guide to the literature. *Journal of Economic Surveys*, 13(3):227–286.

Kneip, A. and Utikal, K. J. (2001). Inference for density families using functional principal component analysis. *Journal of the American Statistical Association*, 96(454):519–542.

Krishna, V. (2002). *Auction Theory*. Academic Press, San Diego, CA.

Ku, G., Malhorta, D., and Murnighan, J. D. (2004). Competitive arousal in live and Internet auctions. Working Paper, Northwestern University.

Kulkarni, S. R., Lugosi, G., and Venkatesh, S. S. (1998). Learning pattern classification: a survey. *IEEE Transactions on Information Theory*, 44(6):2178–2206.

Kulkarni, S. R. and Posner, S. E. (1995). Rates of convergence of nearest neighbor estimation under arbitrary sampling. *IEEE Transactions on Information Theory*, 41(4):1028–1039.

Kullback, S. and Leibler, R. A. (1951). On information and sufficiency. *The Annals of Mathematical Statistics*, 22:79–86.

Kumar, M. and Patel, N. (2008). Clustering data with measurement error. In Jank, W. and Shmueli, G. (eds.), *Statistical Methods in eCommerce Research*, Wiley, New York.

Lambert, D. (1992). Zero-inflated Poisson regression, with an application to defects in manufacturing. *Technometrics*, 34:1–14.

Levinson, N. (1946). The Wiener RMS (root mean square) error criterion in filter design and prediction. *Journal of Mathematics and Physics*, 25:261–278.

Levy, P. S. and Lemeshow, S. (1999). *Sampling of Populations, Methods and Applications*, 3rd edition, Wiley.

Li, L. I. (2006). Reputation, trust, and rebates: how online auction markets can improve their feedback mechanisms. Working Paper.

Liebovitch, L. S. and Schwartz, I. B. (2003). Information flow dynamics and timing patterns in the arrival of email viruses. *Physical Review*, 68:1–4.

Liu, B. and Müller, H.-G. (2008). Functional data analysis for sparse auction data. In Jank, W. and Shmueli, G. (eds.), *Statistical Methods in eCommerce Research*, Wiley, New York, pp. 269–290.

Liu, B. and Müller, H.-G. (2009). Estimating derivatives for samples of sparsely observed functions, with application to on-line auction dynamics. *Journal of the American Statistical Association*, 104:704–714.

Livingston, J. (2005). How valuable is a good reputation? A sample selection model of Internet auctions. *Review of Economics and Statistics*, 87(3):453–465.

Loh, W. Y. (2002). Regression trees with unbiased variable selection and interaction detection. *Statistica Sinica*, 12:361–386.

Lohr, S. L. (1999). *Sampling: Design and Analysis*. Duxbury Press.

Lucking-Reiley, D. (1999). Using field experiments to test equivalence between auction formats: magic on the Internet. *American Economic Review*, 89(5):1063–1080.

Lucking-Reiley, D. (2000) Auctions on the Internet: what's being auctioned, and how? *Journal of Industrial Economics*, 48(3):227–252.

Lucking-Reiley, D., Bryan, D., Prasad, N., and Reeves, D. (2007). Pennies from eBay: the determinants of price in online auctions. *The Journal of Industrial Economics*, 55(2):223–233.

Lucking-Reiley, D., Bryan, D., and Reeves, D. (2000). Pennies from eBay: the determinants of price in online auctions. Working Paper 00-W03, Department of Economics, Vanderbilt University.

Mankiw, N. G., Romer, D., and Weil, D. N. (1992). A contribution to the empirics of economic growth. *Quarterly Journal of Economics*, 107(2):407–437.

Matas, A. and Schamroth, Y. (2008). Optimization of search engine marketing bidding strategies using statistical techniques. In Jank, W. and Shmueli, G. (eds.), *Statistical Methods in eCommerce Research*, Wiley, New York.

McAfee, R. P. and McMillan, J. (1987). Auctions with stochastic number of bidders. *Journal of Economic Theory*, 43:1–19.

McCulloch, C. and Searle, S. (2000). *Generalized, Linear, and Mixed Models*. Wiley.

McGill, R., Tukey, J. W., and Larsen, W. A. (1978). Variations of box plots. *The American Statistician*, 38:12–16.

Menasce, D. A. and Akula, V. (2004). Improving the performance of online auction sites through closing time rescheduling. In *First International Conference on Quantitative Evaluation of Systems (QEST'04)*, pp. 186–194.

Milgrom, P. and Weber, R. (1982). A theory of auctions and competitive bidding. *Econometrica*, 50(5):1089–1122.

Mithas, S., Almirall, D., and Krishnan, M. S. (2006). Do CRM systems cause one-to-one marketing effectiveness? *Statistical Science*, 21(2):223–233.

Müller, H.-G. and Yao, F. (2010). Empirical dynamics for longitudinal data. Working Paper, Department of Statistics, University of California at Davis.

Nalewajski, R. F. and Parr, R. G. (2000). Information theory, atoms in molecules, and molecular similarity. *Proceedings of the National Academy of Sciences of the United States of America*, 97(16):8879–8882.

Ngo, L. and Wand, M. (2004). Smoothing with mixed model software. *Journal of Statistical Software*, 9(1):1–54.

Ockenfels, A. and Roth, A. E. (2002). The timing of bids in Internet auctions market design, bidder behavior, and artificial agents. *AI Magazine*, 23(3):79–87.

Olea, R. A. (1999). *Geostatistics for Engineers and Earth Scientists*. Springer.

Overby, E. and Konsynski, B. (2008). Modeling time-varying relationships in pooled cross-sectional eCommerce data. In Jank, W. and Shmueli, G. (eds.), *Statistical Methods in eCommerce Research*, Wiley, New York.

Pan, X., Ratchford, B. T., and Shankar, V. (2003). The evolution of price dispersion in Internet retail markets. *Advances in Applied Microeconomics*, 12:85–105.

Pan, X., Ratchford, B. T., and Shankar, V. (2004). Price dispersion on the Internet: a review and directions for future research. *Journal of Interactive Marketing*, 18(4):116–135.

Park, Y.-H. and Bradlow, E. (2005). An integrated model for whether, who, when, and how much in Internet auctions. *SSRN eLibrary*.

Pelt, J. (2005). Astronomical time series analysis. Lecture Notes. Available at www.aai.ee/pelt/main.pdf, with software available at www.aai.ee/pelt/soft.htm#ISDA.

Pennock, D. M., Lawrence, S., Giles, C. L., and Nielsen, F. A. (2001). The real power of artificial markets. *Science*, 291(5506):987–988.

Perlich, C. and Rosset, S. (2008). Quantile modeling for wallet estimation. In Jank, W. and Shmueli, G. (eds.), *Statistical Methods in eCommerce Research*, Wiley, New York.

Peters, M. and Severinov, S. (2006). Internet auctions with many traders. *Journal of Economic Theory*, 127(1):220–245.

Pinker, E., Seidmann, A., and Vakrat, Y. (2003). The design of online auctions: business issues and current research. *Management Science*, 49(11):1457–1484.

Quinlan, J. R. (1993). *C4.5: Programs for Machine Learning*. Morgan Kaufmann Publishers, San Mateo, CA.

Ramsay, J. O. (1996). Principal differential analysis: data reduction by differential operators. *Journal of the Royal Statistical Society, Series B*, 58:495–508.

Ramsay, J. O. (1998). Estimating smooth monotone functions. *Journal of the Royal Statistical Society, Series B*, 60:365–375.

Ramsay, J. O. (2000). Differential equation models for statistical functions. *Canadian Journal of Statistics*, 28:225–240.

Ramsay, J. O., Hooker, G., Campbell, D., and Cao, J. (2007). Parameter estimation for nonlinear differential equations: a smoothing-spline approach. *Journal of the Royal Statistical Society, Series B*, 69(5):741–796.

Ramsay, J. O. and Silverman, B. W. (2002). *Applied Functional Data Analysis: Methods and Case Studies*, Springer, New York.

Ramsay, J. O. and Silverman, B. W. (2005). *Functional Data Analysis. Springer Series in Statistics*, 2nd edition, Springer, New York.

Rao, C. (1985). In Weighted distributions arising out of methods of ascertainment: what population does a sample represent?, *A Celebration of Statistics: The ISI Centenary Volume*, Springer, pp. 543–569.

Raubera, T., Braun, T., and Berns, K. (2008). Probabilistic distance measures of the Dirichlet and Beta distributions. *Pattern Recognition*, 41(2):637–645.

Reddy, S. K. and Dass, M. (2006). Modeling on-line art auction dynamics using functional data analysis. *Statistical Science*, 21(2):179–193.

Reithinger, F., Jank, W., Tutz, G., and Shmueli, G. (2008). Smoothing sparse and unevenly sampled curves using semiparametric mixed models: an application to online auctions. *Journal of the Royal Statistical Society, Series C*, 57(2):127–148.

Roth, A. E., Murninghan, J. K., and Schoumaker, F. (1998). The deadline effect in bargaining: some experimental evidence. *American Economic Review*, 78(4):806–823.

Roth, A. E. and Ockenfels, A. (2002). Last-minutes bidding and the rules for ending second price auctions: evidence from eBay and Amazon on the Internet. *American Economic Review*, 92:1093–1103.

Roth, A. E. and Xing, X. (1994). Jumping the gun: imperfections and institutions related to the timing of market transactions. *American Economic Review*, 84:992–1044.

Rubin, D. B. and Waterman, R. P. (2006). Estimating the causal effects of marketing interventions using propensity score methodology. *Statistical Science*, 21(2):206–222.

Ruppert, D., Wand, M. P., and Carroll, R. J. (2003). *Semiparametric Regression*. Cambridge University Press, Cambridge.

Russo, R. P., Shmueli, G., Jank, W., and Shyamalkumar, N. D. (2010). Models for bid arrivals and bidder arrivals in online auctions. In Balakrishnan, N. (ed.), *Handbook of Business, Finance and Management Sciences*, Wiley.

Russo, R. P., Shmueli, G., and Shyamalkumar, N. D. (2008). Models of bidder activity consistent with self-similar bid arrivals. In Jank, W. and Shmueli, G. (eds.), *Statistical Methods in eCommerce Research*, Wiley, New York, pp. 325–339.

Salls, M. (2005). How to harness auction fever (interview with Deepak Malhotra). Harvard Business School Working Knowledge.

Scott, D. W. (1979). On optimal and data-based histograms. *Biometrika*, 66:605–610.

Shekhar, S., Lu, C. T., and Zhang, P. (2003). Unified approach to spatial outliers detection. *GeoInformatica*, 7(2):139–166.

Shmueli, G. (2009). To explain or to predict? Working Paper RHS 06-099, Robert H. Smith School of Business, University of Maryland.

Shmueli, G. and Jank, W. (2005). Visualizing online auctions. *Journal of Computational and Graphical Statistics*, 14(2):299–319.

Shmueli, G. and Jank, W. (2008). Modeling the dynamics of online auctions: a modern statistical approach. In Kauffman, R. and Tallon, P. (eds.), *Economics, Information Systems & Ecommerce Research II: Advanced Empirical Methods. Advances in Management Information Systems Series*, Sharpe, Armonk, NY.

Shmueli, G., Jank, W., Aris, A., Plaisant, C., and Shneiderman, B. (2006). Exploring auction databases through interactive visualization. *Decision Support Systems*, 42(3):1521–1538.

Shmueli, G., Jank, W., and Bapna, R. (2005). Sampling eCommerce data from the web: methodological and practical issues. In *2005 Proceedings of the American Statistical Association*, American Statistical Association, Alexandria, VA.

Shmueli, G., Jank, W., and Hyde, V. (2008). Transformations for semicontinuous data. *Computational Statistics & Data Analysis*, 52(8):4000–4020.

Shmueli, G. and Koppius, O. (2008). The challenge of prediction in information systems research. Working Paper RHS 06-058, Robert H. Smith School of Business, University of Maryland.

Shmueli, G., Russo, R. P., and Jank, W. (2004). Modeling bid arrivals in online auctions. Working Paper RHS 06-001, Robert H. Smith School of Business, University of Maryland.

Shmueli, G., Russo, R., and Jank, W. (2007). The BARISTA: a model for bid arrivals in online auctions. *The Annals of Applied Statistics*, 1(2):412–441.

Shneiderman, B. (1992). Tree visualization with tree-maps: a 2-dimensional space filling approach. *ACM Transactions on Graphics*, 11:92–99.

Short, R. D. and Fukunaga, K. (1981). The optimal distance measure for nearest neighbor classification. *IEEE Transactions on Information Theory*, 27(5):622–627.

Slade, M. E. (2005). The role of economic space in decision making. *Annales d'Economie et de Statistique*, 77(1):1–21.

Snir, E. M. (2006). Online auction enabling the secondary computer market. *Information Technology and Management*, 7(3):213–234.

Solow, R. M. (1956). A contribution to the theory of economic growth. *Quarterly Journal of Economics*, 70(1):65–94.

Spann, M. and Skiera, B. (2003). Internet-based virtual stock markets for business forecasting. *Management Science*, 49(10):1310–1326.

Spitzner, D. J., Marron, J. S., and Essick, G. K. (2003). Mixed-model functional ANOVA for studying human tactile perception. *Journal of the American Statistical Association*, 98:263–272.

Stewart, K., Darcy, D., and Daniel, S. (2006). Opportunities and challenges applying functional data analysis to the study of open source software evolution. *Statistical Science*, 21(2):167–178.

Stone, C. J. (1977). Consistent nonparametric regression. *The Annals of Statistics*, 5:595–645.

Sugar, C. A. and James, G. M. (2003). Finding the number of clusters in a data set: an information theoretic approach. *Journal of the American Statistical Association*, 98:750–763.

Surowiecki, J. (2005). *The Wisdom of Crowds*. Random House Inc., New York.

Telang, R. and Smith, M. D. (2008). Internet exchanges for used digital goods. *SSRN eLibrary*.

Tibshirani, R., Walther, G., and Hastie, T. (2001). Estimating the number of clusters in a data set via the gap statistic. *Journal of the Royal Statistical Society, Series B*, 63:411–423.

Vakrat, Y. and Seidmann, A. (2000). Implications of the bidders' arrival process on the design of online auctions. In *Proceedings of the 33rd Hawaii International Conference on System Sciences*, 110.

Van der Heijden, P. and Böckenholt, U. (2008). Applications of randomized response methodology in eCommerce. In Jank, W. and Shmueli, G. (eds.), *Statistical Methods in eCommerce Research*, Wiley, New York.

Venkatesan, R., Mehta, K., and Bapna, R. (2007). Do market characteristics impact the relationship between retailer characteristics and online prices? *Journal of Retailing*, 83(3):309–324.

Wand, M. P. (1997). Data-based choice of histogram bin width. *The American Statistician*, 51(1):59–64.

Wand, M. P. (2003). Smoothing and mixed models. *Computational Statistics*, 18:223–249.

Wand, M. P., Coull, B., French, J., Ganguli, B., Kammann, E., Staudenmayer, J., and Zanobetti, A. (2005). The SemiPar 1.0. R package.

Wang, S., Jank, W., and Shmueli, G. (2008a). Explaining and forecasting online auction prices and their dynamics using functional data analysis. *Journal of Business and Economic Statistics*, 26(2):144–160.

Wang, S., Jank, W., Shmueli, G., and Smith, P. (2008b). Modeling price dynamics in eBay auctions using principal differential analysis. *Journal of American Statistical Association*, 103(483):1100–1118.

Warren, R., Eiroldi, E., and Banks, D. (2008). Shared knowledge systems with value: statistical aspects of Wikipedia. In Jank, W. and Shmueli, G. (eds.), *Statistical Methods in eCommerce Research*, Wiley, New York.

Wilcox, R. T. (2000). Experts and amateurs: the role of experience in Internet auctions. *Marketing Letters*, 11:363–374.

Yang, I., Jeong, H., Kahng, B., and Barabasi, A.-L. (2003). Emerging behavior in electronic bidding. *Physical Review*, 68:5–8.

Yang, W., Müller, H.-G., and Stadmüller, U. (2010). Functional singular components and their applications. Working Paper, Department of Statistics, University of California at Davis.

Yao, S. and Mela, C. F. (2007). Online auction demand. *SSRN eLibrary*.

Yu, Y. and Lambert, D. (1999). Fitting trees to functional data, with an application to time-of-day patterns. *Journal of Computational and Graphical Statistics*, 8(4):749–762.

Zeileis, A., Hothorn, T., and Hornik, K. (2005). Model-based recursive partitioning. Research Report Series, Report 19, Department of Statistics and Mathematics, Wirtschaftsuniversität Wien.

Zeithammer, R. (2006). Forward-looking bidding in online auctions. *Journal of Marketing Research*, 43(3):462–476.

Zhang, A., Beyer, D., Ward, J., Liu, T., Karp, A., Guler, K., Jain, S., and Tang, H. K. (2002). Modeling the price–demand relationship using auction bid data. Hewlett-Packard Labs Technical Report HPL-2002-202. Available at `http://www.hpl.hp.com/techreports/2002/HPL-2002-202.eps`.

Zhang, S., Jank, W., and Shmueli, G. (2010). Real-time forecasting of online auctions via functional k-nearest neighbors. *International Journal of Forecasting*, in press.

INDEX

STATISTICS IN PRACTICE

Human and Biological Sciences

Brown and Prescott · Applied Mixed Models in Medicine
Ellenberg, Fleming and DeMets · Data Monitoring Committees in Clinical Trials: A Practical Perspective
Lawson, Browne and Vidal Rodeiro · Disease Mapping With WinBUGS and MLwiN
Lui · Statistical Estimation of Epidemiological Risk
*Marubini and Valsecchi · Analysing Survival Data from Clinical Trials and Observation Studies
Parmigiani · Modeling in Medical Decision Making: A Bayesian Approach
Senn · Cross-over Trials in Clinical Research, *Second Edition*
Senn · Statistical Issues in Drug Development
Spiegelhalter, Abrams and Myles · Bayesian Approaches to Clinical Trials and Health-Care Evaluation
Turner · New Drug Development: Design, Methodology, and Analysis
Whitehead · Design and Analysis of Sequential Clinical Trials, *Revised Second Edition*
Whitehead · Meta-Analysis of Controlled Clinical Trials

Earth and Environmental Sciences

Buck, Cavanagh and Litton · Bayesian Approach to Interpreting Archaeological Data
Cooke · Uncertainty Modeling in Dose Response: Bench Testing Environmental Toxicity
Gibbons, Bhaumik, and Aryal · Statistical Methods for Groundwater Monitoring, *Second Edition*
Glasbey and Horgan · Image Analysis in the Biological Sciences
Helsel · Nondetects and Data Analysis: Statistics for Censored Environmental Data
McBride · Using Statistical Methods for Water Quality Management: Issues, Problems and Solutions
Webster and Oliver · Geostatistics for Environmental Scientists

Industry, Commerce and Finance

Aitken and Taroni · Statistics and the Evaluation of Evidence for Forensic Scientists, *Second Edition*
Brandimarte · Numerical Methods in Finance and Economics: A MATLAB-Based Introduction, *Second Edition*
Brandimarte and Zotteri · Introduction to Distribution Logistics
Chan and Wong · Simulation Techniques in Financial Risk Management
Jank · Statistical Methods in eCommerce Research
Jank and Shmueli · Modeling Online Auctions
Lehtonen and Pahkinen · Practical Methods for Design and Analysis of Complex Surveys, *Second Edition*
Ohser and Mücklich · Statistical Analysis of Microstructures in Materials Science

*Now available in paperback.